Micropropagation of Medicinal Plants

(Volume 2)

Edited by

T. Pullaiah
Department of Botany
Sri Krishnadevaraya University
Anantapur 515003
Andhra Pradesh
India

Micropropagation of Medicinal Plants

(Volume 2)

Editor: T. Pullaiah

ISBN (Online): 978-981-5238-30-3

ISBN (Print): 978-981-5238-31-0

ISBN (Paperback): 978-981-5238-32-7

First published in 2024.

Bentham Science Publishers Pte. Ltd.
80 Robinson Road #02-00
Singapore 068898
Singapore
Email: subscriptions@benthamscience.net

CONTENTS

PREFACE

Micropropagation of medicinal plants has become vital in providing high yielding elite genotypes for pharmaceutical purposes, as well as in producing high quality plantlets for conservation. Due to excess demand and injudicious harvesting, deforestation, climate change, pollution, urbanization, and natural calamities, many medicinal plants are under threat in their natural habitat. Variations in different biotic and abiotic environmental conditions severely limit the conventional method of propagation. As a result, micropropagation techniques may provide a better alternative and make it possible for the rapid multiplication of medicinal plants. It plays a significant role in increasing the production of disease-free plants, regardless of the season, with the goal of restoring these plants in their natural habitat and conserving them. These plants also serve as another source of raw materials used for commercial purposes, reducing the stress on plants growing in natural habitats. Tissue culture protocols for a wide variety of medicinal plants have been developed over the years. It also allows the modification and regulation of their genetic information with the goal of producing valuable phytoconstituents in greater quantities or with better properties, or both.

The present book gives the protocols for micropropagation of more than 40 species of medicinal plants. This book smartly combines the scientific principles with the state of the art in tissue culture techniques presented by experienced and authors. I wish to express my gratitude to all the authors who contributed to the review chapters and research papers. I thank them for their cooperation and erudition. I hope that this will be a source book for the cultivation and improvement of medicinal plants. I request that readers give their suggestions to improve in future editions.

T. Pullaiah
Department of Botany
Sri Krishnadevaraya University
Anantapur 515003
Andhra Pradesh
India

List of Contributors

Ayushi Negi	Department of Life Sciences, Graphic Era (Deemed to be University), 566/6, Bell Road, Clement Town, Dehradun-248002, Uttarakhand, India
Avijit Chakraborty	Plant Biotechnology Laboratory, Department of Botany, Ramakrishna Mission Vivekananda Centenary College, Rahara, Kolkata-700118, India
Bince Mani	Department of Botany, St. Thomas College Palai, Kottayam-686574, Affiliated to Mahatma Gandhi University, Kerala, India
B. L. Manjula	Department of Botany, Sri Jagadguru Renukacharya College of Science, Arts and Commerce, # 9 Race course road, Bengaluru 560 009, Karnataka, India
Boddupalli Krishna Jaswanth	Department of Biotechnology, Vikrama Simhapuri University, Nellore, Andhra Pradesh-524324, India
Biswajit Ghosh	Plant Biotechnology Laboratory, Post Graduate Department of Botany, Ramakrishna Mission Vivekananda Centenary College, Rahara, Kolkata - 700118, India
Chinnadurai Immanuel Selvaraj	Department of Genetics and Plant Breeding, School of Agricultural Innovations and Advanced Learning, Vellore Institute of Technology, Vellore, Tamil Nadu, PIN 632 014, India
Delna Joseph	Department of Botany, St. Thomas College Palai, Kottayam-686574, Affiliated to Mahatma Gandhi University, Kerala, India
Diptesh Biswas	Plant Biotechnology Laboratory, Post Graduate Department of Botany, Ramakrishna Mission Vivekananda Centenary College, Rahara, Kolkata - 700118, India
D. Raghu Ramulu	Department of Botany, Government College (A), Anantapur-515001, Andhra Pradesh, India
Govindugari Vijaya Laxmi	Department of Biotechnology, Chaitanya Bharathi Institute of Technology, Hyderabad, Telangana-500075, India
G. Sangeetha	Department of Botany, TKM College of Arts and Science, Kollam, Kerala, India
Harsha V. Hegde	ICMR-National Institute of Traditional Medicine, Nehru Nagar, Belagavi, Karnataka-590010, India
Jaindra Nath Tripathi	International Institute of Tropical Agriculture (IITA), PO Box 30709-00100, Old Naivasha Road, Nairobi, Kenya
Kalpana Agarwal	Department of Botany, IIS (Deemed to be University), Jaipur, Rajasthan, India
Kiranmai Chadipiralla	Department of Biotechnology, Vikrama Simhapuri University, Nellore, Andhra Pradesh-524324, India
K. Dharmalingam	Department of Biotechnology, Chaitanya Bharathi Institute of Technology, Hyderabad, Telangana-500075, India
Kannan Gandhi	International Institute of Tropical Agriculture (IITA), PO Box 30709-00100, Old Naivasha Road, Nairobi, Kenya
K. Sri Rama Murthy	R & D Center for Conservation Biology and Plant Biotechnology, Shivashakti Biotechnologies Limited, S. R. Nagar, Hyderabad-500038, Telangana, India

Leena Tripathi International Institute of Tropical Agriculture (IITA), PO Box 30709-00100, Old Naivasha Road, Nairobi, Kenya

Megha Rawat Department of Life Sciences, Graphic Era (Deemed to be University), 566/6, Bell Road, Clement Town, Dehradun-248002, Uttarakhand, India

Manu Pant Department of Life Sciences, Graphic Era (Deemed to be University), 566/6, Bell Road, Clement Town, Dehradun-248002, Uttarakhand, India

Mala Agarwal Department of Botany, B.B.D. Government College, Chimanpura (Shahpura), Jaipur, Rajasthan, India

M.V. Rao Department of Botany, Bharathidasan University, Tiruchirappalli-620024, Tamil Nadu, India

Nayan Kumar Sishu School of Biosciences and Technology, Vellore Institute of Technology, Vellore, Tamil Nadu, PIN 632 014, India

Parmila Saini Department of Botany, University of Rajasthan, Jaipur-302004, Rajasthan, India

Priya Saini Department of Life Sciences, Graphic Era (Deemed to be University), 566/6, Bell Road, Clement Town, Dehradun-248002, Uttarakhand, India

Parasurama Deepa Sankar Department of Genetics and Plant Breeding, School of Agricultural Innovations and Advanced Learning, Vellore Institute of Technology, Vellore, Tamil Nadu, PIN 632 014, India

Pichili Vijaya Bhaskar Reddy Department of Life Science & Bioinformatics, Assam University Diphu Campus, Diphu-782462, Assam, India

Pradeep Bhat ICMR-National Institute of Traditional Medicine, Nehru Nagar, Belagavi, Karnataka-590010, India

Poornananda Madhava Naik Kanara E-vision Science and Commerce PU College, Chalageri-581145, Ranebennur, Karnataka, India

Ridhi Joshi Department of Botany, University of Rajasthan, Jaipur-302004, Rajasthan, India

Raghunandan Singh Nathawat Department of Botany, University of Rajasthan, Jaipur-302004, Rajasthan, India

Richa Bhardwaj Department of Botany, IIS (Deemed to be University), Jaipur, Rajasthan, India

R. Lavanaya Department of Botany, Bharathidasan University, Tiruchirappalli-620024, Tamil Nadu, India
Department of Botany, Thanthai Periyar Govt. Arts & Science College, Tiruchirappalli-620023, Tamil Nadu, India

Sakshi Juyal Department of Life Sciences, Graphic Era (Deemed to be University), 566/6, Bell Road, Clement Town, Dehradun-248002, Uttarakhand, India

Sinjumol Thomas Department of Botany, Carmel College, (Autonomous), Mala, Thrissur-680732, Affiliated to University of Calicut, Kerala, India

S. Manjula Department of Botany, Government First Grade College, B. M. Road, Ramanagara-562159, Karnataka, India

Sandeep R. Pai Department of Botany, Rayat Shikshan Sanstha's Dada Patil Mahavidyalaya, Karjat-414402, Ahmednagar, Maharashtra, India

Suproteem Mukherjee Plant Biotechnology Laboratory, Post Graduate Department of Botany, Ramakrishna Mission Vivekananda Centenary College, Rahara, Kolkata - 700118, India

S. Asha Department of Biotechnology, VFSTR (Deemed to be University), Vadlamudi, Guntur Dt., A.P., India

Santoshkumar Jayagoudar Department of Botany, G. S. S. College & Rani Channamma University P. G. Centre, Belagavi, Karnataka-590006, India

Sachet Hegde Department of Botany, Bangurnagar Degree College, Ambewadi, Dandeli, Karnataka-581325, India

Savaliram G. Ghane Department of Botany, Shivaji University, Vidyanagar, Kolhapur, Maharashtra-416004, India

S. Parthibhan Department of Botany, Bharathidasan University, Tiruchirappalli-620024, Tamil Nadu, India
Department of Botany, Kalaignar Karunanidhi Govt. Arts College for Women (A), Pudukkottai- 622001, Tamil Nadu, India

T. S. Swapna Department of Botany, TKM College of Arts and Science, Kollam, Kerala, India

Vidya Patni Department of Botany, University of Rajasthan, Jaipur-302004, Rajasthan, India

Varsha S. Dhoran Department of Botany, Sant Gadge Baba Amravati University, Amravati-444602, Dist-Amravati, Maharashtra, India

Vishal P. Deshmukh Department of Botany, Jagadamba Mahavidyalaya, Achalpur city-444806, Dist-Amravati, Maharashtra, India

Varsha N. Nathar Department of Botany, Sant Gadge Baba Amravati University, Amravati-444602, Dist-Amravati, Maharashtra, India

Vinay Kumar Hegde Bhandimane Life Science Research Foundation, Sirsi-581401, Uttara Kannada, Karnataka, India

Vinayak Upadhya Department of Forest Products and Utilization, College of Forestry, (University of Agricultural Sciences Dharwad), Sirsi-581401, Uttara Kannada, Karnataka, India

V. Kumaresan Department of Botany, Bharathidasan University, Tiruchirappalli-620024, Tamil Nadu, India
Department of Botany, Arignar Anna Govt. Arts College, Attur-636121, Tamil Nadu, India

CHAPTER 1

Micropropagation Studies on Genus *Cissus* A Review

Ridhi Joshi[1,*], **Parmila Saini**[1], **Raghunandan Singh Nathawat**[1] and **Vidya Patni**[1]

[1] *Department of Botany, University of Rajasthan, Jaipur-302004, Rajasthan, India*

Abstract: The genus *Cissus* Linn. belongs to the Family Vitaceae (formerly Ampelidaceae) and comprises about 350 species distributed all over the world, having rich phytochemicals with medicinal as well as commercial value. This genus is a storehouse of large varieties of phytochemicals such as alkaloids, flavonoids, and phytosterols, making this genus pharmaceutically important. Some species contain high quantity of calcium ions in their stem extract, which is possibly responsible for their bone healing activity. *In vitro* propagation of plantlets provides the opportunity to conserve endangered species as well as to use the beneficial species without disturbing their natural habitat. The present review comprises *in vitro* protocols used to conserve the species, exploit and enhance useful metabolites. The whole plant, parts and metabolites isolated from *in vitro* cultures of *Cissus* species may be used further for pharmaceutical purposes.

Keywords: *Cissus*, Conserve, *In vitro* propagation, Phytochemicals, Pharmaceutical.

INTRODUCTION

The genus *Cissus* Linn. belonging to the Family Vitaceae (formerly Ampelidaceae) comprises about 350 species [1, 2]. It has been reported by many researchers that the genus has approximately 135 species in Africa, 85 in Asia, 12 in Australia and 65 in the Neotropics [3]. This genus has cosmopolitan distribution across the globe and is characterized by polypetalous flowers having prominent disk-shaped thalamus below the ovary. and represents the largest of the 14 genera of Vitaceae, primarily distributed in tropical and temperate regions of the world [3]. The *Cissus* group of plants have a variety of bioactive properties and are known for their medicinal uses since ages. Diarrhoea, loose stools,

* **Corresponding author Ridhi Joshi:** Department of Botany, University of Rajasthan, Jaipur-302004, Rajasthan, India; E-mail: ridhi.joshi316@gmail.com

T. Pullaiah (Ed.)

coughs, and breast cancer are some common diseases that can be cured by the use of various preparations of *Cissus* species.

PHYTOCHEMISTRY OF THE GENUS

Cissus species are being used in all parts of the world and are implicated in treating various ailments. As reported by several researchers, species of the genus *Cissus* are often used as medicinal plants because they contain vitamins, proteins, carbohydrates, and polyphenols Table **1**.

Table 1. Various phytochemicals reported from different species of genus *Cissus* responsible for their pharmacological properties.

Plant Name	Extract	Chemicals Reported	References
C. quadrangularis	Whole plant	Vitamin C, flavonoids, triterpenoids, stilbene derivatives and several secondary metabolites, *e.g.*, quercetin and kaempferol, quadrangularin A-C, resveratrol, pallidol, perthenocissin piceatannol and phytosterols. Calcium ions, β-sitosterol, d-amyrin, onocer-7-ene-3a, 21beta-diol, d-amyrone and 3,3',4,4'-tetrachloro-1,1'-biphenyl. Anabolic steroidal substances and carotene 7-oxo-onocer-8-ene-3-β 21-α-diol have been reported.	[8-11]
C. vitiginea	Methanolic leaf extract	Diethyl phthalate, 3,7,11,15-tetramethyl-2-hexadecen-1-ol, myristic acid, azelaic acid, 2,6,10-trimethyl 14-ethylene-14-pentadecene, 2-hexadecene, palmitic acid, 2-hydrox--1,3-propanedyl ester, 22-tricosanoic acid, oleic acid, dibutyl ester, L-ascorbyl 2,6-dipalmitate, dibutyl phthalate oleic acid, heptadecanoate, 2-hexadecen-1-ol, 1H-cyclopropan-α-aphthalene, 9,12-octadecadienoic acid (Z,Z)-, nonadecanoic acid, palmitic acid, icosanoic acid, octadecanoic acid, 2,3-dihydroxypropyl ester, stearic acid glycidyl ester, and bis(2-ethylhexyl) phthalate.	[12]
C. pteroclada	Ethanolic extract of the roots with stems	Gallic acid, β-sitosterol, bergenin, 11-O-(4-hydroxy benzoyl) bergenin, 11-O-galloylbergenin and daucosterol	[13]
C. ibuensis	Ethanolic leaf extract	Rutinoside and quercetin.	[14]
C. assamica	Whole plant	Lupeol, β sitosterol, n-hexacosinic acid, daucostenin, isolariciesinol-9-O-beta-D-glucopyranoside, 3,8-Di-O-methylellagic acid and bergenin.	[15]
C. repens	Ethanolic extraction of aerial part	Stilbene C glucosides.	[16]

Good quantity number of Ca^{2+} and P ions, essential for bone growth, have been reported from *Cissus quadrangularis* stem extract [4]. Fracture healing studies on *C. quadrangularis* have also been reported by some Indian laboratories (Bulletin of Department of Pharmacology, Nagpur, 2002). *In vitro* screening and pharmacological studies [5] and chemical components [6] of *C. quadrangularis* have also been reported. *C. quadrangularis* has also been used to synthesize calcite crystals [7].

Advanced biochemical analyses of various species of *Cissus* have revealed the presence of a number of useful phytochemicals summarised in Table **1**.

MICROPROPAGATION

Micropropagation has the upper hand over vegetative propagation methods as it can rapidly multiply valuably genotypes, release improved varieties and disease-free plants at a quick pace. It is also important for off-seasonal production of plantlets, germplasm conservation and secondary metabolites production.

The plant tissue culture technique has been tremendously exploited for *in vitro* propagation of desired genotypes on a mass scale. The results of such studies have practical as well as economical value-producing industrially valuable compounds [17]. This technique is a potential renewable source of obtaining valuable compounds from a particular species, we can also get flavours, fragrances, colourants, which cannot be produced by microbial cells or chemical synthesis [18]. Therefore, plant cell culture is being utilized for extensive production of valuable medicinal plants these days.

Need for *in vitro* Propagation of Genus *Cissus*

All the species belonging to keep it genus *Cissus* are medicinally important as they possess useful antiarthritic, antidiabetic, anticholestrolemic, anticancerous, anticonvulsive, antimicrobial, and anti-inflammatory properties as indicated in Table **2**.

Table 2. The effect of the *Cissus* plant part on target organisms showing various activities.

Plant Name	Useful Plant's Part	Activity	Target Organism	References
Cissus quadrangularis	Stem and root	Bone healing activity,	Murine osteoblastic cell line, Wistar albino rats	[86]
Cissus adnata	Whole plant parts	Antioxidant activity	-	[87]

(Table 2) cont.....

Plant Name	Useful Plant's Part	Activity	Target Organism	References
Cissus javana	Root	Antidiabetic	Rat muscle cells	[88]
Cissus woodrowii	Leaf	Antioxidant activity	-	[89]
Cissus sicyoides	Stem, aerial part	Antidiabetic, anti-convulsant, anxiolytic	Rats	[65, 78, 90]
Cissus cornifolia	Leaf and root	Antioxidant activity, anti-convulsant activity	-	[66, 91]
Cissus populnea	Rhizome	Antioxidant and anthelmintic activities	*Onchocerca ochengi* adult worms	[92]
Cissus trifoliata	Stem	Anticancer activity	Prostate cancer (PC3), lung cancer (A549), breast cancer (MCF7), liver cancer (HepG2, Hep3B), cervix cancer (HeLa)	[53]
Cissus repanda	Stem Root	Anti-inflammatory activity	Rat paw	[92]
Cissus gongylodes	Leaf	Anti-inflammatory and anti-urolithiatic	Albino mice male	[93]
Cissus assamica	Stem	Anticancer and anti-inflammatory activity	Human cancer cell lines, non-small cell lung carcinoma (NCI-H226) and colon cancer cell line (HCT116), nasopharyngeal carcinoma (NPC-TW01), Human neutrophils	[94]
Cissus rotundifolia	Leaf	Antioxidant, antibacterial, antidiabetic activity	Wistar rats *Bacillus cereus, Staphylococcus aureus, Listeria monocytogenes, Escherichia coli* and *Salmonella infantis*	[95 - 97]
Cissus araloides	-	Anti-microbial	*Enterococcus faecalis, Staphylococcus aureus, Pseudomonas aeruginosa, Shigella flexneri, Escherichia coli, Salmonella typhi, Klebsiella pneumoniae, Proteus mirabilis, Candida krusei, Candida albicans.*	[98]
Cissus debilis	-	Anti-cell proliferation	Human colon cancer cell line (CaCo-2)	[49]

To conserve these species and to maximize the benefits for the folklore, it is a present-day need to develop protocols to efficiently grow the medicinally important species *in vitro*. These techniques are helpful because, without disturbing biodiversity, it is possible to extract phytocompounds from them for the development of new drugs.

One of the plant species belonging to this genus which has come into the limelight in recent decades is *Cissus quadrangularis* due to its huge beneficial aspects. Tremendous work reported on this species indicates its medicinal importance and the possible threat it may face due to over-exploitation in the near future. This plant, when given alone or in formulations such as Cylaris, CORE and also combined with *Irvingis gabonensis*, reduces body weight [19 - 21]. It has hypoglycaemic and hypolipidemic activity [22]. It decreases insulin resistance and has antioxidant activity [23]. The presence of antioxidant and antimicrobial activity [24], gastroprotective activity [25 - 27], hepatoprotective properties [28], and suppressed chronic ulcer activity [8, 29] have been reported. It also demonstrated anti-inflammatory and analgesic properties [30]. In rats, *Cissus quadrangularis* extracts reduced oedema of the ears and paws. In addition, it also showed anti-tumour activity [31].

Altogether, it has many biological activities such as antidiabetic, antimalarial, antiallergic, antiasthmatic, and cytotoxic activities, wound healing, bone fractures, reducing cholesterol, and pain during menstruation. *Cissus quadrangularis* has also been reported to be used as body building supplements as an alternative to anabolic steroids [32]. *C. latifolia* Lam. and *C. quadrangularis* L. together are used in the treatment of weaker bones, fractures, cancer, scurvy, haemorrhoids, peptic ulcer disease, malaria, pain, and asthma [33].

So, it is important to develop protocols and multiply such medicinally important species like *C. quadrangularis* and others *in vitro* to conserve and propagate for medicinal uses.

Sterilization Protocol

For *in vitro* shoot propagation, nodal or stem explants with axillary buds have been used and for raising callus, nodal, internodal and leaf segments as explants have been used by various researchers for the genus *Cissus*.

Various explants were cut into 8-10 mm sizes and were washed under tap water to remove dirt and soil particles, they were then washed with detergent (1% labolene) and again rinsed with distilled water to remove every trace of detergents. The explants can thereafter be treated with bavistin (0.1% w/v) (Saraswati Agro Chemicals, Jammu, India) for 10 minutes to remove fungal contamination, if needed.

Under aseptic conditions, they were then treated with mercuric chloride (0.1%) (Ranbaxy Fine Chemicals Ltd, New Delhi, India) for 30 seconds to 3 minutes, depending on the explants taken, with continuous shaking and were repeatedly washed thrice with sterile distilled water after treatment and inoculated onto the respective media.

Other detergents like TEEPOL (Reckitt Benckiser India Pvt Limited, Uttarakhand, India) and Tween-20 (Loba Chemie Pvt Ltd, India) have been used, and their sterilization time was increased or decreased according to the response and fragility of the tissue.

Callus Initiation and Establishment

Callus induction in *Cissus quadrangularis* was observed on MS medium fortified with NAA (2.0 mg/L) (HiMedia, Mumbai, India). This callus was yellow-green in colour, fast growing and fragile in nature, but when auxin in combination with cytokinin was added to the medium (NAA-2.0 mg/L+BAP-0.5 mg/L) (HiMedia, Mumbai India), the callus produced was greenish, fast growing with high regenerative potential [34].

Stem and leaf explants on Murashige and Skoog's medium supplemented with NAA (5.37µm) and Kn (2.32 µm) produced callus. Increased callus growth was observed with 15% CM and 3% glucose in the same study. Root differentiation took place on higher auxin concentration of NAA (10.74-21.48 µm) and Kn (4.65 µm) after 30-35 days. few in place of a smaller number of shoots differentiated on NAA (14.7 µm) and Kn (2.32 µm) after 56 days [35].

For *C. sicyoides* L., leaf segments were inoculated onto a Milk-Tween medium having NAA (1.0 mg/L) after surface disinfection. For induction and establishment, the combination of 1.0 mg/L NAA and BAP (2, 4, 6, 12 mg/L) were used. 6.0 mg/L BAP proved best for callusing in *C. sicyoides* [36].

In *Cissus verticilliata*, leaf explants were inoculated on MS media supplemented with factorial concentration and combinations of 2,4-D and BA (0, 1, 2, and 4 mg/L). The absence of growth regulators produced no callus. The higher percentages of Leaf Area Covered with Callus (LACC) were observed with 1.0 - 4.0 mg/L BA. The maximum fresh weight was obtained with 4.0 mg/L BA in the culture medium [37].

In another study, three species, *viz., Cissus rotundifolia, Cissus repanda* and *Cissus quadrangularis*, were tried and tested for the development of a protocol for callus initiation and establishment. *In vivo* nodal segments (approximately 0.5 cms) were found to be the best for optimum callus production in *Cissus repanda* and *Cissus quadrangularis.* Leaf segments showed callusing response near the cut end of the midrib region in *Cissus rotundifolia.*

Among the various explants (leaf/nodal segments) tried, nodal segments proved to be the best in *Cissus repanda* and *Cissus quadrangularis*, while leaf proved best in the case of *Cissus rotundifolia.* Among the various auxins tried, 2,4-D (0.5mg/L) was best for callus induction in *Cissus rotundifolia.* In *Cissus repanda* and *Cissus quadrangularis*, NAA (1 mg/L) proved best for callus induction.

During this research, various auxins (2,4-D, IBA, IAA, NAA ranging from 0.25-2.0 mg/L) and cytokinins (BAP, Kn from 0.25-2.0 mg/L) were added to the growth medium at different concentrations, used separately and in combination.

Cissus rotundifolia leaf explants when inoculated on MS medium supplemented with 2,4-D (0.5mg/L) gave green, compact and fast-growing callus (Fig. **1A**). In the case of *Cissus repanda* and *Cissus quadrangularis*, compact, fast-growing greenish callus was obtained on MS medium supplemented with NAA (1.0 mg/L) (Fig. **1B** & **C**). IAA (0.25 mg/L) and 2,4-D (0.5 mg/L) produced very scanty brownish callus. IBA showed necrosis of tissue in both plant species.

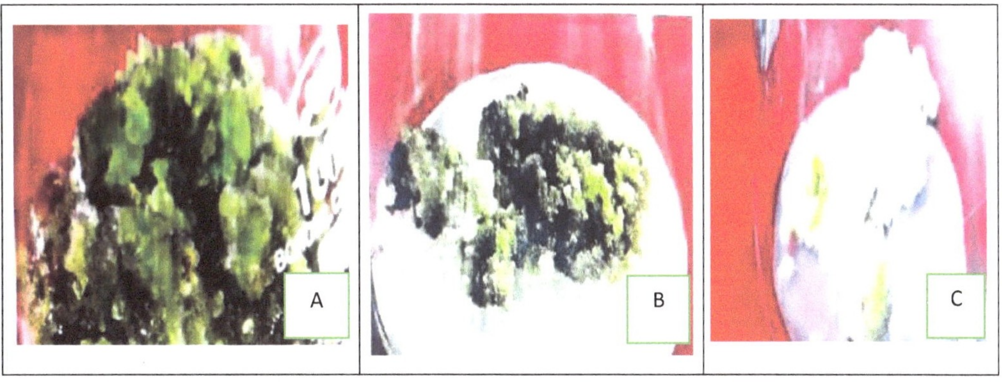

Fig. (1). A) Callus induction from leaf explant on MS medium supplemented with auxin (2,4-D-0.50 mg/L) in *Cissus rotundifolia.* **B**) Callus induction from nodal segment on MS medium supplemented with auxin (NAA-1.0 mg/L) in *Cissus repanda.* **C**) Callus induction from nodal segment on MS medium supplemented with auxin (NAA-1.0 mg/L) in *Cissus quadrangularis.*

In the case of *Cissus rotundifolia,* leaf explants cultured on MS medium supplemented with BAP (0.5mg/L) found to be the best for the production of green, fast-growing callus. In *Cissus repanda* and *Cissus quadrangularis*, MS medium supplemented with BAP (1.0mg/L) gave rise to the best, fast-growing, greenish callus, on the other hand addition of Kn (0.5 mg/L) gave rise to fragile callus.

the combined concentrations of auxin and cytokinins led to the production of a large amount of callus. In *Cissus rotundifolia*, 2,4-D (0.5 mg/L) initiated callus formation at the margins of leaf explants. When the optimum concentration of 2,4-D was combined with BAP (0.5 mg/L) and $AdSo_4$ (0.10 mg/L), it gave a green callus with a morphogenic response. While Kn (1.0 mg/L) in combination with 2,4-D (0.5 mg/L) and $AdSo_4$ (0.10 mg/L) gave green, fast growing and higher amount of callus. In *Cissus repanda* and *Cissus quadrangularis*, optimum concentrations of NAA (1.0mg/L) with BAP (0.5mg/L) and $AdSo_4$ (0.10 mg/L) gave rise to green callus [38].

Morphogenesis/Differentiation in Callus Culture

The regeneration medium contains a balance of auxin and cytokinins (growth regulators) and sucrose, which are involved in the process of organogenesis and differentiation. Manipulation of these growth regulators, *i.e.*, auxin to cytokinin ratio in the medium, leads to the development of shoots, roots or somatic embryos, from which plants can subsequently be produced.

BAP (1.5 mg/L) evoked morphogenetic response in the form of shoot buds from explants, whereas with Kn (0.5 mg/L), the frequency of morphogenetic response was low as compared to BAP. Percent morphogenetic response increased in all experimental plants when callus was subcultured on MS medium containing both BAP (1.5mg/L) and Kn (0.5 mg/L).

Optimum concentrations of various auxins (NAA, IAA, IBA, 2,4-D) were incorporated along with BAP (1.5 mg/L) and Kn (0.5 mg/L) for differentiation of callus. The most favourable response was obtained on MS medium supplemented with BAP (1.5 mg/L), Kn (0.5 mg/L) and 2,4-D (0.5 mg/L) in *Cissus rotundifolia*. In case of *Cissus repanda* and *Cissus quadrangularis* BAP (1.5 mg/L), Kn (0.5 mg/L) and NAA (1.0 mg/L) evoked the most favourable response (Fig. **2**). At higher concentration of NAA (1.0-2.0 mg/L), shoot buds decreased. When these cytokinins were combined at optimum concentrations with IBA, IAA, no morphogenetic response was observed.

Multiple Shoot Induction and Proliferation

selection of the explant in place of the choice of the explants is a critical factor in the success of *in vitro* propagation protocols. Various explants like nodal, stem, leaf, *etc*., were used in several studies.

Fig. (2). A) Shoot bud formation on MS medium supplemented with cytokinin (BAP-1.50 mg/L) in *Cissus repanda*. **B**) Shoot bud formation on MS medium supplemented with cytokinin (BAP-1.50 mg/L) in *Cissus quadrangularis*. **C**) Shoot bud induction on MS medium supplemented with combination of auxin and cytokinin(s) (NAA (1.0 mg/L) + BAP-(1.50 mg/L) + Kn (0.50 mg/L) in *Cissus repanda*. **D**) Shoot bud induction on MS medium supplemented with a combination of auxin and cytokinin(s) (NAA (1.0 mg/L) + BAP-(1.50mg/L) + Kn (0.50 mg/L) in *Cissus quadrangularis*.

Nodal Explants

In vitro propagation of genus *Cissus* has been attempted by various workers. For obtaining multiple shoots, nodal stem segments were inoculated on various media (MS, B5, MSB5) supplemented with different concentrations of phytohormones. Various critical factors, such as the evaluation of seasonal influence on bud proliferation, contamination, concentration effect of sucrose, inorganic and organic nutrients, vitamins and sulfates were also evaluated.

Experiments in series and multiplicates were set up to obtain a wide number of shoots from the nodal explants of *C. quadrangularis*. Shoot buds emerged after 2-3 weeks of incubation under controlled environmental conditions. The highest shoot buds (9.6±0.457) emerged from this callus at a BAP concentration of 3.0 mg/L alone and increase in the number of shoots (18.2± 0.25) was observed at BAP-3.0 mg/L and Kn -1.0 mg/L (HiMedia, Mumbai India) designated as shoot bud differentiation and elongation medium whereas addition of auxins for shoot bud differentiation proved unfruitful [34].

Various synthetic media like MS [39], B_5 and MSB_5 (modified MS-medium) were evaluated for the initiation and development of the optimal shoot buds from a single nodal segment in *C. quadrangularis*. Full strength MS medium with proper inorganic salts concentrations proved best. Inoculation of cultures for four weeks led to the maximum number of shoot buds proliferation on MS-medium without much callus formation (Fig. **3**) [34].

Nathawat *et al.* (2013) in an extensive *in vitro* study reported that in this media, sucrose (3.0%) and plant hormones like cytokinins (BAP/Kn (0.5-5.0 mg/L) and auxins (NAA (1.0- 3.0 mg/L)) were incorporated. Nodal segments were surface sterilized and inoculated on media containing various concentrations and combinations of growth regulators. MS medium with BAP alone produced a fairly good number of shoots (8.24±0.09), but the maximum shoot buds (11.42±0.19) were obtained from nodal explants with BAP (3.0mg/L) in combination with Kn (1.0 mg/L) (HiMedia, Mumbai India). The addition of auxin (NAA/IAA) to this medium showed an inhibitory effect on shoot bud production and the number of emergences reduced [34].

Seasonal effect on *in vitro* propagation response was also observed, and it was found that April-June proved best for maximum induction of shoots in *Cissus quadrangularis* [34]. For shoot induction, nodal explants when inoculated on MS Medium supplemented with 0.5/1.0 mg/L zeatin proved to be the best, producing 5 shoots per explant with a mean length of approximate 2.5 ±0.2 cms [40]. In another study, 10 mm nodal segments of *Cissus sicyoides* were sterilized and placed on MS medium containing combinations of NAA: BAP and NAA: Kn. Shoot induction was best observed at 2.7 µm NAA and 4.64 µm of Kn. Rooting was best recorded at 160 mg/L IBA media supplemented with sucrose and boric acid [41].

Stem Explants

Shoot tip explants were also tested and produced shoots with a lower concentration and combination of various growth regulators (Zeatin, BAP and adenine sulphate). In this study, when *in vitro* grown shoot tips were taken as explants, a maximum of 4 shoots were produced at a concentration of BAP (3.0 mg/L) and Zeatin (1.0 mg/L) of approximately 2.75 cm mean shoot length. However, only 3 shoots were produced when *in vivo* grown shoot tips were taken as explants on MS medium containing 2,4-D 1.0mg/L and Kn 2.0 mg/L with 1.12 cm mean shoot length [40].

Fig. (3). Shoot induction in *C. quadrangularis*. **A**) Effect of kinetin alone after four weeks on axillary bud proliferation from nodal stem explants. **B**) Effect of Kn in combination with BAP after two weeks on axillary bud proliferation from nodal stem explants at a growth hormone concentration of BAP 3.0 mg/L+ Kn 1.0 mg/L. **C**) Effect of Auxin IAA (1.5 mg/L) in combination with cytokinins after four weeks on axillary bud proliferation from nodal stem explants at a growth regulator concentration of BAP 3.0 mg/L+ Kn 1.0 mg/L. **D**) Effect of Auxin NAA (0.5 mg/L) in combination with cytokinins after four weeks on axillary bud proliferation from nodal stem explants at a growth regulators concentration of BAP 3.0 mg/L+ Kn 1.0 mg/L.

Adventitious shoot regeneration from the stem explant directly occurred on MS medium supplemented with 2, 4-D (18.12µM) and the addition of 4.90 µM IBA resulted in best rooting when stem explants were taken, forming 3-5 roots after 20 days of inoculation [42].

Somatic Embryogenesis

Ramar *et al.* [43] developed a simple and effective protocol for somatic embryogenesis in *Cissus quadrangularis*. Somatic embryos were obtained from callus suspension cultures. Leaf explants produced callus on medium fortified with MS salts, B5 vitamins, BAP, 2,4-D, glutamine, polyvinylpyrrolidone, and sucrose. High frequency somatic embryos were produced in a suspension medium containing 4.54 µM Kn along with others used in the callusing medium. When embryos were mature enough, they were transferred onto solid media for their germination. These somatic embryos developed into plantlets (53%) on half strength MS medium containing B5 vitamins, 0.984 µM GA_3, and 0.88 µM BAP [43].

Rooting, Hardening, Acclimatisation

The rooting media comprised of full, half and one–fourth strength MS medium devoid or with various auxins (IBA, IAA, NAA and 2,4-D) at varied concentrations (1.0-5.0 mg/L). Media also contained 2-3 gm Phytagel (Sigma). Phytagel is a clear, colourless and high strength gelling compound composed of glucuronic acid, rhamnose and glucose. Positive rooting response (90-100%) was observed on full strength MS Media containing auxins like NAA, IBA, and IAA (1.0-5.0 mg/L) except 2,4-D, which showed no response. It was concluded that full strength MS medium with NAA (1.0 mg/L) was the best medium for rooting in *Cissus quadrangularis*.

After successful root establishment, *in vitro* developed rooted plantlets were taken out from culture vessels without damaging the delicate root system, washed with distilled water to remove the trace of media, and then were treated with Bavistin (0.05% for 10 Seconds) (Saraswati Agro Chemicals, Jammu, India) and transferred to small pots containing vermiculite and soil in ratio 1:3. To maintain humidity they were covered by glass beakers. After two weeks, they were uncovered and transferred to earthen pots with normal soil. The plantlets were watered daily along with a few drops of MS–inorganic solution [34].

Garg and Malik [40] developed a fast micropropagation and plantlet regeneration method for *C. quadrangularis*. In this finding, rooting in *C. quadrangularis* was observed with IAA (1-2 mg/L) and IBA (1 mg/L) supplementation to MS media. Plantlets were acclimatized and transferred to pots containing 50:50 sterile soil and vermicompost. They were later transferred to greenhouse with a higher survival rate (80%) observance. These were transferred to autoclaved garden soil and vermicompost (1:1) and kept in plant growth chambers at 26±2°C for 3-4 weeks and watered every third day. These were then transferred to pots containing normal soil and kept in greenhouse under normal day length condition.

In the study conducted by Ramar *et al.* [43], the regenerated plantlets from somatic embryos were hardened in pots with sand and soil (1:1), and these pots were covered with plastic bags, which were removed after the eighth leaf formation. Developed plantlets showed a 72% survival rate in greenhouse conditions [43].

In vitro Enhancement of Secondary Metabolites

Profuse callusing was obtained in *C. quadrangularis* by using nodal stem explants on MS medium containing a combination of auxin and cytokinin (NAA-2.0 mg/L and BAP-0.5 mg/L) (HiMedia, Mumbai, India). Identification studies of steroidal compounds were done through colour reactions, TLC and HPLC techniques. Beta sitosterol and stigmasterol were reported to be present in the callus extracts. For the purpose of enhancement, the addition of plant growth regulators was done in the callusing medium, and there was a 10-12 times increase in their concentration with IAA (HiMedia, Mumbai India) at 5PPM concentration and 6-8 times increase with 2,4-D (5PPM) (HiMedia, Mumbai India) in the callus tissues [44]. In this study, the addition of NAA showed negligible effect in increasing the amount of sterol production, but callus content increased considerably.

Nodal segments of *C. quadrangularis* produced callus on MS medium containing NAA (1.5mg/L) and BAP (0.5mg/L) [45]. It was observed that lower concentrations of $CaCl_2$ induced stress conditions and increased production of alkaloids and sterols in the callus culture. Callus treated with 0.1 mM $CaCl_2$ had the highest levels of total Alkaloids (2.495 mg/10 g) and highest levels of sterols (0.252mg/10g), whereas control cultures reported the maximum amount of flavonoids (1.882 mg/10g) and addition of $CaCl_2$ resulted in a decrease in flavonoid content in cultures [45].

Nano based *in vitro* enhancement of phytoestrogen was achieved in a study conducted by Gour [38], in which ZnO particles gave positive results, while in the case of Fe_3O_4, satisfactory results were observed. *In vitro* grown callus (3-4 weeks old) was subcultured on MS medium containing NAA (0.5mg/L) and BAP (0.5mg/L) with different concentrations of various nanoparticles ranging from 0.015ml/l to 0.065 ml/l. After 4-5 weeks, the concentration of Phytoestrogens was determined through RP-HPLC. The results indicated the enhancement of daidzein content after treatment of Zn containing nanomaterial. ZnO (0.025 ml/L) showed two times quantification valuer ($2.437x10^{-1}$ mg/ml) as compared to Fe_3O_4. Zinc has an important role in bone development and remodeling [38].

IN VITRO BIOLOGICAL ACTIVITIES

Antiproliferative and Antioxidant Activity

According to the National Cancer Institute, if plant extracts have IC value ($IC_{50} \leq$ 30 μg/mL), the plant has good anticancerous activity [46].

In the DPPH free radical assay, silver nanoparticles produced with *C. vitiginea* leaf extracts reported maximum inhibition (73.18%) at 80 μg/ml concentration [47] *C. sicyoides* also reported anti-mitotic activity in HEp-2 cells and reduced gastric ulcers in rodents [48, 49].

The antioxidant activity of the stem extract of *Cissus quadrangularis* was assayed with both the β-carotene linoleic acid system and DPPH free radical assay. Ethyl acetate stem extract (Both fresh and dry) of *Cissus quadrangularis* at 100 ppm concentration showed 65% antioxidant activity when assessed through the β-carotene linoleic acid system and 62% through DPPH free radical assay [74]. For the *in vitro* study of the breast cancer cell line, different extracts (acetone, chloroform, ethanol, ethyl acetate, and methanol) of *Cissus quadrangularis* were used. Among all the extracts, ethyl acetate showed the lowest (41.5%) viability, hence the highest anti-cancerous activity [50].

In another study, stem aqueous extract of *Cissus quadrangularis* showed 53.43% and the ethanol extract showed 77.42% scavenging activity [51]. The methanolic extract of the plants *Cissus pallida* (stem and roots) and *Cissus vitiginea* (aerial parts) were used to find out the antioxidant and anticancer properties. *In vitro* studies indicated that at the same methanolic extract dose (20μg/ml), *Cissus pallida* showed 75.4% inhibition, whereas *Cissus vitiginea* showed 66.5% inhibition against the superoxide free radical. *In vitro* anticancer activity of extracts against MCF-7, A549, HT-29, and HeLa cell lines were studied. At some dose (200 ug/ml) of both plant extracts, *Cissus pallida* showed the best result (33.4%) against the MCF-7 cell line and *Cissus vitigenia* showed (23.28%) anticancer activity against the HeLa cell line [52].

The hexane stem extract of the *Cissus trifoliata* has an IC_{50} value of 62 ± 3 μg/ml (SD) against prostate cancer (PC3) [53]. In the case of *C. sicyoides*, hydrochloride extract of leaves inhibited sarcoma-180 to 62% and Ehrlich Carcinoma to 84.4% at the 600m/kg dose [54].

The skin carcinoma cell line (A431) was treated with different extracts of *Cissus quadrangularis in vitro*. The best result came out with dry acetone, which showed the highest anticancer activity with a 50% inhibition value at 8μg/ml. This was the first report of apoptosis in skin cancer due to active biocompounds of *Cissus quadrangularis* [55]. For *in vitro* antioxidant study, green copper oxide nanoparticles (CuONPs) formed from leaf extracts of *Cissus vitiginea* indicated that CuONPs possess excellent antioxidant activity than *Cissus vitiginea* extract and is near to ascorbic acid standard. The value of radical scavenging activity of CuONPs and *Cissus vitiginea* extract was 86.78% and 82.37%, respectively, at 80 μg/ml concentration [56].

Anthelmintic Activity

In a study, *C. rotundifolia* showed an anti-parasitic property. Four concentrations (25, 50, 75, 100mg/ml) of aqueous extract of *Cissus quadrangularis* were studied in comparison with albendazole as standard and saline water as control. Aqueous extracts at 100 mg/ml showed a good effect that paralysed and killed earthworms (*Pheretima posthuma*) model taken because of their similarity with parasites (roundworms) infecting the human gut. The methanolic stem extract of *Cissus quadrangulris* showed excellent anthelmintic activity at 20 mg/ml concentration compared to the standard drug albendazole [57]. The methanol extract of *C. populnea* rhizomes showed better anthelmintic activity against *Onchocerca ochengi* male than ivermectin, respectively [58].

Bone Healing and Antiarthritic Activity

In vitro studies revealed that the alcoholic extract of *Cissus quadrangularis* has potent osteoblastic activity, increased trabecular bone thickness, and enhanced the bone fracture healing process. The different fractions (acetone, benzene, chloroform, diethyl ether, petroleum ether and ethyl acetate) were used for the *in vitro* study. The petroleum ether fraction showed efficient anti-osteoporotic activity compared to other fractions. The 500 mg/kg body weight dose of petroleum ether extract from fresh stems showed beneficial effects for the treatment of osteoporosis on 3-month-old female Wistar rats [59]. The petroleum ether extracts enhanced the mineralization of the bone that made the bone stronger. Its ethanolic extract (100mg/kg) raised the level of serum calcium after 7 weeks in the tested Wistar albino rats. This restoration of the serum calcium led to the speeding up of bone healing and bone strengthening without inducing side effects [60].

An osteoinductive herbal scaffold prepared from ethanolic extract of *Cissus quadrangularis,* alginate solution and O-carboxy methyl chitosan showed excellent osteoinductive property for bone tissue regeneration. The presence of the phytosteroids in this plant enhanced the biomineralization process of human mesenchymal stem cells and elevated osteogenic differentiation on the herbal scaffolds [61]. Another bioactive scaffold *Cissus quadrangularis*/chitosan/Na-carboxymethyl cellulose (CQ/CHI/Na-CMC) was also formed. *In vitro* cell culture study showed that scaffold provided osteoinductive property by enhancing the alkaline phosphatase activity of the Sarcoma osteogenic cells without osteogenic media supplement. Scaffold serves as a promising biomaterial for bone tissue engineering [62].

Alkaline phosphatase is a ubiquitous enzyme. It acts as an early biochemical marker for bone cell formation *via* hydrolysis of phosphate esters at alkaline pH. The hexane and aqueous stem extracts of *Cissus quadrangularis* have bioactive secondary metabolites, which elevate bone tissue regeneration. *In vitro* study suggested that these extracts have unique potential to enhance early osteogenesis [63]. *Cissus quadrangularis* could be a good source for recovering the bones during and after menopause by down-regulating proinflammatory cytokines, which are increased after ovariectomy. These beneficial effects are due to its flavonoids [64].

Anticonvulsive Activity

C. sicyoides showed anti-anxiolytic and anti-convulsive properties in mice [65]. *Cissus cornifolia* leaves are used for the treatment of epilepsy. The methanolic leaf extract of *Cissus cornifolia* delayed the convulsion and death at doses of 600 mg/kg and protected 33.33% of mice induced by 4-aminopyridine. At doses of 150 and 300 mg/kg, protected 1/3rd of mice against convulsion induced by pentylenetetrazole and strychnine, respectively [66].

Antimicrobial Activity

Ethyl acetate fractions and methanolic fraction of *C. quadrangularis* stem extracts exhibited antimicrobial activity against gram positive bacteria such as *Bacillus subtilis, Bacillus cereus, Staphylococcus aureus* and *Streptococcus* [24]. Methanolic and acetone root extracts of *C. vitiginea* species were found effective gram positive bacteria having role in skin infections [67]. Medicinal Gel prepared from *C. vitiginea* leaves mixed with certain ingredients exhibited significant antibacterial activity [68].

In vitro study concluded that copper nanoparticles, which are formed by leaf extract of *C. vitiginea* have antibacterial activity against urinary tract infection pathogens, namely *E. coli, Enterococcus* sp., *Proteus* sp. and *Klebsiella* sp. The result showed the plants have a high zone of inhibition against *E. coli* (22.2 mm diameter) and *Enterococcus* sp. (20.3mm diameter) [69].

Two compounds β-sitosterol and sitosterol-β-D-glucopyranoside isolated from aerial parts of *Cissus sicyoides* showed the highest antibacterial activity against *Bacillus subtilis*. The ethyl acetate fraction of methanol extracts of the aerial parts of the plant showed a 14-17 mm inhibition zone at 0.1 mg/ml concentration against *B. subtilis* [70]. *In vitro* study of *Cissus arnottiana* indicated that the methanolic stem extract showed antimicrobial activity against *K. planticola* bacterial (gram negative) extract (13.20 mm inhibition zone) at 80μL concentration [71].

Antilipidemic Activity

100 g of the powdered plant material was taken and suspended in 500 ml of 99% ethanol. *Cissus quadrangularis,* along with *Tribulus terrestris*, showed anti-hyperlipidemic effect on high fat fed albino Wistar rats when fed at a dose of 583 mg/kg body weight for 30 days [72].

Anti-inflammatory Activity

Methanolic extracts of *C. sicyoides* showed *in vitro* anti-allergic properties [73]. There was a significant ear and paw oedema reduction after its application [74]. In another study, protein denaturation bioassay was selected for assessment of the *in vitro* anti-inflammatory studies. For this study, silver nanoparticles (AgNPs) were formed by using leaf extract of *Cissus vitiginea* [75]. The effects of AgNPs and *Cissus vitiginea* on inhibition of bovine serum albumin protein denaturation, respectively, was 88.78% and 82.52% at 500 μg/ml concentration. This study indicated that the synthesis of AgNPs using *Cissus vitiginea* leaf extract may be a good source for developing green nano-medicine for anti-inflammation [76].

Antidiabetic Activity

Cissus rotundifolia shows antidiabetic activity [76]. The aqueous leaf extracts of *C. sicyoides* decreased blood sugar levels in alloxan-induced diabetic rats both in the short-term (7 days) and long-term (30 days) studies [77]. It might also work by increasing the conversion of circulating glucose to glycogen [78] or inhibiting gluconeogenesis [79].

Anticytotoxic Activity

Dose-dependent cytotoxic and genoprotective effects of plant extract of *C. latifolia* have been studied [80]. *C. populnea* extracts were safer and showed no adverse side effects or toxicity even after long-term administration on Rabbits [81].

Other Activities

Cissus assamica shows anti-venom activity by decreasing endothelin-1 and sarafotoxin 6b and has ethnobotanical importance in Southeast Asia [82].

Methanolic extracts of *C. populnea* showed spermatogenic activity and proliferated Sertoli cells TM4 in *in-vitro* studies [83], but human experiments treated for 72 days gave nil results. Extracts of this plant inhibit the sickling of RBCs [84]. *Cissus ibuensis*, proved to be useful in treating gastrointestinal disorders [85].

Various other biological activities of *Cissus* species are tabulated below Table **2**.

CONCLUSION

This review is an attempt to summarize various protocols available for *in vitro* propagation of *Cissus* species and their potential uses. Moreover, this review can also help in restructuring future research possibilities and endeavours. The most pharmaceutically valuable species belonging to this genus are *C. quadrangularis, C. repanda, C. discolor, C. verticillata, C. arnottiana, C. sicyoides, C. populnea, C. cornifolia, C. repens, C. aralioides, C. rotundifolia* and *C. vitiginea* to name a few. Elaborated research is needed as various species are untouched out of these and some have not produced positive micropropagation results to date. Leaf explants have not produced shoots in many *Cissus* species, which also need to be researched.

ABBREVIATIONS

***Cissus* Linn.**	*Cissus* Linnaeus
C. quadrangularis	*Cissus quadrangularis*
NAA	Naphthalene Acetic Acid
BAP	6-Benzyl Amino Purine
2,4-D	2,4 Dichlorophenoxy Acetic Acid
IAA	Indole Acetic Acid
IBA	Indole Butyric Acid
Kn	Kinetin
$AdSo_4$	Adenine Sulphate
mg/Kg	Milligram per Kilogram
mg/L	milligram per litre
µg/L	Microgram per litre
ng/L	Nanogram per Litre
µM	Micromolar
cm	Centimeter
ml	Millilitre
µl	Microlitre

PPM	Parts Per Million
MS	Murashige and Skoog Medium
B_5	Gamborg'S B-5 Basal Medium
MSB_5	Murashige & Skoog Medium Including B5 Vitamins
$Ca\ Cl_2$	Calcium Chloride
RP-HPLC	Reverse Phase High Performance Liquid Chromatography
DPPH	2,2-Diphenyl-1-Picrylhydrazyl.
ZnO	Zinc Oxide
Fe_3o_4	Iron Oxide
IC_{50}	Half Maximal Inhibitory Concentration
CuONP's	Cupric Oxide Nanoparticles
AgNP's	Silver Nanoparticles
pH	Power of Hydrogen
Sp.	Species
RBC's	Red Blood Cells

ACKNOWLEDGEMENTS

The authors would like to thankfully acknowledge the Department of Botany, University of Rajasthan for providing an infrastructure facility to carry out callus induction and *in vitro* propagation work on Genus *Cissus*.

REFERENCES

[1] Airy Shaw HK. A dictionary of the flowering plants and ferns. London: Cambridge University Press 1985.

[2] Eggli UR. Illustrated Handbook of succulent plants: Dicotyledons. Germany: Springer 2002.

[3] Wen J. Vitaceae, The Families and genera of vascular plants. Germany: Springer-Verlag-Berlin 2007.

[4] Deka DK, Lahon LC, Saikia J, Mukit A. Effect of *Cissus quadrangularis* in accelerating healing process of experimentally fractured Radius-Ulna of dog: A preliminary study. Indian J Pharm 1994; 26: 44-8.

[5] Austin A, Jegadeesan M, Gowrishankar R. *In-vitro* screening of *Cissus quadrangularis* L. Variant ii against *Helicobacter pylori*. Anc Sci Life 2003; 23(1): 55-60.
[PMID: 22557114]

[6] Singh G, Rawat P, Maurya R. Constituents of *Cissus quadrangularis*. Nat Prod Res 2007; 21(6): 522-8.
[http://dx.doi.org/10.1080/14786410601130471] [PMID: 17497424]

[7] Sanyal A, Ahmad A, Sastry M. Calcite growth in *Cissus quadrangularis* plant extract, a traditional Indian bone healing aid. Curr Sci 2005; 82(10): 1742-8.

[8] Jainu M, Devi CSS. Effect of *Cissus quadrangularis* on gastric mucosal defensive factors in experimentally induced gastric ulcer : A comparative study with sucralfate. J Med Food 2004; 7(3): 372-6.

[PMID: 15383234]

[9] Enechi OC, Odonwodo I. An assessment of the phytochemical and nutrient composition of the pulverized root of *Cissus quadrangularis* Linn. Bio-Research 2003; 1(1): 63-68p.

[10] Shirley DA, Sen SP. High resolution X-ray photoemission studies on the active constituents of *Cissus quadrangularis* Linn. Curr Sci 1966; 35: 317.

[11] Gupta MM, Verma RK. Lipid constituents of *Cissus quadrangularis.* Phytochemistry 1991; 30(3): 875-8.
[http://dx.doi.org/10.1016/0031-9422(91)85270-A]

[12] Selvan SP, Velavan S. Analysis of bioactive compounds in methanol extract of *Cissus vitiginea* leaf using GC-MS technique. RASAYAN 2015; 8: 443-7.

[13] Chi CY, Wang F, Lei T, Xu SY, Hong AH, Cen YZ. Studies on the chemical constituents from *Cissus pteroclada.*. Zhong Yao Cai 2010; 33(10): 1566-8.
[PMID: 21355191]

[14] Ahmadu AA, Onanuga A, Aquino R. Flavonoid glycosides from the leaves of *Cissus ibuensis* Hook (Vitaceae). Afr J Tradit Complement Altern Med 2010; 7(3): 225-30.
[http://dx.doi.org/10.4314/ajtcam.v7i3.54780] [PMID: 21461150]

[15] Xie YH, Deng P, Zhang YQ, Yu WS. Studies on the chemical constituents from *Cissus assamica.*. Zhong Yao Cai 2009; 32(2): 210-3.
[PMID: 19504963]

[16] Wang YH, Zhang ZK, He HP, *et al.* Stilbene *C* -glucosides from *Cissus repens.* J Asian Nat Prod Res 2007; 9(7): 631-6.
[http://dx.doi.org/10.1080/10286020600979548] [PMID: 17943557]

[17] Celiktas OY, Gurel A, Sukan FV. Large scale cultivation of plant cell and tissue culture in bioreactors. Transworld Research Network 2010; 1: 54.

[18] Sree NV, Udayasri PVV, Kumar YA, Babu BR, Kumar YP, Varma MV. Advancements in the production of secondary metabolites. J Nat Prod 2010; 3: 112-23.

[19] Oben J, Kuate D, Agbor G, Momo C, Talla X. The use of a *Cissus quadrangularis* formulation in the management of weight loss and metabolic syndrome. Lipids Health Dis 2006; 5(1): 24.
[http://dx.doi.org/10.1186/1476-511X-5-24] [PMID: 16948861]

[20] Oben JE, Enyegue D, Fomekong GI, Soukontoua YB, Agbor GA. The effect of *Cissus quadrangularis (*CQR-300) and a *Cissus* formulation (CORE) on obesity and obesity-induced oxidative stress. Lipids Health Dis 2007; 6(1): 4.
[http://dx.doi.org/10.1186/1476-511X-6-4] [PMID: 17274828]

[21] Oben JE, Ngondi JL, Momo CN, Agbor GA, Sobgui C. The use of a *Cissus quadrangularis/Irvingia gabonensis* combination in the management of weight loss: A double-blind placebo-controlled study. Lipids Health Dis 2008; 7(1): 12.
[http://dx.doi.org/10.1186/1476-511X-7-12] [PMID: 18377661]

[22] Balasubramanian P, Rajasekaran A, Prasad SN. Folk medicine of the Irulas of Coimbatore forests. Anc Sci Life 1997; 16(3): 222-6.
[PMID: 22556796]

[23] Chidambaram J, Carani Venkatraman A. *Cissus quadrangularis* stem alleviates insulin resistance, oxidative injury and fatty liver disease in rats fed high fat plus fructose diet. Food Chem Toxicol 2010; 48(8-9): 2021-9.
[http://dx.doi.org/10.1016/j.fct.2010.04.044] [PMID: 20450951]

[24] Chidambara Murthy KN, Vanitha A, Mahadeva Swamy M, Ravishankar GA. Antioxidant and antimicrobial activity of *Cissus quadrangularis* L. J Med Food 2003; 6(2): 99-105.
[http://dx.doi.org/10.1089/109662003322233495] [PMID: 12935320]

[25] Jainu M, Devi CS. *Invitro* and *In vivo* evaluation of free radical scavenging potential of *Cissus quadrangularis*. Afr J Biomed Res 2005; 8: 95-9.

[26] Jainu M, Devi CSS. Gastroprotective action of *Cissus quadrangularis* extract against NSAID induced gastric ulcer: Role of proinflammatory cytokines and oxidative damage. Chem Biol Interact 2006; 161(3): 262-70.
[http://dx.doi.org/10.1016/j.cbi.2006.04.011] [PMID: 16797507]

[27] Jainu M, Vijai Mohan K, Shyamala Devi CS. Gastroprotective effect of *Cissus quadrangularis* extract in rats with experimentally induced ulcer. Indian J Med Res 2006; 123(6): 799-806.
[PMID: 16885602]

[28] Swamy AHMV, Kulkarni R, Thippeswamy AHM, Koti BC, Gore A. Evaluation of hepatoprotective activity of *Cissus quadrangularis* stem extract against isoniazid-induced liver damage in rats. Indian J Pharmacol 2010; 42(6): 397-400.
[http://dx.doi.org/10.4103/0253-7613.71920] [PMID: 21189914]

[29] Jainu M, Mohan KV. Protective role of ascorbic acid isolated from *Cissus quadrangularis* on NSAID induced toxicity through immunomodulating response and growth factors expression. Int Immunopharmacol 2008; 8(13-14): 1721-7.
[http://dx.doi.org/10.1016/j.intimp.2008.08.005] [PMID: 18773975]

[30] Panthong A, Supraditaporn W, Kanjanapothi D, Taesotikul T, Reutrakul V. Analgesic, anti-inflammatory and venotonic effects of *Cissus quadrangularis* Linn. J Ethnopharmacol 2007; 110(2): 264-70.
[http://dx.doi.org/10.1016/j.jep.2006.09.018] [PMID: 17095173]

[31] Nalini G, Vinoth PV, Chidambaranathan N, Jeyasundari K. Evaluation of anti-tumour activity of *Cissus quadrangularis* L. against Dalton's ascitic lymphoma and Ehrlich ascetic induced carcinoma in mice. Int J Pharm Sci Rev Res 2011; 8: 75-9.

[32] Jadhav A, Rafiq M, Devanathan R, *et al.* Ketosteroid standardized *Cissus quadrangularis* L. extract and its anabolic activity: Time to look beyond ketosteroid? Pharmacogn Mag 2016; 12(46) (Suppl. 2): 213.
[http://dx.doi.org/10.4103/0973-1296.182177] [PMID: 27279709]

[33] Poornima RK, Vidya SM. Ethnomedicinal importance of rare family member of Vitaceae, *Cissus elongata*. Roxb Conference on Conservation and sustainable management of ecologically sensitive regions in Western Ghats. 409-16.

[34] Nathawat RS. *In vitro* propagation and phytochemical studies on fracture healing plants: *Cissus quadrangularis* L. and *Lepidium sativum* L. Ph.D. Thesis submitted to University of Rajasthan Jaipur. 2013.

[35] Anand M, Kaur T. Effect of growth regulators on callus induction and organ formation in *Cissus quadrangularis* Linn. A valuable medicinal plant. Intl J Pharma and Clin Res 2019; 11(4): 106-11.

[36] Rebouças FS. *In vitro* cultivation of medicinal plants: *Ocimum basilicum* L. i.e. *Cissus sicyoides* l. (Dissertation) (Federal University of Recôncavo da Bahia Center for Environmental and Biological Agricultural Sciences Postgraduate Program in Agricultural Sciences Master's Dissertation,) Brazil 2013.

[37] Dos Santos MR, da Rocha JF, Paz ES, Smozinski CV, de Oliveira Nogueira W, Guimarães MD. Callus induction in leaf explants of *Cissus verticillata* (L.) Nicolson & CE Jarvis. Plant Cell Culture & Micropropagation 2014; 10(2): 41-6.

[38] Gour K. Biochemical investigations on some bone healing plants of semi-arid lands belonging to Vitaceae. Jaipur: Thesis submitted to University of Rajasthan 2013.

[39] Murashige T, Skoog F. A revised medium for rapid growth and bioassay with tobacco cultures. Physiol Plant 1962; 15(3): 473-97.
[http://dx.doi.org/10.1111/j.1399-3054.1962.tb08052.x]

[40] Garg P, Malik CP. Multiple shoot formation and efficient root induction in *Cissus quadrangularis.* Int J Pharma Cli Res 2012; 4(1): 4-10.

[41] Abreu IN, Pinto JE, Bertolucci SK, Geromel C, Castro EM. Evaluation of different concentrations of auxins and explant types in the induction and growth of callus of (*Cissus sicyoides* L.) a medicinal plant. Revista Brasileira de plantas Medicinais 2003; 5(2): 83-9.

[42] Anand M, Kaur T. A micropropagation system for the cloning of *Cissus quadrangularis* : A valuable medicinal plant. Medicinal Plants : Int J Phytomed Rel Indus 2018; 10(2): 125-32. [http://dx.doi.org/10.5958/0975-6892.2018.00020.5]

[43] Ramar K, Ganesan M, Lakshmi Prabha A, Nandagopalan V. *in vitro* clonal propagation of wild *Cissus quadrangularis* by suspension culture mediated somatic embryogenesis. Int J Appl Biotechnol Biochem 2011; 1(1): 1-14.

[44] Sharma N, Nathawat RS, Gour K, Patni P. Establishment of callus tissue and effect of growth regulators on enhanced sterol production in *Cissus quadrangularis* L. Int. J Pharm 2011; 7(5): 653-8.

[45] Cholke PB, Bhalerao DV, Shete AM, Bhor AK. The effect of $CaCl_2$ induced stress on callus of *in-vitro* propagated *Cissus quadrangularis* Linn. Shodhasamhita 2022; 9(5): 1.

[46] Grever MR, Schepartz SA, Chabner BA. The national cancer institute: Cancer drug discovery and development program. Semin Oncol 1992; 19(6): 622-38. [PMID: 1462164]

[47] Gnanasundaram I, Balakrishnan K. *In vitro* antioxidant activity of *Cissus vitiginea* leaves and its silver nanoparticles. World J Pharm Res 2018; 7: 997-1006.

[48] Sáenz MT, Garcia MD, Quilez A, Ahumada MC. Cytotoxic activity of *Agave intermixta* L. (Agavaceae) and *Cissus sicyoides* L. (Vitaceae). Phytother Res 2000; 14(7): 552-4. [http://dx.doi.org/10.1002/1099-1573(200011)14:7<552::AID-PTR639>3.0.CO;2-U] [PMID: 11054850]

[49] Ferreira MP, Nishijima CM, Seito LN, *et al.* Gastroprotective effect of *Cissus sicyoides* (Vitaceae): Involvement of microcirculation, endogenous sulfhydryls and nitric oxide. J Ethnopharmacol 2008; 117(1): 170-4. [http://dx.doi.org/10.1016/j.jep.2008.01.008] [PMID: 18304768]

[50] Ruskin RS, Kumari VP, Gopukumar ST, Praseetha PK. Evaluation of phytochemical, antibacterial and anti-cancerous activity of *Cissus quadrangularis* from South Western Ghats regions of India. Int J Pharm Sci Rev Res 2014; 28(1): 12-5.

[51] Abinaya P, Sampathkumar P, Sumathi S, Poornima A, Parthasarathy S, Parthasarathy S. Analyses of *In-vitro* antioxidant and anticancer activity of *Cissus quadrangularis* stem extract in osteoblastic cell line -UMR- 106. Int J Res Pharmac Sci 2020; 11(4): 5293-300. [http://dx.doi.org/10.26452/ijrps.v11i4.3148]

[52] Parimala S, Selvan TA. Anticancer and antioxidant activity of *Cissus pallida* and *Cissus vitiginea.* J Pharmacogn Phytochem 2017; 6(4): 1521-6.

[53] Méndez-López LF, Garza-González E, Ríos MY, *et al.* Metabolic profile and evaluation of biological activities of extracts from the stems of *Cissus trifoliata.* Int J Mol Sci 2020; 21(3): 930. [http://dx.doi.org/10.3390/ijms21030930] [PMID: 32023823]

[54] Lucena FRS, Almeida ER, Aguiar JS, Silva TG, Souza VMO, Nascimento SC. Cytotoxic, antitumor and leukocyte migration activities of resveratrol and sitosterol present in the hidroalcoholic extract of *Cissus sicyoides* L., Vitaceae, leaves. Rev Bras Farmacogn 2010; 20(5): 729-33. [http://dx.doi.org/10.1590/S0102-695X2010005000002]

[55] Bhujade A, Gupta G, Talmale S, Das SK, Patil MB. Induction of apoptosis in A431 skin cancer cells by *Cissus quadrangularis* Linn stem extract by altering Bax–Bcl-2 ratio, release of cytochrome c from mitochondria and PARP cleavage. Food Funct 2013; 4(2): 338-46.

[http://dx.doi.org/10.1039/C2FO30167A] [PMID: 23175101]

[56] Thakar MA, Saurabh Jha S, Phasinam K, Manne R, Qureshi Y, Hari Babu VV. X ray diffraction (XRD) analysis and evaluation of antioxidant activity of copper oxide nanoparticles synthesized from leaf extract of *Cissus vitiginea.* Mater Today Proc 2022; 51: 319-24. [http://dx.doi.org/10.1016/j.matpr.2021.05.410]

[57] Pathak AK, Kambhoja S, Dhruv S, Singh HP, Chand H. Anthelimintic activity of *Cissus quadrangularis* Linn stem. Pharmacologyonline 2010; 3: 15-8.

[58] Nyemb J, Ndoubalem R, Talla E, *et al.* DPPH antiradical scavenging, anthelmintic and phytochemical studies of *Cissus poulnea* rhizomes. Asian Pac J Trop Med 2018; 11(4): 280. [http://dx.doi.org/10.4103/1995-7645.231468]

[59] Potu BK, Rao MS, N GK, Bhat KMR, Chamallamudi MR, Nayak SR. Petroleum ether extract of *Cissus quadrangularis* (LINN) stimulates the growth of fetal bone during intra uterine developmental period: a morphometric analysis. Clinics 2008; 63(6): 815-20. [http://dx.doi.org/10.1590/S1807-59322008000600018] [PMID: 19061006]

[60] Ramachandran S, Fadhil L, Gopi C, Amala M, Dhanaraju MD. Evaluation of bone healing activity of *Cissus quadrangularis* (Linn), *Cryptolepis buchanani*, and *Sardinella longiceps* in Wistar rats. Beni Suef Univ J Basic Appl Sci 2021; 10(1): 1-9.

[61] Soumya S, Sajesh KM, Jayakumar R, Nair SV, Chennazhi KP. Development of a phytochemical scaffold for bone tissue engineering using *Cissus quadrangularis* extract. Carbohydr Polym 2012; 87(2): 1787-95. [http://dx.doi.org/10.1016/j.carbpol.2011.09.094]

[62] Tamburaci S, Kimna C, Tihminlioglu F. Novel phytochemical *Cissus quadrangularis* extract–loaded chitosan/Na-carboxymethyl cellulose–based scaffolds for bone regeneration. J Bioact Compat Polym 2018; 33(6): 629-46. [http://dx.doi.org/10.1177/0883911518793913]

[63] Nair PR, Sreeja S, Sailaja GS. *In vitro* biomineralization and osteogenesis of *Cissus quadrangularis* stem extracts: An osteogenic regulator for bone tissue engineering. J Biosci 2021; 46(4): 88. [http://dx.doi.org/10.1007/s12038-021-00206-x] [PMID: 34544907]

[64] Guerra JM, Hanes MA, Rasa C, *et al.* Modulation of bone turnover by *Cissus quadrangularis* after ovariectomy in rats. J Bone Miner Metab 2019; 37(5): 780-95. [http://dx.doi.org/10.1007/s00774-018-0983-3] [PMID: 30756174]

[65] de Almeida ER, de Oliveira Rafael KR, Couto GBL, Ishigami ABM. Anxiolytic and anticonvulsant effects on mice of flavonoids, linalool, and α-tocopherol presents in the extract of leaves of *Cissus sicyoides* L. (Vitaceae). J Biomed Biotechnolo. 2009.

[66] Yaro AH, Musa AM, Magaji MG, Nazifi AB. Anticonvulsant potentials of methanol leaf extract of *Cissus cornifolia* Planch (Vitaceae) in mice and chicks. Intern J Herbs and Pharmacol Res 2015; 4(2): 25-32.

[67] Kumar RB, Suryanarayana B. Antimicrobial screening of some selected tribal medicinal plants from Sriharikota Island. Andhra Pradesh, India: Ethnobot. Leaflets 2008; pp. 1269-82.

[68] Kavitha K, Kavitha T, Fathima KR. Preparation and evaluation of herbal gel formulation of *Cissus vitiginea* leaf. J Pharma Res 2018; 7: 179-81.

[69] Wu S, Rajeshkumar S, Madasamy M, Mahendran V. Green synthesis of copper nanoparticles using *Cissus vitiginea* and its antioxidant and antibacterial activity against urinary tract infection pathogens. Artif Cells Nanomed Biotechnol 2020; 48(1): 1153-8. [http://dx.doi.org/10.1080/21691401.2020.1817053] [PMID: 32924614]

[70] Beltrame FL, Pessini GL, Doro DL, Dias Filho BP, Bazotte RB, Cortez DAG. Evaluation of the antidiabetic and antibacterial activity of *Cissus sicyoides.* Braz Arch Biol Technol 2002; 45(1): 21-5. [http://dx.doi.org/10.1590/S1516-89132002000100004]

[71] Selvi P, Murugesh S, Yuvarajan RY, Rajasekar AR. Screening the therapeutic potential of methanolic stem extract of *Cissus arnottiana.* Biomed Pharmacol J 2021; 14(3): 1405-13. [http://dx.doi.org/10.13005/bpj/2243]

[72] Jiji MJ, Visalakshi S, Meenakshi P, *et al.* Antilipidemic activity of *Cissus quadrangularis* and *Tribulus terrestris* on obesity in high fat fed rats. Pharmacologyonline 2009; 2: 1250-8.

[73] Quílez AM, Saenz MT, García MD, de la Puerta R. Phytochemical analysis and anti-allergic study of *Agave intermixta* Trel. and *Cissus sicyoides* L. J Pharm Pharmacol 2010; 56(9): 1185-9. [http://dx.doi.org/10.1211/0022357044102] [PMID: 15324488]

[74] García MD, Quílez AM, Sáenz MT, Martínez-Domínguez ME, de la Puerta R. Anti-inflammatory activity of *Agave intermixta* Trel. and *Cissus sicyoides* L., species used in the Caribbean traditional medicine. J Ethnopharmacol 2000; 71(3): 395-400. [http://dx.doi.org/10.1016/S0378-8741(00)00160-4] [PMID: 10940576]

[75] Gnanasundaram I, Balakrishnan K. Synthesis and evaluation of anti-inflammatory activity of silver nanoparticles from *Cissus vitiginea* leaf extract. J Nanoscience and Technology 2017; 3(3): 266-9.

[76] Onyechi UA, Judd PA, Ellis PR. African plant foods rich in non-starch polysaccharides reduce postprandial blood glucose and insulin concentrations in healthy human subjects. Br J Nutr 1998; 80(5): 419-28. [http://dx.doi.org/10.1017/S0007114598001482] [PMID: 9924263]

[77] Viana GSB, Medeiros ACC, Lacerda AMR, Leal LKAM, Vale TG, Matos FJ. Hypoglycemic and anti-lipemic effects of the aqueous extract from *Cissus sicyoides.* BMC Pharmacol 2004; 4(1): 9. [http://dx.doi.org/10.1186/1471-2210-4-9] [PMID: 15182373]

[78] Salgado JM, Mansi DN, Gagliardi A. *Cissus sicyoides*: Analysis of glycemic control in diabetic rats through biomarkers. J Med Food 2009; 12(4): 722-7. [http://dx.doi.org/10.1089/jmf.2008.0157] [PMID: 19735170]

[79] Pepato MT, Baviera AM, Vendramini RC, Perez MPMS, Kettelhut IC, Brunetti IL. *Cissus sicyoides* (princess vine) in the long-term treatment of streptozotocin-diabetic rats. Biotechnol Appl Biochem 2003; 37(1): 15-20. [http://dx.doi.org/10.1042/BA20020065] [PMID: 12578546]

[80] Rubeena M, John ET. Genotoxicity evaluation of *Cissus latifolia* lam. and its genoprotective effect on oxidative damage induced by hydrogen peroxide. Asian J Pharma CliRes 2020; 13(7): 185-91.

[81] Ojekale AB, Lawal OA, Lasisi AK, Adeleke TI. Phytochemisty and spermatogenic potentials of aqueous extract of *Cissus populnea* (Guill. and Per) stem bark. ScientWorldJ 2006; 6: 2140-6. [http://dx.doi.org/10.1100/tsw.2006.343] [PMID: 17370009]

[82] Yang LC, Wang F, Liu M. A study of an endothelin antagonist from a Chinese anti-snake venom medicinal herb. J Cardiovasc Pharmacol 1998; 31(1) (1): S249-50. [http://dx.doi.org/10.1097/00005344-199800001-00070] [PMID: 9595451]

[83] Osibote E, Noah N, Sadik O, McGee D, Ogunlesi M. Electrochemical sensors, MTT and immunofluorescence assays for monitoring the proliferation effects of *cissus populnea* extracts on Sertoli cells. Reprod Biol Endocrinol 2011; 9(1): 65. [http://dx.doi.org/10.1186/1477-7827-9-65] [PMID: 21575213]

[84] Moody JO, Ojo OO, Omotade OO, Adeyemo AA, Olumese PE, Ogundipe OO. Anti-sickling potential of a Nigerian herbal formula (ajawaron HF) and the major plant component (*Cissus populnea* L. CPK). Phytother Res 2003; 17(10): 1173-6. [http://dx.doi.org/10.1002/ptr.1323] [PMID: 14669251]

[85] Irvine FR. Woody Plants of Ghana. Oxford University Press 1961; pp. 300-1.

[86] Banu J, Varela E, Bahadur AN, Soomro R, Kazi N, Fernandes G. Inhibition of bone loss by *Cissus quadrangularis* in mice: A preliminary report. J osteoporosis 2012; 2012: 1-10.

[87] Ghosh T, Zarif Morshed M, Islam N, Al Masud KN, Akter M, Islam R. *In-vitro* investigation of antioxidant activity of *Cissus adnata* in different fractions. J Pharmacogn Phytochem 2018; 7(2): 2625-8.

[88] San HT, Boonsnongchee P, Putalun W, Sritularak B, Likhitwitayawuid K. Bergenin from *Cissus javana* DC. (Vitaceae) root extract enhances glucose uptake by rat L6 myotubes. Trop. J Pharma Res 2020; 19(5): 1081-6.

[89] Kolap RM, Gulave AB, Zimare SB. Bioprospecting of the underutilized endemic taxon *Cissus woodrowii* (Stapf ex Cooke) Santapau for its antioxidant activity and phenolic profiling. Indian J Nat Prod Resour 2011; 11(4): 250-9.

[90] Beltrame F, Ferreira A, Cortez D. Coumarin glycoside from *Cissus sicyoides.* Nat Prod Lett 2002; 16(4): 213-6.
[http://dx.doi.org/10.1080/10.575630290015736] [PMID: 12168753]

[91] Chipiti T, Ibrahim MA, KOoorbanally NA, Islam MS. *In vitro* antioxidant activity and GC-MS analysis of the ethanol and aqueous extracts of *Cissus cornifolia* (Baker) Splanch (Vitaceae) parts. Acta Pol Pharm 2015; 72(1): 119-27.
[PMID: 25850207]

[92] Kumar VS, Satyanarayana T, Mathew A, Chandrasekhar S, Rajendra G. *In-vivo* anti-inflammatory activity of methanolic extract of *Cissus repanda.* Indian J Res Pharmacy and Biotechnology 2013; 1(5): 668.

[93] Salem PPO, Vieira NB, Garcia DA, *et al.* Anti-urolithiatic and anti-inflammatory activities through a different mechanism of actions of *Cissus gongylodes* corroborated its ethnopharmacological historic. J Ethnopharmacol 2020; 253: 112655.
[http://dx.doi.org/10.1016/j.jep.2020.112655] [PMID: 32045681]

[94] Chan YY, Wang CY, Hwang TL, *et al.* The constituents of the stems of *Cissus assamica* and their bioactivities. Molecules 2018; 23(11): 2799.
[http://dx.doi.org/10.3390/molecules23112799] [PMID: 30373325]

[95] Al-Mehdar AA, Al-Battah AM. Evaluation of hypoglycemic activity of *Boswellia carterii* and *Cissus rotundifolia* in streptozotocin/nicotinamide-induced diabetic rats. Yemeni J Med Sci 2016; 10(1): 30-8.
[http://dx.doi.org/10.20428/yjms.v10i1.959]

[96] Alzoreky NS, Nakahara K. Antibacterial activity of extracts from some edible plants commonly consumed in Asia. Int J Food Microbiol 2003; 80(3): 223-30.
[http://dx.doi.org/10.1016/S0168-1605(02)00169-1] [PMID: 12423924]

[97] Assob JCN, Kamga HLF, Nsagha DS, *et al.* Antimicrobial and toxicological activities of five medicinal plant species from Cameroon Traditional Medicine. BMC Complement Altern Med 2011; 11(1): 70.
[http://dx.doi.org/10.1186/1472-6882-11-70] [PMID: 21867554]

[98] Line-Edwige M, Raymond Ft, François E, Edouard NE. Antiproliferative effect of alcoholic extracts of some Gabonese medicinal plants on human colonic cancer cells. Afr J Tradit Complement Altern Med 2009; 6(2): 112-7.
[PMID: 20209001]

CHAPTER 2

Micropropagation of *Juglans regia* L.

Sakshi Juyal[1], **Megha Rawat**[1], **Priya Saini**[1], **Ayushi Negi**[1] and **Manu Pant**[1,*]

[1] *Department of Life Sciences, Graphic Era (Deemed to be University), 566/6, Bell Road, Clement Town, Dehradun-248002, Uttarakhand, India*

Abstract: *Juglans regia* L., commonly called walnut, is a nutrient-rich fruit. Besides many therapeutic properties, the plant is highly valued for its timber, which along with the fruit-nut, fetches a very high demand in the domestic and international markets. The ever increasing demand for these plant products is not being sufficed by the existing supplies. This is owing to the fact that conventional methods of walnut propagation are time and space-consuming and show limited responsiveness. Walnut cuttings are also difficult to root, making large-scale propagation a challenge. Consequently, walnut micropropagation has become extremely important to ensure rapid mass production of selected cultivars in a small space, and for an indefinite time period. The tissue-culture-raised products are robust, disease-free, and have desirable characteristics. The aim of this chapter is to compile information on tissue culture studies on *Juglans regia* with a special focus on the latest developments in the field. The chapter covers various pathways employed for the *in vitro* propagation of walnut, hardening, and acclimatization of tissue culture raised plantlets to ensure better quality, quantity, and sustainability of walnut trees to meet the demand of the growing global population.

Keywords: Callus Culture, Embryo Culture, Hardening, *In vitro* Rooting, *Juglans regia*, Nodal Explant Culture, Somatic Embryos, Walnut.

INTRODUCTION

Juglans regia L. (family Juglandaceae) is one of the most popular fruit-nut trees. Commonly known as Persian walnut, English walnut or walnut, is native to the region that runs eastward from the Balkans to the western Himalayan Mountains [1]. The edible seeds of the drupe on any Juglans tree are known as walnuts. The grain is encircled by the shell. It typically has a fibrous shell barrier that breaks as it ages, dividing it into two halves. The brown seed coat that surrounds the seeds, which are often sold as walnuts without the shell, contains antioxidants. This nut is referred to as the King of Nuts because of its ultimate benefits.

[*] **Corresponding author Manu Pant:** Department of Life Sciences, Graphic Era (Deemed to be University), 566/6, Bell Road, Clement Town, Dehradun-248002, Uttarakhand, India; E-mail: manupant@geu.ac.in

T. Pullaiah (Ed.)

Walnuts contain significant amounts of fats, proteins, vitamins, and minerals, making them nutrient-dense food and accounting for their extensive usage in the traditional system of medicine. The average amount of protein in walnuts is around 18%, wherein glutelins make up around 70% of the total protein in seeds, and have 18% globulins, 7% albumins, and 5% prolamins [2 - 6]. Potassium, magnesium, and calcium are known to be the most prominent nutrients in walnuts abundance in walnut cultivars [7 - 9]. They are also rich in phenolic acids, related polyphenols, pectic compounds, sterols, flavonoids, and sterols. The nutritional composition varies between cultivars and varies with factors like genotype, cultivator, environment, and soil type [10 - 12].

The phytochemically active ingredients of the species are responsible for its wide range of therapeutic properties like antioxidant, anti-hypertensive, antihistamine, bronchial relaxant, analgesic, neuroprotective, immuno-modulator, antiulcer, antidiabetic, hepatoprotective, antibacterial, anti-inflammatory, lipolytic, dental-cavity preventive, and many others. Walnuts have high amounts of omega-6 and omega-3 polyunsaturated fatty acids, which prevent heart-related ailments [13]. The high protein and oil content of the walnut kernels has led to the FAO (Food And Agriculture Organization) designating the walnut as a priority plant and a key species for human nutrition [14]. Besides, the wood is robust and solid, making it a top choice for high-end furniture and woodwork. Due to its significant commercial value and great health advantages, it is a crop that is in high demand in both domestic and international markets.

Conventional methods of walnut propagation by cutting or grafting are cumbersome and less productive, as is the case with many other woody plant species. One of the cutting-edge and important techniques for plant multiplication is the tissue culture technique, sometimes referred to as micropropagation [15 - 18]. A lot of plants may be produced quickly using the micropropagation method, which is especially advantageous when compared to vegetative propagation or when regeneration species like walnuts are difficult to regenerate naturally.

The present chapter discusses the different methods of mass propagation of *Juglans regia* using different techniques of plant tissue culture.

PLANT TISSUE CULTURE GROWTH MEDIUM FOR WALNUT

One of the most crucial elements of plant cell and tissue culture is the growth media. Different culture media have been employed for the micropropagation of walnut, including Murashige and Skoog medium (MS), Driver and Kuniyuki medium (DKW) and Woody plant medium (WPM) with varying success. The most used media for walnut tissue culture is the DKW medium, which was created for the cultivar paradox and has proven useful for a number of

Juglandaceae species, including *J. regia* [19 - 22]. However, a number of researchers have successfully cultured Persian walnut using MS media [23 - 26]. The suitability of DKW medium for walnut propagation is attributed to the fact that walnut requires high salt media for shoot multiplication, a condition which is provided by DKW medium [27 - 30].

Countering Metabolic Exudates

Growing walnuts has been significantly hampered by the phytotoxic exudates that are produced by recently cultivated explants [31]. Metabolic byproducts frequently cause the culture media underneath the explant to become darker. To regulate these exudes, a number of techniques have been explored. These include soaking the seed explants in water for twelve hours [32], use of antioxidants like cPIBA (cyclo-pentano-isobutyric acid), TIBA (2,3,5 tri-iodobenzoic acid), ascorbic acid, phloroglucinol, sodium thiosulfate, dithiothreitol and charcoal [33, 34]. A study with mature tissues (glabrous shoot tips, internodal segments) of field-grown *J. regia* highlighted the use of fungicide solution (Captan + Benomil 1g/l each) and antioxidant solution (20mg/l cysteine + 5mg/l ascorbic acid) to establish healthy cultures [35].

In general, periodic transfer of explants to new media has proven to be more effective in eliminating the exudate problem. The process is continued till the release of exudates from the cultures subsides.

APPROACHES FOR *IN VITRO* PROPAGATION OF *JUGLANS REGIA*

Nodal Explant Culture

Juglans regia and hybrids are often micropropagated from shoots, branches, zygotic embryos, or adult plant scions [29]. In one of the initial studies, incubation of explants in a modified MS medium supplemented with BAP for 4 weeks resulted in longer axillary buds from cultured explants [36]. A study [23] was carried out to improve walnut micropropagation procedures using embryonic and juvenile nodal explants. MS medium supplemented 1.0 mg/l IBA was most optimal for *in vitro* shoot development. The study showed that higher levels of cytokinin BAP(6-Benzylaminopurine) induced deformation of leaves and shoots, eventually causing vitrification, while GA_3(Gibberellic acid) promoted elongation of shoots. Another study [16] showed that a modified MS medium with BAP proved to be most optimal for shoot induction from immature herbaceous shoots of *Juglans regia.* However, for further development, DKW medium containing IBA,2.0mg/l and riboflavin1.0mg/l proved to be most optimal when cultures were given a ten-day darkness period. Through this system, *in vitro* regenerated plantlets could be transplanted after 4-5 weeks. A study on *in vitro* propagation of

J. regia [30] confirmed the superiority of DKW medium solidified with Phytagel over MS and WPM. The same study also showed that a combination of BAP and IBA (Indolebutyric acid) was most suitable for shoot elongation.

The use of IBA in *J. regia* shoot induction and multiplication has also been confirmed by other species. Heile-Sudholt *et al.* [37] reported that the use of IBA in DKW medium produced the greatest number of axillary shoots per explant, which is consistent with the findings of Gruselle *et al.* [15], who found that IBA at 0.1 mg/1 was more beneficial for Persian walnut shoot proliferation.

In a contrasting study, Bosela and Michler [38] zeatin was shown to give improved results over BAP at levels ranging from 2.5 to 25 M for fast shoot elongation of seedling black walnut nodal sections. The study also showed that higher concentrations of zeatin and BAP in the medium were responsible for shoot necrosis and subsequent *in vitro* rooting. Lately, a study [39] on elite black walnut genotypes showed that the best system for nodal explant-based micropropagation was culturing on DKW medium containing BAP, IBA, and additives casein hydrolysate, adenine hemisulfate. The study also showed that *in vitro* produced shoots were cultivated in liquid initiation media in 3-L polycarbonate on a rotary shaker (100 rpm) under a 16-hr photoperiod at 25^{0}C, long-term survival, and proliferation of microshoots could be accomplished. Another micropropagation of an elite genotype (Mj209xRa) of walnut was performed by Licea-Moreno *et al.* [40], through a nodal explant. They obtained desired results by introducing explants to a corrected formulation DKW-C medium supplemented with 4.4 μM of BAP, 0.5 μM of IBA and 60mM of Phloroglucinol (PG), they also followed an advanced procedure of temporary immersion of microshoots in a self-designed bioreactors, these had a liquid culture medium provided with 4.4 μM of BAP, 0.5 μM of IBA and 40mM of PG. They aimed for a better clonal propagation of walnut plantlets of high timber-producing potential.

Another study [41] determines that DKW medium when supplemented with 3% sucrose and varied combination of BAP, IBA and GA_3 has a positive effect on bud proliferation and also serves well for other purposes such as embryo germination and callus development. Yegizbayeva *et al.* [42] also confirmed the superiority of DKW medium for walnut shoot culture while working on four different *J. regia* varieties (Fig. **1**).

Fig. (1). (a-d) *In vitro* establishment of shoot cultures and shoot multiplication in *J. regia* [42].

Callus Culture

Refers to the *in vitro* culture of de-differentiated plant cells that are induced on media that often include relatively high auxin concentrations or a combination of auxin and cytokinin. It is particularly challenging to sustain continuous cultivation in the *Juglans* genus due to the browning that occurs after the callus emerges from the primary explant. However, callus production in walnut has frequently been observed when embryogenic explants are grown to acquire other morphogenetic responses. The amount of BAP supplied is shown to have a stronger impact on callus production in mature embryos grown on MS or WP media, generating several shoots simultaneously [43]. When cytokinin and kinetin were added to MS medium containing CH and NAA (Napthalene acetic acid), optimal callus development was observed with mature and immature endosperm [44]. A study investigated the effect of culture medium on the caulogenic response of walnut leaves [45]. Out of the different media tried, broadleaf tree medium (BTM) supplemented with NAA and BAP proved to be the most efficient, followed by WPM.

The effect of different plant growth regulators on callus induction from *in vitro* generated leaves has been studied [46]. The study revealed that cytokinin TDZ (N-phenyl-1, 2, 3-thidiazole-5yl urea) and auxins IBA, NAA and 2,4-D were capable of callus induction when used at low concentrations, while high concentrations were inhibitory for healthy callus development.

The efficiency of DKW medium on callus induction from shoots of different genotypes of walnut has also been established [47]. The study showed that both BAP and IBA have the capacity to induce callus, however, the effect varies with the genotype. Similar observations have been reported by Nikolova *et al.* [48] where DKW medium was suitable to initiate callus from leaves when supplemented with BAP and NAA. Fig. (**2**) shows the effect of different plant growth regulators on the morphology of callus induced from leaf explants of *J. regia*.

Fig. (2). Morphological variations in walnut callus with different plant growth regulators in the culture medium.

Embryo Culture

Embryo culture is the sterile isolation and *in vitro* development of a developing or mature embryo with the goal of yielding a healthy plant. It involves aseptically removing embryos and cultivating them on a medium, primarily aiding in lowering post-fertilization hurdles to hybridization [49] and helping plants to multiply more effectively. Additionally, it shortens the generation cycle in plants like walnut, where seeds need to be stratified for two to three months before germination [29, 50, 51]. The germination of walnut embryos is regulated by a number of variables, including physical aspects, media cultures, plant development regulators, *etc*. Kaur *et al*. [52] showed that MS medium fortified with kinetin, gibberellic acid and BAP proved to be the best for embryo germination, wherein the cultured explants responded within 15 days of culture. A study Payghamzadeh and Kazemitabar [26, 53] showed that modified DKW medium with gibberellic acid can be used for effective embryo germination in walnut under *in vitro* cold and dark conditions. Another study [54] showed that BAP and IBA have a positive impact on *in vitro* embryo germination. A study [55] showed that embryos from a selected *J. regia* cultivar exhibited the best germination and proliferation rate on WPM medium supplemented with BAP. Toosi and Dilmagani [18] showed that as compared to other media (MS, WPM, DKW, B5(Gamborg medium, Lyod and Mc Cown, Cheng), NGE media proved to be most optimal for embryo germination in Persian walnut, when tried with BAP and IBA combination. Myrselaj *et al*. [56] in their study on walnut regeneration from mature zygotic embryos showed that although germination percentage on DKW and MS media was not statistically different, MS medium containing Kn (kinetin) was found to give a better organogenic response.

Conclusively, media requirements for embryo culture in *J. regia* seem to vary with the plant genotype, and all the different types of nutrient media have been reported to show better responses in experiments conducted in different parts of the world.

Somatic Embryogenesis

Somatic embryogenesis is the process of developing embryos from cultured somatic tissues. The generated embryos are called somatic embryos and parallel the developmental path of zygotic embryos. The pathway is advantageous over organogenic differentiation as it produces entire plants rather than microshoots that have to be subsequently rooted. Tulecke and McGranahan [57] created the first walnut somatic embryos from immature cotyledon explants, while another study showed the ability of immature zygotic embryos to be converted to somatic embryos [58]. As per these studies, basal DKW medium proves to be suitable for somatic embryogenesis [59]. Chenevard *et al*. [60] showed that embryogenic reactivity can be observed within 4 weeks of culturing and developed somatic embryos can be germinated within three weeks on DKW medium under dark conditions. Another study [61] established that somatic embryos can be obtained from immature zygotic embryos of English walnut. They showed that the desiccation of somatic embryos by pretreatment with calcium chloride or calcium nitrate in dark conditions drastically improved somatic embryo germination. Further, a lower concentration of nutrients in the medium favored plantlet development. Breton *et al.* [62] showed that somatic embryogenesis and *in vitro* flowering can be induced in early mature walnut varieties from central Asia using zygotic embryos as the starting material and DKW medium supplemented with sucrose, myoinositol, L-glutamine, BAP, Kinetin, IBA and phytagel as the nutrient medium. A study conducted by Baya *et al.* [63] investigated the effect of abscisic acid (ABA) and sucrose on somatic embryo maturation in *J. regia*. Their research established that 2mg/l ABA significantly increased the maturation of somatic embryos and their conversion rate into plantlets. Somatic embryogenesis can be envisaged as a better option for the rapid production of superior genotypes of otherwise hard-to-root walnuts.

In vitro Rooting and Hardening

Rooting in walnut is highly labor intensive when conventional methods like grafting are used. Even under *in vitro* conditions, microshoots fail to root unless appropriate media and growth regulator conditions are provided. Additionally, there is no standard protocol for *in vitro* rooting. Just like organogenic differentiation, rooting method and media requirements vary with the walnut genotype. MS, WPM and DKW media have been tried by different workers to obtain *in vitro* rooted plantlets of *J. regia*. Of the different auxins that have been tried, IBA has proven to be the most optimal for root regeneration in tissue culture raised shoots of walnut (Table **1**).

Table 1. Different nutrient media standardized for *in vitro* rooting in *J. regia*.

S. No.	Rooting Medium for *J. regia* microshoots
1	½ strength DKW medium + full calcium nitrate [64]
2	¼ MS medium + 3.0 mg/l IBA [65]
3	¼ strength DKW medium + 25 µM IBA [66]
4	½ DKW medium + 10 µM IBA [67]
5	½ DKW medium + vermiculite 250/200 ml (v/v) + 24.6 µM [62]
6	MS medium + 15 µM IBA [68]
7	DKW medium + 4.0 mg/l BAP + NAA [69]
8	WPM medium + 0.5 mg/l BAP + 0.5mg/l IAA [55]
9	½ strength MS medium + 4 mg/l IBA [35]
10	DKW medium + 2 mg/l ABA (abscisic acid) [63]
11	1/4-strength DKW medium + (25-50 µM) IBA [70]
12	MS + 3 µM IBA [71]

(Table 1) cont.....

S. No.	Rooting Medium for *J. regia* microshoots
13	½ strength MS + 2.5 μM IBA [72]
14	DKW + 5-10 mg/l IBA [73]
15	½ strength of DKW + 5 mg/l IBA [74]
16	NGE +5mg/l IBA [75]
17	1/4 DKW medium +12 μM IBA + vermiculite [76]
18	1/2 DKW medium + 50 μM IBA [39]
19	DKW medium + 69 mg/l salicylic acid ; DKW medium +1 mg/l paclobutrazol [77]
20	1/2 DKW medium + 3 mg/l IBA [78]

The success of the transfer of a plant from laboratory to nursery or field conditions is a crucial phase. This stage of micropropagation often witnesses maximum plant mortality as the tissue culture raised plantlets are heterotrophic and lack a well-developed root system to be able to survive under natural conditions. Consequently, appropriate hardening and acclimatization methods need to be standardized for each plant species being propagated by tissue culture technique.

It has been observed that walnut is particularly susceptible to the many stimuli like thermal stress or plantation shocks that limit the successful establishment of micropropagated plantlets in the field [79]. Several workers have standardized hardening methods for such plantlets. The use of jiffy pots for improved growth and survival of rooted plantlets has been suggested [64]. It has also been shown that in the initial phase of hardening, the plantlets need to be kept in a greenhouse under humid conditions for around 10 days before shifting them to field conditions [65]. Vermiculite has been reported to be a better choice as rooting and hardening medium [66]. An interesting study [68] showed that mixtures comprising fir bark:oyster lime and peat:volcanic rock aid in better root development and acclimatization of rooted plantlets under high humidity conditions. The study also suggested that promalin spray can be used to induce bud break in case resetting is observed. The use of well-drained soil as growing material during hardening has also been suggested [80]. A study showed that a mixture of soil:fir, bark:fine and sand:peat moss irrigated with Hoagland solution during 2 weeks of acclimatization under high humidity proves to be the best for survival of *J. regia* plantlets. Perlite: vermiculite mix [35], and sorbarod [72], peat: perlite mix [76] peat:perlite:vermiculite [39] have also been suggested as suitable hardening medium under greenhouse conditions. Rohr *et al.* [81] suggested that the use of microbial inoculant during acclimatization may improve field performance of *J. regia* plantlets. Another study [82] reported that rooted plantlets of *J. regia* need to be covered for not less than a month after transplantation to maintain high humidity. Thereafter, around three to four months are required to gradually decrease the moisture content to enable the plantlet to survive under nursery conditions.

CONCLUSION

The primary motive behind the tissue culture (micropropagation) of walnuts has been the necessity to produce superior trees, rootstocks, and selections for both academic and industrial interests. By using axillary bud culture and the rooting of microshoots, walnut cultivars can be preserved, and new clones with desirable genotypes can be produced. The recalcitrance of explants, long propagation cycles and hybridization barriers can be successfully overcome by zygotic embryo culture. There is a considerable prospect for walnut improvement and rapid proliferation by somatic embryogenesis *in vitro*. This method is also very beneficial for biotechnological procedures, including transformation and the production of synthetic seeds. Callus-mediated propagation of walnuts can be successfully employed for rapid plant generation and also for secondary metabolite production. These *in vitro* methods have the potential to raise disease-free, healthy plants with desirable traits that can be rapidly provided for field transfer. Once the challenges of microbial contamination, phenolic release, microenvironment control and generation of rooted plantlets are overcome, a tissue culture system can significantly impact the production and contribute to the walnut improvement programs.

REFERENCES

[1] Fernandez-Lopez J, Aleta N, Alıas R. Forest genetic resources conservation of *Juglans regia*. Rome, Italy: L. IPGRI 2000; pp. 20-5.

[2] Amaral JS, Casal S, Pereira JA, Seabra RM, Oliveira BPP. Determination of sterol and fatty acid compositions, oxidative stability, and nutritional value of six walnut (*Juglans regia* L.) cultivars grown in Portugal. J Agric Food Chem 2003; 51(26): 7698-702. [http://dx.doi.org/10.1021/jf030451d] [PMID: 14664531]

[3] Muradolu F. Selection of promosing genotypes in native walnut (*Juglans regia* L.) populations of Hakkari central and Ahlat (Bitlis) districht, and genetic diversity.

[4] Mitrovic M, Stanisavljevic M, Gavrilovic-Danjanovic J. Biochemical composition of fruits of some important walnut cultivars and selections. In III International Walnut Congress. 442 1995, 13 pp. 205-208.

[5] Muradoglu F, Oguz HI, Yildiz K, Yilmaz H. Some chemical composition of walnut (*Juglans regia* L.) selections from Eastern Turkey. Afr J Agric Res 2010; 5(17): 2379-85.

[6] Savage GP. Chemical composition of walnuts (*Juglans regia* L.) grown in New Zealand. Plant Foods Hum Nutr 2001; 56(1): 75-82. [http://dx.doi.org/10.1023/A:1008175606698] [PMID: 11213171]

[7] Ravai M. Quality characteristics of California walnuts. (USA): Cereal foods world 1992.

[8] Payne T. California walnuts and light foods. (USA): Cereal foods world 1985.

[9] Souci SW. Garching Deutsche Forschungsanstalt für Lebensmittelchemie. Food composition and nutrition tables 5,. rev. and completed ed. Medpharm Scientific Publ.; 1994.

[10] Çağlarırmak N. Biochemical and physical properties of some walnut genotypes (*Juglans regia L.*). Food/Nahrung 2003; 47(1): 28-32.

[11] Crews C, Hough P, Godward J, *et al.* Study of the main constituents of some authentic hazelnut oils. J Agric Food Chem 2005; 53(12): 4843-52. [http://dx.doi.org/10.1021/jf047836w] [PMID: 15941325]

[12] Martínez ML, Labuckas DO, Lamarque AL, Maestri DM. Walnut (*Juglans regia* L.): Genetic resources, chemistry, by-products. J Sci Food Agric 2010; 90(12). [http://dx.doi.org/10.1002/jsfa.4059] [PMID: 20586084]

[13] Davis L, Stonehouse W, Loots DT, *et al.* The effects of high walnut and cashew nut diets on the antioxidant status of subjects with metabolic syndrome. Eur J Nutr 2007; 46(3): 155-64. [http://dx.doi.org/10.1007/s00394-007-0647-x] [PMID: 17377830]

[14] Gandev S. Budding and grafting of the walnut (*Juglans regia* L.) and their effectiveness in Bulgaria. Bulg J Agric Sci 2007; 13(6): 683.

[15] Gruselle R, Boxus P. Walnut micropropagation. Acta Hortic 1990; (284): 45-52. [http://dx.doi.org/10.17660/ActaHortic.1990.284.3]

[16] Lopez JM. Walnut tissue culture: Research and field applications. United States Department of agriculture forest service general technical report NC 2004; 243-146.

[17] Vahdati K, Razaee R, Mirmasoomi M. Micropropagation of some dwarf and early mature walnut genotypes. Biotechnol 2009; pp. 171-5.

[18] Toosi S, Dilmagani K. Proliferation of *Juglans regia* L. by *in vitro* embryo culture. J Cell Biol Genet 2010; 1(1): 12-9.

[19] Cornu D, Jay-Allemand C. Micropropagation of hybrid walnut trees (Juglans nigra x Juglans regia) through culture and multiplication of embryos. In: Annales des sciences forestières 1989; 46: 113s-6s.

[20] Bonga JM. Clonal propagation of mature trees: Problems and possible solutions.Cell and tissue culture in forestry. Dordrecht: Springer 1987; pp. 249-71. [http://dx.doi.org/10.1007/978-94-017-0994-1_15]

[21] McGranahan G, Leslie CA, Driver JA. *In vitro* propagation of mature Persian walnut cultivars. HortScience 1988; 23(1): 220. [http://dx.doi.org/10.21273/HORTSCI.23.1.220]

[22] Payghamzadeh K, SK K. Comparison effects of MT novel medium with modified DKW basal medium on walnut micropropagation. Proceeding book of the 1st national conference of student biology and modern world 2008; 204.

[23] Revilla MA, Majada J, Rodriguez R. Walnut (*Juglans regia* L.) micropropagation. Ann Sci For 1989; 46 (Suppl.): 149s-51s. [http://dx.doi.org/10.1051/forest:19890533]

[24] Garavito C, Penuela R, Sanchez-Tames R, Rodriguez R. Multiple shoot-bud stimulation and rhizogenesis induction of embryogenic and juvenile explants of walnut. In: International Symposium on Vegetative Propagation of Woody Species. 227: 457-9.

[25] Kornova K, Stephanova A, Terzijsky D. *In vitro* culture of immature embryos and cotyledons of *Juglans regia* L., morphological and anatomical analyses of some regenerants. Acta Hortic 1993; (311): 125-33. [http://dx.doi.org/10.17660/ActaHortic.1993.311.17]

[26] Payghamzadeh K. An investigation on asexual propagation of adult Persian walnut (*Juglans regia L.*)

via single nod culture and apical shoot culture. In: Abstract book of the 15th national and 3rd international conference of biology 2008; 208.

[27] Driver JA, Kuniyuki AH. *In vitro* propagation of Paradox walnut rootstock. HortScience 1984; 19(4): 507-9. [http://dx.doi.org/10.21273/HORTSCI.19.4.507]

[28] Lee MH, Ahn CY, Park CS. *In vitro* propagation of *Juglans sinensis* Dode from bud culture. Research report of the Institute of Forest Genetics 1986.

[29] Leslie C, McGranahan G. Micropropagation of persian walnut (*Juglans regia L.*). In: Bajaj YPS, Ed. Biotechnology in Agriculture and Forestry High Technology and Micropropagation II. Berlin Heidelberg: Springer-Verlag 1992; pp. 137-50.

[30] Saadat YA, Hennerty MJ. Factors affecting the shoot multiplication of Persian walnut (*Juglans regia L.*). Sci Hortic 2002; 95(3): 251-60. [http://dx.doi.org/10.1016/S0304-4238(02)00003-1]

[31] Preece JE, Van Sambeek JW, Huetteman CA, Gaffney GR. Biotechnology: *In vitro* studies with walnut (*Juglans*) species. In: Phelps JE, Ed. The continuing quest for quality Proc 4th Black walnut Symposium. Carbondale.IL 1989; pp. 159-80.

[32] Rodriguez R. Callus initiation and root formation from *in vitro* culture of walnut cotyledons. Science Horticulture 1982; 17: 195-6.

[33] Jay-Allemand C. *In vitro* from Walnut Experimental study on the seeding of isolated embryos and buds. France: PEA in agromy University of Sciences and Techniques of Languedoc in Montpellier 1982; p. 125.

[34] Liu S, Han B. *In vitro* propagation of walnut (*Juglans regia L.*). Acta Agric Univ Pekinensis 1986; 12: 143-51.

[35] Leal DR, Sánchez-Olate M, Avilés F, *et al.* Micropropagation of *Juglans regia* L. Protocols for micropropagation of woody trees and fruits. Dordrecht: Springer 2007; pp. 381-90. [http://dx.doi.org/10.1007/978-1-4020-6352-7_35]

[36] Chalupa V. Clonal propagation of broad-leaved forest trees *in vitro.* Commun Inst for Cech 1981; 12: 255-71.

[37] Heile-Sudholt C, Huetteman CA, Preece JE, Van Sambeek JW, Gaffney GR. *In vitro* embryonic axis and seedling shoot tip culture of *Juglans nigra L.* Plant Cell Tissue Organ Cult 1986; 6(2): 189-97. [http://dx.doi.org/10.1007/BF00180804]

[38] Bosela MJ, Michler CH. Media effects on black walnut (*Juglans nigra* L.) shoot culture growth *in vitro*: Evaluation of multiple nutrient formulations and cytokinin types. *In Vitro* Cell Dev Biol Plant 2008; 44(4): 316-29. [http://dx.doi.org/10.1007/s11627-008-9114-5]

[39] Stevens ME, Pijut PM. Rapid *in vitro* shoot multiplication of the recalcitrant species *Juglans nigra L.* In Vitro Cell Dev Biol Plant 2018; 54(3): 309-17. [http://dx.doi.org/10.1007/s11627-018-9892-3]

[40] Ricardo JLM, Alexandru F, Georgi C. Micropropagation of valuable walnut genotypes for timber production: New advances and insights. Ann Silvic Res 2020; 44(1): 5-13.

[41] Pejić S, Mikropropagacija O. (*Juglans regia* L.) *in vitro.* Diss. Josip Juraj Strossmayer University of Osijek. Faculty of Agrobiotechical Sciences Osijek. Department of Plant Production and Biotechnology 2021.

[42] Yegizbayeva TK, García-García S, Yausheva TV, *et al.* Unraveling factors affecting Micropropagation of four Persian Walnut varieties. Agronomy 2021; 11(7): 1417. [http://dx.doi.org/10.3390/agronomy11071417]

[43] Somers PW, Van Sambeek JW, Preece JE, Gaffney G. Myers 0. In vitro micropropagation of black

walnut.Proc 7th North Am For Bioi. Lexington, KY: University Kentucky Press 1982; pp. 224-30.

[44] Cheem GS, Mehra PN. Morphogenesis in endosperm cultures.Plant tissue culture. Tokyo: Maruzen 1982; pp. 111-2.

[45] Avilés F, Ríos D, González R, Sánchez-Olate M. Effect of culture medium in callogenesis from adult walnut leaves (*Juglans regia L.*). Chil J Agric Res 2009; 69(3): 460-7. [http://dx.doi.org/10.4067/S0718-58392009000300020]

[46] Li HX, Cao SY. Callus induction from *in vitro* leaves of *Juglans regia* 'Lvbo'. Acta Hortic 2014; (1050): 131-8. [http://dx.doi.org/10.17660/ActaHortic.2014.1050.15]

[47] Zarghami R, Salari A. Effect of different hormonal treatments on proliferation and rooting of three persian Walnut (*Juglans regia* L.) genotypes. Pak J Biol Sci 2015; 18(6): 260-6. [http://dx.doi.org/10.3923/pjbs.2015.260.266]

[48] Nikolova V, Dimanov D, Gandev S. Investigating the possibilities for callus induction from walnut (*Juglans regia*) *in vitro*. Rastenievadni nauki. Rastenievudni Nauki 2016; 53(4): 15-9.

[49] Kornova K, Stephanova A, Terzijsky D. *In vitro* culture of immature embryos and cotyledons of *Juglans regia L.* Morphological and anatomical analysis of some regenerants. Acta Hortic 1993; (311): 125-33. [http://dx.doi.org/10.17660/ActaHortic.1993.311.17]

[50] Bonga JM, Aderkas P, von Aderkas P. In vitro culture of trees. Springer science & Business Media 1992; 38. [http://dx.doi.org/10.1007/978-94-015-8058-8]

[51] Ramming DW. The use of embryo cultura in fruti breeding. HortScience 1990; 25(4): 393-8. [http://dx.doi.org/10.21273/HORTSCI.25.4.393]

[52] Kaur R, Sharma N, Kumar K, Sharma DR, Sharma SD. *In vitro* germination of walnut (*Juglans regia L.*) embryos. Sci Hortic 2006; 109(4): 385-8. [http://dx.doi.org/10.1016/j.scienta.2006.05.012]

[53] Payghamzadeh K, Kazemitabar SK. Assessment of effects of physical factors and its interaction with GA3 on walnut local cultivar micropropagation. Abstract book of the 10th Iranian genetic congress 2008.

[54] Payghamzadeh K, Kazemitabar SK. The effects of BAP and IBA and genotypes on *in vitro* germination of immature walnut embryos. Inter. J Plant Production 2010; 4(4): 309-22.

[55] Sánchez-Zamora MÁ, Cos-Terrer J, Frutos-Tomás D, García-López R. Embryo germination and proliferation *in vitro* of *Juglans regia* L. Sci Hortic 2006; 108(3): 317-21. [http://dx.doi.org/10.1016/j.scienta.2006.01.041]

[56] Myrselaj M, Sota V, Bitri I, Zekaj Z. *In vitro* regeneration of walnut (*Juglans regia*) from embryo culture and histological analysis of leaf epidermal structures in *in vitro* derived plantlets. J Multidis Eng Sci Stud 2021; 7(2): 3680-4.

[57] Tulecke W, McGranahan G. Somatic embryogenesis and plant regeneration from cotyledons of walnut, *Juglans regia L.* Plant Sci 1985; 40(1): 57-63. [http://dx.doi.org/10.1016/0168-9452(85)90163-3]

[58] Cornu D. Walnut somatic embryogenesis: Physiological and histological aspects. Ann Sci For 1989; 46 (Suppl.): 133s-5s. [http://dx.doi.org/10.1051/forest:19890529]

[59] Preece JE, McGranahan GH, Long LM, Leslie CA. Somatic embryogenesis in walnut (*Juglans regia*). In: Somatic embryogenesis in woody plants Springer. Dordrecht 1995; pp. 99-116.

[60] Chenevard D, Jay-Allemand C, Frossard JS, Genduard M. Morphological and biochemical factors affecting the survival rate of microcuttings of two hybrid walnut (*Juglans nigra*×*Juglans regia*) clones

during their acclimatization. Ann Sci For 1995; 52: 147-56. [http://dx.doi.org/10.1051/forest:19950205]

[61] Tang H, Ren Z, Krczal G. Improvement of English walnut somatic embryo germination and conversion by desiccation treatments and plantlet development by lower medium salts. *In Vitro* Cell Dev Biol Plant 2000; 36(1): 47-50. [http://dx.doi.org/10.1007/s11627-000-0011-9]

[62] Breton C, Cornu D, Chriqui D, *et al.* Somatic embryogenesis, micropropagation and plant regeneration of "Early Mature" walnut trees (*Juglans regia*) that flower *In Vitro* Tree Physiol 2004; 24(4): 425-35. [http://dx.doi.org/10.1093/treephys/24.4.425] [PMID: 14757582]

[63] Baya SH, Ebrahimzadeh H, Vahdati K, Mirmasoumi M. Somatic embryo maturation and germination of Persian walnut (*Juglans regia L.*). III International Symposium on Acclimatization and Establishment of Micropropagated Plants . 812. 2007, pp. 313-318.

[64] Saadat Y, Hennerty M. Factors affecting microshoot rooting of Persian walnut (*Juglans regia* L.). Iran J Hortic Sci Technol 2000; 1(3-4): 135-46.

[65] Bourrain L, Navatel JC. Plant production of walnut Juglans regia L. by *in vitro* multiplication. In: IV International Walnut Symposium 544. . 465-71.

[66] Tetsumura T, Tsukuda K, Kawase K. Micropropagation of Shinano Walnut(*Juglans regia* L.). Engei Gakkai Zasshi 2002; 71(5): 661-3. [http://dx.doi.org/10.2503/jjshs.71.661]

[67] Caboni E, Delia G, Starza SL. *in vitro* rooting of walnut (*Juglans regia* L.). Italy: Italus Hortus 2003.

[68] Vahdati K, Leslie C, Zamani Z, McGranahan G. Rooting and acclimatization of *in vitro* grown shoots from mature trees of three Persian walnut cultivars. HortScience 2004; 39(2): 324-7. [http://dx.doi.org/10.21273/HORTSCI.39.2.324]

[69] Saadat Y, Javid Tash I. Micropropagation of Mature Persian Walnut (*Juglans regia L.*) Trees. 2006. Available from: https://agris.fao.org/agris-search/search.do?recordID=IR2009000187

[70] Dong P, Lichai Y, Qingming W, Ruisheng G. Factors affecting rooting of *in vitro* shoots of walnut cultivars. J Hortic Sci Biotechnol 2007; 82(2): 223-6. [http://dx.doi.org/10.1080/14620316.2007.11512223]

[71] Monireh C, Hassan E, Ali MN, Kourosh V, Charles L. Effect of endogenous phenols and some antioxidant enzyme activities on rooting of Persian walnut (*Juglans regia* L.). Afr J Plant Sci 2010; 4(12): 479-87.

[72] Payghamzadeh K, Kazemitabar SK. *In vitro* propagation of walnut : A review. Afr J Biotechnol 2011; 10(3): 290-311.

[73] Moreno RJ, Morales AV, Gradaille MD, Gómez L. Towards scaling-up the micropropagation of *Juglans major* (Torrey) Heller var. 209 x *J. regia* L., a hybrid walnut of commercial interest. Integrating vegetative propagation, biotechnologies and genetic improvement for tree production and sustainable forest management. Czech Republic: IUFRO 2012; pp. 80-91.

[74] Licea-Moreno RJ, Contreras A, Morales AV, Urban I, Daquinta M, Gomez L. Improved walnut mass micropropagation through the combined use of phloroglucinol and FeEDDHA. Plant Cell Tissue Organ Cult 2015; 123(1): 143-54. [http://dx.doi.org/10.1007/s11240-015-0822-3]

[75] Kepenek K, Kolağasi Z. Micropropagation of Walnut (*Juglans regia L.*). Acta Phys Pol A 2016; 130(1): 150-6. [http://dx.doi.org/10.12693/APhysPolA.130.150]

[76] Tuan PN, Meier-Dinkel A, Höltken AM, Wenzlitschke I, Winkelmann T. Factors affecting shoot multiplication and rooting of walnut (*Juglans regia* L.) *in vitro.* Acta Hortic 2017; (1155): 525-30. [http://dx.doi.org/10.17660/ActaHortic.2017.1155.77]

[77] Gentile A, Frattarelli A, Urbinati G, Caboni E. Effect of CaCl2, paclobutrazol and salicylic acid on *in vitro* rooting of walnut (*Juglans regia* L.). XXX International Horticultural Congress IHC2018: II International Symposium on Micropropagation and *in vitro* Techniques. 2018; 1285: 1-8.

[78] Ribeiro H, Ribeiro A, Pires R, *et al. Ex Vitro* Rooting and Simultaneous Micrografting of the Walnut Hybrid Rootstock 'Paradox' (*Juglans hindsi* × *Juglans regia*) cl. 'Vlach'. Agronomy 2022; 12(3): 595. [http://dx.doi.org/10.3390/agronomy12030595]

[79] Höltken AM, Wenzlitschke I, Winkelmann T, Tuan PN, Meier-Dinkel A. Factors affecting shoot multiplication and rooting of walnut (*Juglans regia* L.) *in vitro.* InVI International Symposium on Production and Establishment of Micropropagated Plants. 1155 2015; pp. 525-530.

[80] Britton MT, Leslie CH, McGranahan GH, Dandekar AM. II 1 Walnuts. Edited by T. Nagata (Managing Editor) H. Lörz 2007; p. 349.

[81] Rohr R, Iliev I, Scaltsoyinnes A, Tsoulpha P. Acclimatization of micropropagated forest trees. I International Symposium on Acclimatization and Establishment of Micropropagated Plants. 59-69.

[82] Quintero-García OD, Jaramillo-Villegas S. Rescue and *in vitro* germination of immature embryos of black cedar (*Juglans neotropica* Diels). Acta Agron 2012; 61(1): 50-8.

CHAPTER 3

Micropropagation and *in vitro* studies in *Alpinia* Roxb. (Zingiberaceae)

Delna Joseph[1], **Sinjumol Thomas**[2] and **Bince Mani**[1,*]

[1] *Department of Botany, St. Thomas College Palai, Kottayam-686574, Affiliated to Mahatma Gandhi University, Kerala, India*

[2] *Department of Botany, Carmel College, (Autonomous), Mala, Thrissur-680732, Affiliated to University of Calicut, Kerala, India*

Abstract: A tropical and subtropical Asian genus called *Alpinia* is used for both horticultural and medicinal purposes. Species having ornamental uses are now distributed widely all over the world. Different species of *Alpinia* are widely used in traditional medicine for treating many diseases. Several *Alpinia* species have now been experimentally demonstrated to have medicinal properties. Excess trade of many species of *Alpinia*, such as *A. calcarata*, *A. galanga etc.*, as well as habitat loss and urbanization demands its mass propagation. Therefore, one of the best methods for its mass propagation and conservation is micropropagation. *In vitro* studies of medicinal taxa such as *A. calcarata*, and *A. galanga* and ornamental species such as *A. purpurata* has been well established. Different *in vitro* approaches such as direct organogenesis, callogenesis and indirect organogenesis, somatic embryogenesis (SEs), and multiplication using inflorescence buds were generally tried for the successful micrpropagation of different species of *Alpinia*. Genetic and phytochemical fidelities of the *in vitro* raised plants were also studied in many instances to enhance the commercial use of it.

Keywords: *Alpinia calcarata*, *A. galanga*, *A. purpurata*, *A. zerumbet*, *Alpinia malaccensis*, Callus, Clonal Propagation, Explant, Genetic Fidelity, Hormones, *In vitro* Propagation, Meristemoids, Organogenesis, Plant Growth Regulators, Rhizome Buds, Somatic Embryogenesis.

INTRODUCTION

Alpinia is the largest genus in the family Zingiberaceae, and comprises approximately 245 species worldwide and it is native to Tropical and Subtropical Asia to West Pacific [1]. The species of *Alpinia* have been widely used for various purposes for centuries. *Alpinia* is widely used as traditional medicines in China,

* **Corresponding author Bince Mani:** Department of Botany, St. Thomas College Palai, Kottayam-686574, Affiliated to Mahatma Gandhi University, Kerala, India; E-mail: binsnm@gmail.com

T. Pullaiah (Ed.)

India, and Japan to treat various diseases such as gastralgia, indigestion, vomiting, *etc*. Research on *Alpinia* species revealed its pharmaceutical and pharmacological properties scientifically, including antianxiety, antibacterial, antiemetic, antifungal, antitumor, antiulcer, cardioprotective, hypoglycemic, and neuroprotection activities. Many species are cultivated for its ornamental uses such as *A. calcarata*, *A. purpurata*, *A. zerumbet*, *etc*. *Alpinia blepharocalyx* is used as a natural dye [2]. The species namely *A. galanga* is a multipurpose plant has variety of uses. It is widely used by the traditional medicinal practitioners due to its ethnomedicinal properties such as anti-inflammatory, antioxidant, antimicrobial, anticancerous, spasmolytic, properties [3 - 5]. The therapeutic uses of *A. galanga* are due to its medicinally active constituents such as α-fenchol, α-fenchyl acetate, myrcene, 1, 8-cineole, camphor, camphene, *etc* [6]. Compounds namely acetoxyeugenol acetate, acetoxychavicol acetate (ACA) and p-coumaryl diacetate characterised from *A. galanga* are well-known for its anti-HIV, anti-parasitic, anti-tumour, and antituberculous activities [7 - 9]. Besides, the rhizome has also been used as condiment in many parts of the world [5, 10]. The pharmacological and pharmaceutical properties of other species of *Alpinia* such as *A. katsumadai*, *A. oxyphylla*, *A. calcarata*, *A. purpurata*, *A. zerumbet* has also been proved recently [2].

Due to deforestation, climate change, natural calamities, indiscriminate harvesting, commercial exploitation, and pollution, plants in natural habitats are in great decline [11]. Consequently, these plants are under threat or become endangered day by day, hence demanding their conservation. Moreover, the conventional method of propagation either could not be accomplished in many plant species or may not be adequate to fulfill the demand of many species, especially medicinal plants. These impediments can be avoided through biotechnological methods of propagation such as *in vitro* propagation [12]. Micropropagation of *Alpinia* species is quite occasional, with only about less than 10 species studied so far. The *in vitro* method that was most extensively studied was direct organogenesis from axillary meristems of rhizomes, though callus induction and somatic embryogenesis were also established in some species. This chapter outlined the different methods used for the micropropagation of the medicinal herb *Alpinia*.

MICROPROPAGATION OF *ALPINIA* SPECIES

Alpinia calcarata Roscoe

Alpinia calcarata is a species that can be found in moderate distributions, primarily originating from South India and extending to China (Guangdong) and Indo-China. This plant is a rhizomatous geophyte and thrives in wet tropical biomes [1]. With its diverse medicinal properties, clonal propagation of *A. calcarata* has been carried out to some extent.

Clonal Propagation from Axillary Rhizome/Shoot Buds

Sudha and co-workers [13] studied the requirements of optimum concentration and composition of plant growth regulators (PGRs) in clonal propagation of *A. calcarata.* The explant used were axillary shoot buds taken from a 12-month-old rhizome of field-grown plants. After proper disinfection, the explants were cultured on Murashige and Skoog (MS) medium [14] fortified with thidiazuron (TDZ), 6-benzylaminopurine (BAP) and kinetin (Kn) at a concentration of 0.1-5.0 mg/L for the initiation of shoot buds. For the standardization of ideal concentrations of auxins (0.05-5.0 mg/L), such as 1-naphthylacetic acid (NAA), indole-3-butyric acid (IBA) and indole-3-acetic acid (IAA) with 2.0 mg/L BAP were tried. After 8 weeks of initiation, the regenerated axillary shoot buds were subcultured on MS medium augmented with a combination of BAP and 0.2 mg/L IAA or BAP alone (1.0-3.0 mg/L) for multiple shoot induction. Solitary axillary shoot buds with meristemoids and groups of multiple shoot initials were separately sub-cultured on MS medium augmented with the mixture of 0.1 mg/L IAA and BAP or BAP only (0.5-1.0 mg/L) at the next stage of propagation. Meristemoids and groups of three to four multiple shoot initials were sub-cultured on MS medium augmented with 0.1 mg/L BAP for the maintenance of cultures. ½ strength MS (full-strength sucrose and myo-inositol) liquid and solid medium with varied levels of auxins (NAA, IBA, IAA) alone or in combinations were used for the initiation of roots from well-grown shoots (3.0-4.0 cm) of six-week-old cultures. The role of Kn, BAP, and TDZ on the initiation of the axillary bud was evaluated after 8 weeks of culture initiation. Maximum length and number of shoots were obtained in MS medium augmented with 1 mg/L Kn, 2 mg/L BAP, and 3 mg/L TDZ in the first stage of *in vitro* cultures. A mixture of 2.0 mg/L BAP + 0.5 mg/L NAA gave a significantly higher rate of callusing (39%). The mean number of shoots/explants observed was 5.2, whereas the highest mean length of shoots was 4.8 cm in the combination of 2.0 mg/L BAP + 0.1 mg/L IAA. Maximum multiple shoot initials (12.1) and meristemoids (4.0) were produced on MS medium fortified with 2.0 mg/L BAP and 0.2 mg/L IAA (Table **1**). Once the same propagules were transferred to the third subculture on a medium containing 1.0 mg/L BAP alone, the rate of multiplication was further increased to 32 shoots. Within two weeks of culturing, the meristemoids produced an average of 10 shoot initials at each stage. The best PGR combination for the rooting was 0.2 mg/L IAA and IBA with a response rate of 100%, 9.3 roots/shoot and 6.85 cm mean root length. After 7 months, 500 shoots were produced from a single axillary shoot bud explant by employing the above *in vitro* culture method. When root induction and shoot multiplication were carried out in liquid media, a substantial and expensive component of the gelling agent (Agar) was omitted. The rooted plantlets were hardened (survival rate of 95%) and well grown in the field.

Using rhizome bud explants, an effective strategy was developed for the clonal propagation of *A. calcarata* by optimizing the levels of BAP and also a combination of BAP (0.5 mg/L) with different levels of Kn for the successful multiplication of shoots. Myo-inositol (100 mg/L), 3% (w/v) sucrose, various quantities of BAP (0.5-3.0 mg/L), and mixture of Kn (0.5-3 mg/L) and BAP (0.5 mg/L) were supplemented to the basal solid medium to promote shoot multiplication. Root induction was performed by the sub-culturing of individual regenerated shoots using the same medium and full and ½ strength solid medium augmented with IBA (0.5 mg/L) (Table **1**). Rooted individual shoots were then subjected to acclimatization in the field. Variation in regeneration percentage, percentage of rooting and number of shoots per explant were evaluated among various levels of BAP alone and BAP (0.5 mg/L) in combination with Kn (0.5-3 mg/L). MS medium augmented with 0.5 mg/L of BAP and 1.5 mg/L of Kn gave enhanced regeneration rate (90%), number of shoots/explant (13.6) and rate of rooting percentage (82%) in four weeks after inoculation. After three weeks of subculture, shoots were rooted in the same medium. Shoots transferred to the ½ MS augmented with 0.5 mg/L IBA was found best for healthy rooting. The plantlets were hardened and established well in the field [15].

An effective method for the *in vitro* proliferation of *A. calcarata* was developed by Mathew and co-workers [16]. For the *in vitro* propagation, axillary shoot buds (1-1.5 cm in length) that were actively growing and emerged from the 12-month-old rhizomes were chosen. The explants were disinfected and inoculated on a shoot proliferation medium. Kn (0.5-3 mg/L), BAP (0.53 mg/L), combination of Kn (0.5-3 mg/L) with NAA (0.5 mg/L), combination of BAP (0.5-3 mg/L) with Kn (2.5 mg/L) and combination of BAP (0.5-3 mg/L) with 0.5 mg/L NAA were employed to promote shoot proliferation. Effectively regenerated multiple shoots were maintained for thirty days for succeeding experiments on solid MS medium (basal) with a different set of growth hormones, as mentioned above. The shoots were moved to ½ strength MS augmented with various levels of NAA and IAA. The finest medium for shoot regeneration was basal MS medium augmented with Kn (2.5 mg/L) and BAP only or combined with 0.5 mg/L NAA (Table **1**). A promising result in multiple shoot proliferation was not exhibited by MS basal media without a plant growth regulator. Maximum rooting was accomplished in ½ strength MS medium augmented with IAA (2 mg/L) and NAA (2 mg/L). The rooted plantlets were successfully hardened, and field survival of 90% was observed.

Table 1. A detailed account of the media used, as well as the optimum concentrations of PGRs, growth additives, and organic supplements used for the micropropagation of *Alipnia*.

Species	Explant Used	Basal Culture Medium, PGRs and Supplements				References
		Callusing	Shooting	Rooting	Somatic Embryogenesis	
Alpinia Calcarata	Axillary shoot buds	MS + BAP 2.0 + NAA 0.5 mg/L	MS + BAP 2.0 mg/L + IAA 0.2 mg/L	½MS + IAA 0.2 mg/L + IBA 0.2 mg/L	-	[13]
A. calcarata	Axillary shoot buds	-	MS + BAP 0.5 mg/L + Kn 1.5 mg/L	½ MS + IBA 0.5 mg/L	-	[15]
A. calcarata	Axillary shoot buds	-	MS + BAP 2.5 mg/L + Kn 2.5 mg/L	½MS + IAA 2 mg/L + NAA 2 mg/L	-	[16]
A. calcarata	Axillary rhizome buds	-	MS + BAP 10 µM + Kn 15 µM + NAA 2.5 µM	MS + BAP 5 µM + Kn 10 µM + NAA 2.5 µM	-	[17]
A. galanga	Axillary buds	-	MS + BAP 3 mg/L + NAA 1 mg/L + Kn 3 mg/L	MS + BAP 3 mg/L + NAA 1 mg/L + Kn 3 mg/L	-	[12]
A. galanga	Rhizome buds	-	MS + Kn 3 mg/L	MS + Kn 3 mg/L or MS + Kn 3 + NAA 0.5 mg/L	-	[18]
A. galanga	Rhizome buds	-	MS + BAP 5 mg/L + IAA 2 mg/L or BAP 3 mg/L + IAA 0.5 mg/L	MS + BAP 5 mg/L + IAA 2 mg/L	-	[19]
A. galanga	Rhizome buds	-	MS + BAP 0.75 mg/L	MS + IBA 1 mg/L	-	[20]
A. galanga	Rhizome buds	MS + BAP 2 mg/L + 2,4 D 2 mg/L + NAA 2 mg/L	MS + Z 2 mg/L	MS + IBA 2 mg/L	-	[21]

(Table 1) cont.....

Species	Explant Used	Basal Culture Medium, PGRs and Supplements				References
		Callusing	Shooting	Rooting	Somatic Embryogenesis	
A. galanga	Axillary buds	-	MS + Kn 3 mg/L + BAP 3 mg/L + NAA 1 mg/L	BAP 1 mg/L + IBA 1 mg/L	-	[22]
A. galanga	Rhizome buds	-	MS + BAP 13.32 µM + TDZ 0.45 µM	MS + Kn 13.94 µM	-	[23]
A. galanga	Rhizome discs	MS + Kn (1-5 mg/l) or BAP (1-5 mg/l) or 2,4-D (0.5-3 mg/l)	-	-	MS + BAP 3 mg/l + 2,4-D 2.5 mg/l	[24]
A. malaccensis	Rhizome buds	MS + BAP 4.4 µM + 2,4-D 2.26 µM	MS + BAP 2.22 µM or BAP 6.66 µM	½MS + BAP 4.44 µM + IAA 2.85 µM	-	[25]
A. officinarum	Rhizome buds	-	MS+Kn 3 mg/L + NAA 1 mg/L	½MS + IBA 0.5 mg/L	-	[26]
A. purpurata	Rhizome apical buds	-	MS + BAP 5 mg/l	MS + BAP 1 mg/l + NAA 1.5 mg/l	-	[27]
A. purpurata	Rhizome buds	-	MS + BAP 3 mg/L +Kn 2 mg/L	MS + BAP 3 mg/L + Kn 2 mg/L	-	[28]
A. purpurata	Inflorescence buds	-	MS + BAP 10µM + IAA or NAA 5 µM	MS + BAP 10µM + IAA or NAA 5 µM	-	[29]
A. purpurata	Rhizome buds	MS + 2,4-D 2 ppm + Kn 2 ppm	MS + NAA 0.1 ppm + BAP 3.0 ppm	MS + IAA 3 ppm	-	[30]

(Table 1) cont.....

Species	Explant Used	Basal Culture Medium, PGRs and Supplements				References
		Callusing	Shooting	Rooting	Somatic Embryogenesis	
A. purpurata	Inflorescence buds	-	MS + IAA 2 mg/L MS + TDZ 2/4 mg/L MS + BAP 2 mg/L MS + IAA 2 mg/L + TDZ 2 mg/L MS + IAA 2 mg/L + BAP 2 mg/L	MS + IAA 2 mg/L MS + TDZ 2/4 mg/L MS + BAP 2 mg/L MS + IAA 2 mg/L + TDZ 2 mg/L MS + IAA 2 mg/L + BAP 2 mg/L	-	[31]
A. zerumbet	Rhizome buds	-	MS + TDZ 8 mg/L	MS + TDZ 2 mg/L	-	[32]
A. zerumbet	Rhizome buds	-	MS + IAA 2 mg/L + TDZ 2 mg/L	MS + Kn 2 mg/L	-	[33]
A. zerumbet	Rhizome buds	-	MS+BAP 1.5 mg/L + Kn 0.5 mg/L	½ MS + IBA 0.5 mg/L	-	[34]
A. zerumbet	Primordial leaves	MS + 2,4-D 18.2 µM + BAP 0.5 µM + Kn 9.3 µM	MS + Kn 2.3 µM + BAP 6.65 µM	MS + Kn 2.3 µM + BAP 6.65 µM	-	[35]
A. zerumbet	Seeds	MS + BAP 1.5 mg/l + 2,4-D 0.3 mg/l	MS + TDZ 2 mg/L + 2,4-D 0.1 mg/L	½MS + ABT1 1 mg/L	-	[36]

For direct *in vitro* shoot proliferation and plantlet redevelopment, a quick and effective approach was developed by Bhowmik and colleagues [17]. Explants were the axillary buds (2.5 × 2.5 mm) obtained from the rhizome segments pre-induced *in vitro* and grown for 4 weeks on MS medium (HiMedia, India) with BAP and Kn (5.0-20 µM). The synergic effect of auxin (2.5 µM NAA) and cytokinin, as well as the efficacy of BAP and Kn only and their combinations, were evaluated. After six weeks of culture, the average number of roots and shoots per explant, the percentage of explants developing multiple shoot buds and the average length of shoots were evaluated. The MS media augmented with 5 mM BAP, 10 mM Kn, and 2.5 mM NAA generated an average of 6.2 shoots per explant. Highest percentage of regeneration (100%) is shown by combinations of

15 µM BAP + 15 µM Kn, BAP (10 µM) + Kn (5µM) + NAA (2.5), BAP (5 µM), and Kn (10µM) + NAA (2.5µM). PGR formulations such as BAP, Kn and NAA in 15, 10 and 2.5 µM concentrations, respectively, gave the best outcome in mean shoot length (8.1). A mean number of shoot/explant (6.2) and optimum number of roots per explant (10.0) resulted in treatments with 5, 10, 2.5 µM of BAP, Kn and NAA, respectively (Table **1**). Assessment of the morphogenetic potential of *A. calcarata* shoot culture for four subculture passages using shoot multiplication medium showed that the third subculture exhibited the maximum shoot multiplication (5.4). Inter simple sequence repeats (ISSR) and random amplified polymorphic DNA (RAPD) techniques revealed the genetic fidelity of regenerants of *A. calcarata*.

Alpinia galanga (L.) Willd.

Alipnia galanga, a rhizomatous geophyte native to South China to West & Central Malesia. It grows chiefly in the wet tropical biomes [1]. Micropropagation studies using various explants and different methodologies are available in this species.

Direct Regeneration from Rhizome Buds

Borthakur *et al.* [18] established a single-step procedure for the direct regeneration *of A. galanga* from rhizome buds. MS medium (basal) augmented with cytokinins, Kn (1-4 mg/L) and BAP (2-3 mg/L) only or combined with NAA (0.5 mg/L) were tried. The explants cultured on MS medium augmented with 3.0 mg/L Kn were found to be best for shoot induction (7.1 shoots) (Table **1**), number of leaves per shoot (5.0) and number of roots/shoots (4.5). Within eight weeks, both shoots and roots regenerated from the explant simultaneously. The shoots (2-3) were subcultured on the same fresh media after 8 weeks of culture. Plantlets were successfully acclimatized and 80% of the transferred plantlets persisted in the field.

Singh *et al.* [19] established an effective methodology for the *in vitro* proliferation of *A. galanga* using rhizome bud explants. Explants of 1 mm were transferred on MS media (basal) augmented with combinations of Kn (3 mg/L), BAP (1-3 mg/L), IAA (0.5-1.0 mg/L), IBA (1.0 mg/L), NAA (0.5-2.0 mg/L) and adenine sulphate (ADS) (100 mg/l) for the development of *in vitro* cultures. The *in vitro* formed shoots were transferred to MS medium augmented with BAP (3 mg/L) + Kn (3 mg/L) + NAA (1 mg/L) for shoot proliferation (Table **1**). Further, the shoots were transferred to rooting media containing BAP (1 mg/L) and IBA (1 mg/L). Media combinations of 2 mg/L IAA and 5 mg/L BAP (5.65) or 0.5 mg/L IAA and 3 mg/L BAP (5.70) produced the maximum number of shoots/ explant when related to other dosages. However, the maximum shoot length was found (9.64 cm) on the medium fortified with 1.0 mg/L of BAP and 2.0 mg/L IAA.

Because of the combined effects of cytokinins and auxins, the proliferation and growth of the shoots were greatly improved. MS mediums augmented with 5.0 mg/L of BAP and 2 mg/L of IAA produced a maximum number of roots. On the other hand, 1.0 mg/L BAP produced the lengthiest roots from shoots. The proliferated shoots appeared healthy, and 80% of plantlets were hardened, and successfully established in the field.

Shamsudheen *et al.* [20] comprehend high frequency multiple shoot development and plant regeneration from *A. galanga* by adding various plant growth regulators (PGRs), in MS medium. Sterilized rhizome buds cultured on basal MS medium augmented with various concentrations of TDZ and BAP (0.25 to 1.0 mg/L) for the initiation of multiple shoots. Shoot length, percentage of responsive explants, and number of shoots per explant were recorded during 5th week of culture. In order to facilitate rooting, the elongated shoots (>3.0 cm in length) with completely expanded leaves were shifted to MS media augmented with IBA or IAA at varying levels (0.5 to 1.5 mg/L). The average number of roots/shoots and the percentage of rooting were measured during 5th week. When cultivated on a phytohormone-free medium (control), the rhizome explants did not exhibit any sign of bud break and shoot initiation even after 4th week of culture. However, explants inoculated on MS media with different doses of TDZ and BAP showed bud breakage. The maximum number of shoots (7.33) and the best shoot induction (93%) were obtained on MS media fortified with 0.75 mg/L BAP (Table **1**). The number of shoots and shoot length were found to be more influenced by BAP than TDZ. After 4-5 weeks of growth, IBA (1.0 mg/L) provided the highest root development (94%), optimal root length (11.66 cm), and maximum number of roots per plantlet (15.66). The rooted plantlets having four to six fully grown leaves were taken out of the culture media and subjected to hardening and acclimatization, and finally, a survival rate of 75% was reported (Fig. **1**).

Direct and Indirect Regeneration and Phytochemical Analysis

Rao *et al.* [21] established an effective procedure for the *in vitro* regeneration of *A. galanga*. Explants such as leaf, leaf sheath, rhizome and roots were tried in this study. Explants were inoculated on semi-solid MS media fortified with different concentrations (0.5-5 mg/L) of cytokinins, namely Kn, zeatin (Z), TDZ, and BAP, for direct shoot induction, and shoots were transferred to media augmented with auxins, namely, NAA, IAA and IBA for root induction (chemicals were purchased from Himedia, India). Explants such as leaf, root (1 cm) and rhizome were tried for callus initiation and the explants were inoculated on MS media fortified with different combinations of auxins [2,4-Dichlorophenoxyacetic acid (2,4-D) and NAA] and cytokinins (BAP). The calli formed were moved to MS augmented with different levels (0.5-5 mg/L) of cytokinins (BAP, Kn and TDZ) for shoot induction and the regenerated shoots transferred to media fortified with IBA (2 mg/L) and BAP (2 mg/L) for root regeneration. Multiple shoots were formed during 3-5 weeks of the commencement of the culture from rhizome explants on all the media combinations tried. However, 2 mg/L Z gave high frequency regeneration (Table **1**) with 15.66 shoots per explant. MS medium fortified with IBA at a level of 2 mg/L was most effective in inducing a maximum number of roots (7.66). The calli initiation was achieved only from rhizome explants, and promising results were obtained only on media fortified with 2 mg/L of each BAP, 2,4-D, and NAA. BAP was the most effective cytokinin among the different cytokinins tried for shoot regeneration from calli. BAP at a level of 2 mg/L was the best concentration for shoot regeneration (18.33 shoots) and also for the elongation of the regenerated shoots. The root regeneration was detected on shoots transferred to rooting media. The calli were extracted for testing the presence of acetoxychavicol acetate (ACA), an important phytoconstituent with a wide range of biological properties. The ACA content in the regenerated (indirect) plants (1.253%) was 1.6 times greater than those in the control plants (0.783%).

Fig. (1). Micropropagation and hardening of *A. galanga*. (**A**) Rhizome bud explant (**B**) Rhizome bud cultured on medium (**C**) Shoot development on shoot induction medium (**D**) Multiple shooting (**E** & **F**) Root development (**G**) Plantlets removed for hardening (**H**) Primary hardening in paper cups (**I**) Secondary hardening in earthen pots. © Shamsudheen *et al*. [20].

Direct Regeneration and Assessment of Molecular and Phytochemical Fidelity

Sahoo and co-workers [12] established a successful technique for the micropropagation of *A. galanga* and the micropropagated plantlets were stable over the long term on both molecular and phytochemical characteristics. Axillary rhizome buds were used as explants. The surface disinfected explants were inoculated on MS medium fortified with different concentrations and combinations of PGRs, namely BAP (1–4 mg/L), Kn (1-3 mg/L), IAA (0.5–1 mg/L) and NAA (0.5 mg/L). The same media combinations were tried for both shoot induction and root regeneration. The results were noted after two months of *in vitro* culture. For the best results in terms of plantlet regeneration and multiplication (12.4 shoots /explant and 11.2 roots/shoots), Murashige and Skoog (MS) media augmented with 3 mg/L each of BAP and Kn, and 1 mg/L NAA were found to be ideal (Table **1**). The plantlets were moved to fresh media every seventy-five days intervals. Plantlets micropropagated were profiled using molecular markers (RAPD and ISSR) at intervals of every six months for a period of six years in order to ascertain genetic stability. Similar to the mother plants' banding patterns, monomorphic banding was obtained for *in vitro* plantlets. After 6 years of *in vitro* preservation, the regenerants were shifted to the field, and their ability for producing drugs was assessed using phytoconstituent analyses and bioactivity investigations. Comparative gas chromatography (GC) and mass spectroscopy (MS) analyses of essential oils revealed no appreciable variations in phytoconstitution between field-grown and *in vitro* grown plants. When compared to mother plants, *in vitro* plants' bioactivities, such as antioxidant, antibacterial, and anticancer properties, as well as their total phenolic and total flavonoid content, were shown to be constant with only little variations. Therefore, the study has considerable relevance for the economic utilization of *A. galanga* with stable phytoconstitution.

Direct Regeneration from Rhizome Buds and Assessment of Genetic Fidelity

In vitro regeneration of axillary buds obtained from unsprouted rhizomes of *A. galanga* was achieved by Parida *et al.* [22] who evaluated the genetic fidelity of *in vitro* regenerants. Surface disinfected explants were inserted on MS medium augmented with combinations of BAP (1-3 mg/L), Kn (3 mg/L), IAA (0.5-1.0 mg/L), NAA (0.5-2.0 mg/L), IBA (1.0 mg/L), and ADS (100 mg/L). Further, the *in vitro* formed shoots were transferred on MS medium fortified with BAP (3 mg/L) + Kn (3 mg/L) + NAA (1 mg/L) for multiplication. The shoots were transferred to MS media augmented with BAP (1 mg/L) and IBA (1 mg/L) for root regeneration (Table **1**). The multiple shoot initiation was observed on all media combinations tried, and the highest number of shoots (15.6) was observed on MS augmented with Kn (3 mg/L), BAP (3 mg/L) and NAA (1.0 mg/L). The shoot attained a length of 10.4 cm in this combination, and 4.3 roots per shoot were also induced. The shoots transferred to rooting media formed 7.8 roots/shoots, which was observed to be the maximum. The plantlets were acclimatized and effectively grown in the field. The clonal fidelity of the *in vitro* regenerants was assessed using ISSR and RAPD markers. The banding profiles of both RAPD and ISSR markers of micropropagated plants were monomorphic and similar to that of the mother plant, hence, the genetic fidelity of the regenerants was 100%.

Baradwaj *et al.* [23] established a procedure for the micropropagation of *A. galanga* and assessed the genetic fidelity of *in vitro* regenerated plantlets. Two experimental conditions were set by the authors for the multiplication of *A. galanga*. Rhizome buds were inoculated on MS basal medium augmented with BAP (4.44-14.76 μM), Kn (4.44-14.76 μM), TDZ (0.22-2.25 μM) and 2-iP (4.44-14.76 μM) for multiple shoot initiation in the first set of experimental conditions. MS medium augmented with 0.45 μM TDZ combined with BAP (04.44, 08.87, and 13.32 μM) was studied in the second set of experiments. The shoots were shifted to MS medium augmented with IBA, IAA, and NAA at levels of 4.90 to 17.13 μM for rooting. MS medium augmented with 13.32 μM BAP and 0.45 μM TDZ produced the highest number of shoots (9.4). Media augmented with BAP and TDZ showed considerably higher shoot induction related to cultures on basal MS (control), 2-iP and Kn, after 8 weeks. All the cultures on cytokinins augmented media concurrently formed roots, with a highest of 6.4 in Kn (13.94 μM) (Table **1**) and a lowest of 1.6 in TDZ (2.25 μM). The shoots transferred on rooting medium produced considerably more number of roots (20) amongst the highest was on media with10.74 μM NAA. Micropropagated plants were hardened and 80% of plants survived in the field. The genetic fidelity of the *in vitro* regenerated plantlets was assessed by ISSR markers. The banding profiles of the plantlets displayed a 100% resemblance to those of the mother plant.

Somatic Embryogenesis

Micropropagation *via* somatic embryogenesis was developed in *A. galanga* by Mustafaanand [24]. Rhizome explants were inoculated on MS basal medium fortified with several strengths of Kn (1-5 mg/L), BAP (1-5 mg/L) or 2,4-D (0.5-3 mg/L) singly or in various combinations such as 0.25 mg/L 2,4-D and 0.5-3 mg/L of either BAP or Kn. The explant responded to produce calli on all the tested strengths and combinations of PGRs. However, embryogenesis was detected only on greater concentrations of cytokinin medium (4-5 mg/L BAP & Kn) and cytokinin and auxin combinations tried. These embryogenic calli produced a maximum of 4.0 well-developed globular embryos on medium augmented with 3 mg/L BAP and 0.25 mg/L 2,4-D (Table **1**). Further, the somatic embryos differentiated into plantlets (having shoots and roots) when transferred to auxin free media. Plantlets were hardened, and 40% plantlets survived in the field.

Alpinia malaccensis (Burm.f.) Roscoe

Alpinia malaccensis is a rhizomatous geophyte that chiefly grows in wet tropical biomes. Its native range is Assam to China (Yunnan) and Indo-China (Assam, Bangladesh, Cambodia, China South-Central, Myanmar, Tibet, and Vietnam) [1]. Micropropagation works are scanty in this taxon.

Direct Organogenesis and Callogenesis from Rhizome Buds

Basal MS medium augmented with several concentrations of filter-sterilized IAA, Kn, NAA, BAP and IBA (chemicals were purchased from Merk India, HiMedia India, and Qualigens) was tested in the *in vitro* propagation of *A. malaccensis*. The concentrations of BAP, such as 2.22 µM and 6.66 µM, promoted shoot proliferation among the tested concentrations, whereas shoot initiation and elongation were superior in the 5.37 µM concentration of NAA among the various concentrations tested. Significant progression in calli development was observed on MS medium augmented with 2,4-D and BAP in 2.26 µM and 4.44 µM concentrations, respectively. Rhizome bud explants were most suitable for callogenesis. ½ strength MS medium fortified with 4.44 µm BAP and 2.85 µm IAA were tried for the root induction. *In vitro* rooted plants were removed from the medium, thoroughly cleaned through running water to eliminate any remaining medium, and then transplanted into a plastic cup filled with soil and sand [25].

Alpinia officinarum Hance

This species is native to South-East China to Indo-China (Cambodia, China Southeast, Hainan, Myanmar, and Vietnam). *Alpinia officinarum* is a rhizomatous geophyte, that grows mainly in wet tropical biomes [1]. *In vitro* studies are rare in this taxon.

Multiplication Through Rhizome Bud Culture

The optimum concentration and composition of plant growth hormones in MS medium for the maximum formation of shoots and roots of *A. officinarum in vitro* were determined through rhizome bud cultures [26]. After proper sterilization, rhizome buds cultured on basal MS medium fortified with various levels of BAP (1-4 mg/L) and Kn (1-4 mg/L) singly or in combination with NAA (3 mg/L BAP or Kn + 0.5-2.5 mg/L NAA) were evaluated for the rate of multiple shoot induction. ½ strength MS medium augmented with various concentrations of auxins such as NAA (0.1-1.0 mg/L), IBA (0.1-1.0 mg/L) and IAA (0.1-1.5 mg/L) were tried for root initiation from individual regenerated shoots. Medium devoid of PGRs were used as control groups and the rooted plants were subjected to hardening in the greenhouse with soil and vermiculite in a 1:1 ratio for three to four weeks. The highest number of shoots (10.9), the average length of shoot per culture (7.5 cm) and 100% response were shown by a combination of Kn (3 mg/L) and NAA (1 mg/L). Apart from this combination, 3 mg/L Kn, 1.5 mg/L NAA, 3 mg/L BAP and 0.5 mg/L NAA also showed 100% response. The least number of shoots/node and length of shoot were obtained from hormone free MS medium and combination of 3 mg/L Kn and 0.5 mg/L NAA. All tested combinations of auxin for root induction were effective and maximum roots per shoots (7.2) were obtained on ½MS fortified with 0.5 mg/L IBA (Table **1**). However, media devoid of auxins failed to develop roots. It was also noted that 93% of the plantlets were acclimatized and showed positive response of survival in the field.

Alpinia purpurata (Vieill.) K. Schum.

Alpinia purpurata has environmental uses, as medicine, ornamental and food. Its native range of distribution is Maluku to South-West Pacific (Bismarck Archipelago, Maluku, New Caledonia, New Guinea, Solomon Island, and Vanuatu). It is a rhizomatous geophyte that grows chiefly in the wet tropical biomes. This species has been distributed all over the world for its ornamental uses [1]. There are only few *in vitro* works in this taxon.

Multiplication Through Rhizome Apical Bud

In vitro shoot and root generation in *A. purpurata* was achieved on an MS medium fortified with several concentrations of BAP and NAA. For regeneration of shoots, various concentrations of BAP, such as 0, 1, 5, 10 and 15 mg/L, and for root regeneration, various concentrations of NAA, such as 0.5, 1.0 and 1.5 mg/L, along with BAP, were tried. The results showed that treatments of 5 mg/L BAP and a combination of 1 mg/L BAP and 1.5 mg/L NAA significantly enhanced the shoot and root initiation, respectively. The plantlets were effectively acclimatized and persisted in the field [27].

Multiple Shoot Induction and Plantlet Regeneration from Rhizome Bud

An efficient protocol was developed for the determination of the best concentration of cytokinin for multiple shoot induction of *A. purpurata* using rhizome bud explants by Kochuthressia *et al.* [28]. BAP (0.5-3.5 mg/L) alone and in combination with Kn (0.5-2.5 mg/L) fortified with basal MS medium for multiple shoot initiation and for rooting. After proper sterilization, the explants were transferred to MS medium augmented with BAP and Kn. After 45 days of initial culture, the influence of PGRs on shoot induction in various levels of BAP (0.5-3.5 mg/L) and Kn (0.5-2.5 mg/L) were recorded. The best combination of medium with the highest shoot development (6.4 shoots/explant) exhibited by MS fortified with 3 mg/L BAP and 2 mg/L Kn. Noticeable results in maximum shoot length (7.9 cm), maximum number of roots/shoot (8.4), maximum number of leaves on shoots (4.8), maximum length of leaves (4.6 cm), were recorded after 45 days of culture on the same medium (3 mg/L BAP + 2 mg/L Kn) (Table **1**). The plantlets were successfully acclimatized after hardening.

Multiple Shoot Induction from Inflorescence Bud

Young inflorescence buds were used as explants for the *in vitro* multiplication of ornamental *A. purpurata* by Illg *et al.* [29]. The cultures were initiated and maintained on media with MS salts, but macronutrients were in half-strength, and augmented with 2 mg/L thiamine, 2 mg/L pyridoxine, 1 mg/L nicotinic acid, 100 mg/L myo-inositol, 500 mg/L casein hydrolysate, 2% sucrose, and 2 g/L Gelrite. The media was also fortified with various concentrations of plant growth regulators such as BAP (5, 10, 15 μM), 10 μM BAP + 5 μM IAA, 10 μM BAP + 10 μM IAA, 10 μM BAP + 0.5 μM NAA, 10 μM BAP + 5 μM NAA, 10 μM BAP + 0.5 μM NAA + 163 μM ADS and 163 μM ADS only. Inflorescence buds on Murashige and Skoog medium (MS), which contains 10 μM BAP + 5 μM NAA and 10 μM BAP + 5 μM IAA, resulted in the formation of many shoots, with a mean growth of 15 to 20 new shoots after 2-3 months of the culture. 10 μM BAP along with 5 μM IAA or 5 μM NAA (Table **1**) was the ideal plant growth regulator to use during explant induction and establishment of the culture with 98% response frequency, highest number of shoots/explant on induction medium (8.2), and shoot per cluster on proliferation medium (18.6). The shoots were rooted with or without augmenting auxins in the media. The plantlets were hardened and successfully survived in the field.

Plantlet Regeneration and Quantification of Rutin

Kale and Namdeo [30] successfully established a simple method and cheaper alternatives for the micropropagation of *A. purpurata* in order to examine rutin synthesis *in vitro*. Surface sterilized rhizome buds were transferred to MS medium augmented with low cost additives such as coconut water (15%), commercially available sugar (3%), and corn flour (100 g/l) as alternative gelling agents. The medium was augmented with NAA or 2,4-D along with BAP or Kn or both for callus induction. At the same time, BAP (1-3 ppm) and NAA (0.1 ppm) were fortified with media for shoot regeneration and full or half strength MS augmented with IAA (1-4 ppm) for root initiation. The best results were obtained on 2,4-D (2 ppm) and Kn (2 ppm) for callus initiation, while the greatest number of shoots (9-11) were detected on medium with NAA (0.1 ppm) and BAP (3.0 ppm). At a concentration of 3 ppm IAA, the best root initiation was obtained (Table **1**). Leaf extract of micropropagated plants was fractionated and analyzed by high-performance thin-layer chromatography (HPTLC). The results revealed that leaves of *in vitro* grown plants possessed higher rutin content than naturally grown plants. It suggests that a low-cost medium could be a useful alternative and efficient method for micropropagation and rutin synthesis *in vitro* in *A. purpurata*.

Direct Organogenesis from Inflorescence Buds and Estimation of Phenolics

Victorio *et al.* [31] studied the *in vitro* regeneration of *A. purpurata* and its phytochemical responses to exogenous plant growth regulators. Buds from inflorescence were used as explants, and MS basal medium was augmented with PGRs, namely BAP, TDZ, IAA either alone or in combinations such as 2 mg/L IAA, 2 & 4 mg/L TDZ, 2 mg/L BAP, 2 mg/L IAA + 2 mg/L TDZ, and 2 mg/L IAA + 2 mg/L BAP (Table **1**). Liquid medium was used for all treatments with PGRs. Solid and liquid media devoid of PGRs were used as controls. The same media were used for both shoot as well as root regenerations. Total phenolic compounds of the *in vitro* regenerated and field-grown plants were determined using the Folin-Ciocalteau method. The direct shoot organogenesis process was fast and effective and produced roots in a short period. 100% rooting was reached for all treatments within 3 months. PGRs, namely cytokinins and auxins played an important role in the increase in the leaves number (up to 5.1) and shoot elongation (up to 4.0 cm). Liquid MS was found most effective for fast micropropagation and higher production of shoots. The micropropagated plantlets were acclimatized successfully. HPTLC data profile showed a high level of phenolics in *in vitro* regenerated *A. purpurata* (93%) compared with field-grown plants (100%).

***Alpinia zerumbet* (Pers.) B.L. Burtt & R.M. Sm.**

Alpinia zerumbet is a rhizomatous geophyte that grows mainly in the subtropical biomes. This species is native to Bangladesh, Cambodia, China South-Central, China Southeast, Hainan, Japan, Laos, Malaya, Myanmar, Nansei-shoto, Taiwan, Thailand, and Vietnam. It has medicinal and environmental uses and is also used for food [1]. There are only a few *in vitro* studies in this taxon.

In vitro Organogenesis and Evaluation of Secondary Metabolites

A study on the influence of PGRs in the morphological development and production of flavanols and total phenolics in the *in vitro* grown *A. zerumbet* and field grown donor plants was first evaluated by Victório and co-workers [32]. Rhizome buds were cultured on MS medium devoid of any growth regulators, and the basal segments of pseudo-stems excised from the *in vitro*-grown plantlets were cultured on MS medium without any supplements. The influence of plant growth regulators on the organogenesis of *in vitro* grown plantlets was investigated in the subculture (third stage) of above-grown three-month-old plantlets. After getting augmented with 2 mg/L IAA, 2 mg/L BAP, TDZ (2, 4 & 8 mg/L), 2 mg/L IAA + 2 mg/L BAP, and 2 mg/L IAA + 2 mg/L TDZ, the number of leaves, shoot length, proliferation rate and percentage of rooting were recorded (Chemicals were purchased from Tedia®). Among the observed groups, the maximum rate of proliferation was shown in MS media supplemented with 8 mg/L TDZ. Further, medium augmented with 2 mg/L IAA, 2 & 4 mg/L TDZ, 2 mg/L IAA + 2 mg/L BAP, and 2 mg/L IAA + 2 mg/L TDZ also exhibited significantly higher rates of proliferation. Maximum shoot length was achieved on IAA at 2 mg/L (5.2 cm) and on control (5.1 cm) medium. Among the tested groups, a higher number of leaves was found in the MS medium augmented with IAA (23.6) followed by TDZ (83.4). Whereas a reduced number of leaves was observed in the control group (2.2). Explants inoculated on media comprising 2 mg/L TDZ (Table **1**) simultaneously developed roots and shoots *in vitro*. The analysis of flavonoid and phenolic contents of three months old *in vitro* grown plantlets and field grown donor plants were evaluated by hydro-alcoholic extract of the leaves. HPLC-DAD analysis showed a higher percentage of phenolics shown by the control group. MS media fortified with a combination of 2 mg/L IAA and TDZ displayed better results in Kaempferol glucuronide, Kaempferol rutinoside and rutin, whereas 4 mg/L TDZ 4 showed minimum content of rutin and Kaempferol rutinoside. Apart from phenolic content, flavonoid composition was greater in *in vitro* formed plantlets grown on various concentrations of PGRs compared to donor plants.

The influence of different plant growth regulators on the *in vitro* development of *A. zerumbet* and its efficacy in the production of volatile compounds were examined by Victório and co-workers in 2011 [33]. Rhizome buds were carefully removed and cultured in a glass bottle under aseptic conditions on MS medium. Five different compositions of MS medium were evaluated for the subculturing, such as basal MS (control), MS + 2 mg/L IAA, MS + 2 mg/L Kn, MS + 2 mg/L BAP and MS + 2 mg/L IAA + 2 mg/L TDZ. Length and weight of roots and shoots, proliferation rate and number of leaves were recorded for 4-month-old plantlets. Interpretation of the results revealed a higher rate of proliferation and leaf numbers (4.0) in MS medium augmented with 2 mg/L IAA + 2 mg/L TDZ compared to the rest of the combinations of PGR and control group. MS medium fortified with 2 mg/L IAA exhibited maximum shoot length (13.6 cm), whereas MS medium without any supplements (control) showed superior outcomes in the case of root length (32.4 cm) and shoot weight (100%). The MS medium, along with 2 mg/L Kn, exhibited the maximum root weight (270%). A significant reduction in proliferation rate and number of leaves was showed by MS medium augmented with 2 mg/L IAA (1.6: 1). Reduced shoot length and root length were found on MS medium added with 2 mg/L Kn. Shoot weight and the number of leaves in MS medium added with 2 mg/L BAP was remarkably low. Control media showed low root weight in comparison to other media tried. Volatile analysis was conducted for 4-month-old *in vitro* grown plantlets. Leaf and root extracts were obtained by simultaneous distillation-extraction (SDE) and evaluated by GC-MS, with the quantification of various phytoconstituents by GC-FID (GC-flame ionization detection). A comparative study was also done among the roots and leaves of *in vitro* grown plantlets, donor plants and control groups. Leaf volatiles from *in vitro* grown plants of all the PGR combinations showed a remarkable increase in major constituents such as α-pinne, β-pinene, caryophyllene *etc*., as compared with essential oil from donor plants.

Multiple Shoot Induction from Rhizome Bud

A protocol for the *in vitro* culture of *A. zerumbet* from rhizome bud explants on MS medium fortified with combinations of BAP and Kn and the variation in growth rate according to the hormone composition was performed by Rakkimuthu and colleagues in 2011 [34]. After proper surface sterilization, the explants were cultured on MS medium augmented with a combination of 0.5 mg/L Kn and different concentrations of BAP (0.5, 1.0, 1.5, 2.0, and 2.5 mg/L). Evaluation of the percentage of regeneration and number of shoots/explant showed that a composition of 0.5 mg/L Kn and 1.5 mg/L BAP displayed the best result in the case of both the regeneration percentage and shoot growth (7.9 shoots/explant). Among the tested combinations of media, the least successful result was obtained on the culture media (0.5 mg/L BAP and Kn) with 3.5 shoots/explant with 55% regeneration. Multiple shoots were formed within 6-7 weeks of inoculation on the media augmented with 1.5 mg/L of BAP and 0.5 mg/L of Kn. ½ strength MS augmented with 0.5 mg/L of IBA showed better rooting (Table **1**). The plantlets were successfully hardened and adapted to the field conditions.

Direct Organogenesis, Callogenesis and Antiproliferative Assay

A promising protocol for the callogenesis of *A. zerumbet* from primordial leaf explants of 8-week-old *in vitro* grown plantlets from rhizome explants was established by Shahin *et al.* [35]. Sterile rhizome explants were cultured on MS medium fortified with a combination of 2.3μM Kn and 6.65 μM BAP for rapid direct regeneration. During the subculture, the biggest aerial shoots were removed every 4 weeks. Investigation on the role of growth supplements in callogenesis was observed by the culture of primordial leaf explants of 8-week-old from *in vitro* grown shoots. For callogenesis, various concentrations and combinations of 2,4-D (9.1, 13.6 & 18.2 µM), BAP (0.5 µM BAP+9.1, 13.6 & 18.2 µM 2,4-D), and Kn (9.3 µM Kn+0.5 µM BAP+18.2 µM 2,4-D) were augmented with MS media. Among the tested seven combinations, superior performance in the fresh calli weight (3.87 g/explant) was obtained from MS + 9.1µM 2,4-D + 0.5 µM BAP, whereas MS medium along with 9.1 µM 2,4-D showed the least calli mass. The antiproliferative activity of a crude methanol extract of the root, leaf, and calli of *A. zerumbet* cultured on various media compositions was determined using the MTT assay and the human cervical carcinoma (HeLa) cell line. A direct correlation was observed between the concentration of extract and antiproliferative efficacy. Callus extract on MS medium along with 9.1 µM 2,4-D + 0.5 µM BAP and MS+ 9.3 µM Kn + 0.5 µM BAP+ 9.1µM 2,4-D inhibited HeLa cells in a dose-dependent manner.

Callogenesis and Plantlet Regeneration

Large scale production of *A. zerumbet* was achieved by the establishment of an efficient *in vitro* regeneration system. A study on callus induction, callus differentiation, and rooting was achieved from aseptic cultures of *A. zerumbet* seeds. Analysis of the study demonstrates that MS media augmented with composition of 2,4-D (0.3 mg/L) and BAP (1.5 mg/L) was the most suitable medium for callus induction. The desirable composition of media for the shoot differentiation from callus was found on MS medium augmented with a combination of 2,4-D and TDZ at 0.10 mg/L and 2.00 mg/L, respectively, with a maximum differentiation constant of 10.03. The maximum rate of rooting (98.00%) was exhibited in mediums containing ½ MS supplemented with 1.0 mg/L ABT1 rooting powder (Table **1**). The result of the study showed that an effective *in vitro* regeneration of *A. zerumbet* was possible by callus culture [36].

CONCLUSION

Alpinia is a tropical and sub-tropical Asian genus, having various usages. Many species are used by traditional medicinal practitioners for curing different ailments. The species also has environmental uses and ornamental values. *Alpinia calcarata*, a medicinal species, is chiefly propagated *in vitro* through direct organogenesis using axillary rhizome buds as explants. *Alpinia galanaga* is one of the medicinal and ornamental species studied in detail. Direct regeneration, callogenesis, indirect organogenesis and SEs (somatic embryogenesis) were well established in this taxon. Genetic and phytochemical fidelities of the regenerants have also been well studied in *A. galanga. In vitro* propagation protocols for *A. malaccensis* and *A. officinarum* are also available. Micropropagation and genetic fidelity of the *in vitro* regenerated plantlets were frequently studied in *A. purpurata* and *A. zerumbet*. However, in general, micropropagation and *in vitro* studies are scanty in *Alpinia*. Therefore, micropropagation techniques have yet to be developed for more than 95% of the species in this genus.

REFERENCES

[1] POWO. Plants of the World Online.Facilitated by the Royal Botanic Gardens, Kew. Available from: http://www.plantsoftheworldonline.org/ (Retrieved 01 July 2022.).

[2] Ma XN, Xie CL, Miao Z, Yang Q, Yang XW. An overview of chemical constituents from *Alpinia* species in the last six decades. RSC Adv 2017; 7(23): 14114-44.
[http://dx.doi.org/10.1039/C6RA27830B]

[3] Nam J-W, Kim S-J, Han A-R, Lee SK, SEO E-K. Cytotoxic Phenylpropanoids from the rhizomes of *Alpinia galanga*. J Appl Pharmacol 2005; 13: 263-6.

[4] Tachakittirungrod S, Chowwanapoonpohn S. Comparison of antioxidant and antimicrobial activities of essential oils from *Hyptis suaveolens* and *Alpinia galanga* growing in Northern Thailand. C J Nat Sci 2007; 6: 31-42.

[5] Mahae N, Chaiseri S. Antioxidant activities and antioxidative components in extracts of *Alpinia galanga* (L.) Sw. Witthayasan Kasetsat Witthayasat 2009; 43: 358-69.

[6] Jirovetz L, Buchbauer G, Shafi MP, Leela NK. Analysis of the essential oils of the leaves, stems, rhizomes and roots of the medicinal plant *Alpinia galanga* from southern India. Acta Pharm 2003; 53(2): 73-81.
[PMID: 14764241]

[7] Ye Y, Li B. 1′S-1′-Acetoxychavicol acetate isolated from Alpinia galanga inhibits human immunodeficiency virus type 1 replication by blocking Rev transport. J Gen Virol 2006; 87(7): 2047-53.
[http://dx.doi.org/10.1099/vir.0.81685-0] [PMID: 16760408]

[8] Uchiyama N. Antichagasic activities of natural products against *Trypanosoma cruzi*. J Health Sci 2009; 55(1): 31-9.
[http://dx.doi.org/10.1248/jhs.55.31]

[9] Kaur A, Singh R, Dey CS, Sharma SS, Bhutani KK, Singh IP. Antileishmanial phenylpropanoids from *Alpinia galanga* (Linn.) Willd. Indian J Exp Biol 2010; 48(3): 314-7.
[PMID: 21046987]

[10] Pompimon W, Jomduang J, Prawat U, Mankhetkorn S. Anti-*Phytopthora capsici* activities and potential use as antifungal in agriculture of *Alpinia galanga* Swartz, *Curcuma longa* Linn, *Boesenbergia pandurata* Schut and *Chromolaena odorata*: Bioactivities guided isolation of active ingredients. Am J Agric Biol Sci 2009; 4(1): 83-91.
[http://dx.doi.org/10.3844/ajabssp.2009.83.91]

[11] Chen SL, Yu H, Luo HM, Wu Q, Li CF, Steinmetz A. Conservation and sustainable use of medicinal plants: problems, progress, and prospects. Chin Med 2016; 11(1): 37.
[http://dx.doi.org/10.1186/s13020-016-0108-7] [PMID: 27478496]

[12] Sahoo S, Singh S, Sahoo A, *et al.* Molecular and phytochemical stability of long term micropropagated greater galanga (*Alpinia galanga*) revealed suitable for industrial applications. Ind Crops Prod 2020; 148: 112274.
[http://dx.doi.org/10.1016/j.indcrop.2020.112274]

[13] Sudha CG, George M, Rameshkumar KB, Nair GM. Improved clonal propagation of *Alpinia calcarata* Rosc., a commercially important medicinal plant and evaluation of chemical fidelity through comparison of volatile compounds. Am J Plant Sci 2012; 3(7): 930-40.

[http://dx.doi.org/10.4236/ajps.2012.37110]

[14] Murashige T, Skoog F. A revised medium for rapid growth and bio assays with Tobacco tissue cultures. Physiol Plant 1962; 15(3): 473-97. [http://dx.doi.org/10.1111/j.1399-3054.1962.tb08052.x]

[15] Asha KI, Devi AI, Dwivedi NK, Nair RA. *In vitro* propagation of Lesser Galangal (*Alpinia calcarata* Rosc.) : A commercially important medicinal plant through rhizome bud culture. Res Plant Biol 2012; 2: 13-7.

[16] Mathew S, Britto SJ, Thomas S. *In vitro* Conservation strategies for the propagation of Alpinia calcarata Roscoe (Zingiberaceae) : A valuable medicinal plant. Am J Pharmacol Pharmacother. 2014; pp. 59-67.

[17] Das Bhowmik SS, Basu A, Sahoo L. Direct shoot organogenesis from rhizomes of medicinal zingiber *Alpinia calcarata* Rosc. and evaluation of genetic stability by RAPD and ISSR markers. J Crop Sci Biotechnol 2016; 19(2): 157-65. [http://dx.doi.org/10.1007/s12892-015-0119-4]

[18] Borthakur M, Hazarika J, Singh RS. A protocol for micropropagation of *Alpinia galanga.* Plant Cell Tissue Organ Cult 1998; 55(3): 231-3. [http://dx.doi.org/10.1023/A:1006265424378]

[19] Singh NM, Chanu LA, Devi YP, Singh WRC, Singh HB. Micropropagation- an *in vitro* technique for the conservation of *Alpinia galanga.* Pelagia Res Libr Adv Appl Sci Res 2014; 5(3): 259-63.

[20] Shamsudheen KM, Mehaboob VM, Thiagu G, Shajahan A. High frequency shoot multiplication of *Alpinia galanga* (L.) Willd. using rhizome buds. Res J Life Sci Bioinform Pharm Chem Sci 2018; 4(4): 579-85. [http://dx.doi.org/10.26479/2018.0404.51]

[21] Rao K, Chodisetti B, Gandi S, Mangamoori LN, Giri A. Direct and indirect organogenesis of *Alpinia galanga* and the phytochemical analysis. Appl Biochem Biotechnol 2011; 165(5-6): 1366-78. [http://dx.doi.org/10.1007/s12010-011-9353-5] [PMID: 21892666]

[22] Parida R, Mohanty S, Nayak S. Evaluation of genetic fidelity of *in vitro* propagated greater galangal (*Alpinia galangal* L.) using DNA based markers. Int J Plant, AnimEnviron Sci 2011; 1(3).

[23] Rg B, Mv R, T SK. Curbing Actinomycetes and Thidiazuron enhanced micropropagation in the rare *Alpinia galanga* : A medicinal Zingiber. Asian J Pharm Clin Res 2017; 10(7): 153. [http://dx.doi.org/10.22159/ajpcr.2017.v10i7.17734]

[24] Mustafaanand PH. *In vitro* propagation of *Alpinia galanga* L. ,*via* somatic embryogenesis- A medicinal plant. Eur J Pharm Med Res 2019; 6: 607-10.

[25] Benjamin S, Preethi TP, Shinija K, Rakhi KP, Sabu M. Micropropagation and chemical profiling of *Alpinia malaccensis* and *Hedychium coccineum.* J Trop Med Plants 2009; 10: 95-9.

[26] Selvakkuma C, Balakrishn A, Lakshmi BS. Rapid *in vitro* micropropagation of *Alpinia officinarum* Hance, an important medicinal plant, through rhizome bud explants. Asian J Plant Sci 2007; 6(8): 1251-5. [http://dx.doi.org/10.3923/ajps.2007.1251.1255]

[27] Arunyanart S. Tissue culture of red ginger (*Alpinia purpurata*). Warasan Kaset Phrachomklao 1994; 12: 63-72.

[28] Kochuthressia KP, Britto SJ, Raj LJM, Jaseentha MO, Senthilkumar SR. Efficient regeneration of *Alpinia purpurata* (Vieill.) K. Schum. plantlets from rhizome bud explants. Int Res J Plant Sci 2010; 12: 043-7.

[29] Illg RD, Faria RT. Micropropagation of *Alpinia purpurata* from inflorescence buds. Plant Cell Tissue Organ Cult 1995; 40(2): 183-5. [http://dx.doi.org/10.1007/BF00037673]

[30] Kale VM, Namdeo AG. Micropropogation of *Alpinia purpurata* using low cost media for quantification of rutin. Pharm Lett 2015; 7: 50-7.

[31] Victorio CP, Cruz IPDA, Sato A, Kuster RM, Lages CLS. Effects of auxins and cytokininson *in vitro* development of *Alpinia purpurata* (Vieill) K.Schum and phenolic compounds production. In: Paiva R, Torres AC, Pasqual M, Paiva R, Eds. Plant cell culture & micropropagation. Lavras, MG 2008; pp. 92-8.

[32] Victório CP, Arruda RCO, Lage CLS, Kuster RM. Production of flavonoids in organogenic cultures of Alpinia zerumbet. Nat Prod Commun 2010; 5(8): 1934578X1000500. [http://dx.doi.org/10.1177/1934578X1000500815] [PMID: 20839623]

[33] Victório CP, Kuster RM, Lage CLS. Leaf and root volatiles produced by tissue cultures of *Alpinia zerumbet* (pers.) Burtt & Smith under the influence of different plant growth regulators. Quim Nova 2011; 34(3): 430-3. [http://dx.doi.org/10.1590/S0100-40422011000300012]

[34] Rakkimuthu R, Jacob J, Aravinthan KM. *In vitro* micropropagation of *Alpinia zerumbet* Variegate, an important medicinal plant, through rhizome bud explants. Res Biotechnol 2011; 2: 7-10.

[35] Shahin H, Nasr MY, Nasr GMI. *In vitro* production of callus cultures of *Alpinea zerumbet* and their cytotoxicity on cancer cells. Biosci Res 2019; 16: 870-6.

[36] Mao L, JinHua L, QiuMing T, *et al.* Establishment of *in vitro* propagation technology of *Alpinia zerumbet* Variegata. J South Agric 2016; 47: 1909-13.

CHAPTER 4

Micropropagation of Ginger (*Zingiber officinale* Roscoe)

Nayan Kumar Sishu[1], **Parasurama Deepa Sankar**[2] and **Chinnadurai Immanuel Selvaraj**[2,*]

[1] *School of Biosciences and Technology, Vellore Institute of Technology, Vellore, Tamil Nadu, PIN 632 014, India*

[2] *Department of Genetics and Plant Breeding, School of Agricultural Innovations and Advanced Learning, Vellore Institute of Technology, Vellore, Tamil Nadu, PIN 632 014, India*

Abstract: *Zingiber officinale*, belonging to the family of Zingiberaceae, is commonly known as ginger and is commercially grown as a spice and for culinary purposes. It is a potential Ayurvedic herb with many medicinal properties. A small section of the plant's rhizome is widely used for micropropagation. Besides rhizome explants, callus induction, shoot induction, and meristem culture are used to propagate the plant. For the production of ginger's pest-resistant and disease-free planting material, micropropagation is regarded as the best method. Various classes of bioactive entities, such as flavonoids, alkaloids, glycosides, phenols, tannins, terpenoids, steroids, saponins, and oils, have been identified in the plant. Phenolic bioactives such as gingerols and shogaols are primarily responsible for their therapeutic properties. Various pharmacological activities have been investigated in ginger. This review concentrates on different advanced methods for ginger propagation, especially micropropagation.

Keywords: Ginger, Micropropagation, Multiplication, Plant regeneration, Shoot induction, Zingiber officinale, Zingiberaceae.

INTRODUCTION

Zingiber officinale Roscoe is a perennial herb which is believed to be indigenous to India and China and is widely cultivated in tropical regions such as Southeast Asia, the Caribbean, Central and South America, Australia, and Africa [1]. *Z. officinale* is widely used as Ayurvedic and traditional Chinese medicine because of its promising health benefits. It is widely used to treat numerous health disorders, namely fever, bronchitis, sore throat, digestive issues, nausea, vomiting,

* **Corresponding author Chinnadurai Immanuel Selvaraj:** Department of Genetics and Plant Breeding, School of Agricultural Innovations and Advanced Learning, Vellore Institute of Technology, Vellore, Tamil Nadu, PIN 632 014, India; E-mail: immanuelselvaraj@vit.ac.in

T. Pullaiah (Ed.)

common cold, abdominal pain, and rheumatism [2]. Raw ginger is dried and powdered and used to make tea, which is usually taken to get relief from cold and cough. *Z. officinale* rhizomes are often chewed raw, and they help in reducing abdominal pain. The rhizomes are usually boiled with other herbal ingredients like basil, turmeric, black pepper, and bay leaf to make *kadha* or decoction, which has many health benefits and helps improve body immunity. During the pandemic of COVID-19, the consumption of ginger-based decoction and tea was recommended by health experts to prevent various respiratory disorders such as cold, flu, and cough. Raw ginger rhizome is a home remedy to prevent nausea and constipation. Ginger has stimulant, carminative, and diaphoretic properties; used to flavour food, beer, and other beverages; used in curries as a condiment. Ginger rhizome possesses diuretic, anti-inflammatory, anti-emetic, and sialagogic properties. Rhizome juice is used to prevent migraine, colic, and catarrh. Moreover, it is also believed to provide relief from pain due to menstrual cramps. Ginger oil is used externally to relieve toothache, headache and pain due to swelling and boils; leaves are consumed in Malaysia to reduce stomach pain and rheumatic diseases. Fresh leaf juices are used externally to reduce the symptoms of ague in children. In China and Indonesia, the ginger rhizome paste is used as an antispasmodic and applied externally as an antidote against snake, fish and crab stings.

Ginger is generally propagated through the rhizome. Using rhizome as a starting material affects the yield and supply in the market as the cultivators have to store a high proportion of rhizome for the next growing season. Moreover, the rhizome is easily affected by bacteria and fungi, the newly grown leaves also get infested by pests easily, and all these aspects adversely affect ginger production. This method of cultivation is costly and even needs more human resources; therefore *in vitro* propagation or micropropagation of the crop will be fruitful for producing healthy planting material, which can help achieve the growing demand for ginger. Various types of explants, such as axillary buds from rhizomes, vegetative buds, shoot tips, young buds, root tips, sprouting buds, and rhizome buds, were used for micropropagation of different ginger varieties [3 - 5]. Micropropagation of ginger in a microbial-free controlled condition aids in producing high-frequency multiplication of disease-free clones of ginger.

The pungent smell of ginger is due to the homologous series of phenolic compounds called gingerols. Studies reported the presence of a wide array of phytochemicals in ginger, such as essential oils, phenolic compounds, flavonoids, carbohydrates, proteins, alkaloids, glycosides, saponins, steroids, terpenoids and tannin. It is rich in geraniol, α-curcumene, (E, E)- α-farnesene, α-zingiberene, and β-sesquiphellandrene [6]. Phytochemicals present in ginger exhibit good antioxidant, antimicrobial, antiserotonergic, antispasmodic, anticonvulsant, analgesic, anti-inflammatory, antiulcer, larvicidal, and immunomodulatory activities [7, 8]. Ginger extracts are found to have a potential role in treating Alzheimer's disease and improving the function of the nerve cells [9, 10]. It was reported that in the scopolamine-induced memory deficit paradigm, dried ginger extract at three different doses, containing 6-shogaol (11.7%) and 6-gingerol (5.52%), restored cognitive abilities and delay in the time for the passive avoidance of learning exercise in mice [11]. It was investigated that 10-gingerol (30 M) inhibited 50% of the growth of HCT116 cells with abnormalities characteristic of programmed cell death [12].

Clinical studies among 45 randomized 20-60 age group patients with Type-2 diabetes melitus were conducted where the patients had not been given any insulin treatment for three months. They are subjected to oral administration of powdered ginger (3g/day), which exhibited a significant decrease in blood glucose, triglyceride, highly sensitive-C-reactive protein, and malondialdehyde, and an increase in paraoxonase-1 and total antioxidant capacity (TAC); this indicates that ginger has a potential antidiabetic effect by maintaining blood glucose homeostasis and lowering oxidative stress in the body [13]. Furthermore, it was investigated that the antimicrobial activity of ginger in bacteria forms a zone of inhibition; *Citrobacter* spp. (14 mm), *Escherichia coli* (9.5 mm), *Salmonella* spp. (11.1 mm), *Shigella* spp. (12 mm), and *Enterobacter* spp. (0.66 mm) [14].

MICRO PROPAGATION OF GINGER

Micro-propagation is a method of propagating crops that are infertile and lack natural seeds or reproduce through vegetative propagation. Ginger rhizomes are easily affected by fungi (*Pythium* spp. and *Fusarium oxyporium*) and bacteria (*Ralstonia solanacearum*), which cause rot disease. It spreads from infected rhizomes to whole crop fields and leads to a significant loss in the production of ginger [15, 16]. To protect the plant from pathogens and microbes, applying tissue culture or micropropagation with an effective protocol can help increase ginger yield. The first and vital step for micropropagation is to select disease-free explants, sterilize and establish the culture and then induction of shoot and root multiplication, followed by acclimatization. *In vitro* propagation's most crucial function is to preserve genetic diversity and the evolutionary process in populations of ecologically and economically viable varieties and genotypes to prevent them from extinction [17].

Source and Type of Explants

According to the stages of a plant's development and environmental changes, the plant's physiological state changes spontaneously. Axillary buds and active shoot tips are most commonly used for micropropagation. Different hormone concentrations may be needed during the early establishment of the culture if the ginger rhizome is used as an explant, which is less active or dormant to encourage the sprouting of buds for the formation of shoots. Since rhizome interaction with the soil is a common cause of contamination of field-grown rhizomes in *in vitro* settings, caution should be taken when disinfecting it. Rhizome explants from ginger plants raised in glasshouses could be used to lower the level of contamination. Explants from strong, robust plants are better suited to produce efficient *in vitro* cultures [18].

Explants of various forms, including axillary buds, meristems, aerial pseudostems, and shoot tips, are employed in the micropropagation of ginger and other related species. However, shoot tips and rhizome buds are frequently utilized as explants as they are pathogen-free propagules and effective for micropropagation on a large scale [19]; this is consistent with the results that axillary buds are the most suitable explants for successful clonal multiplication of *Z. officinale* [20]. *Fusarium* and nematode-free rhizomes of *Z. officinale* were used as explants for *in vitro* cultures. The rhizomes were surface-sterilized with 1% sodium hypochlorite for 2 minutes. Furthermore, it was kept at room temperature in the lab until it sprouted. Emerging buds (10 mm^3) were taken out, surface-sterilized for 15 minutes in 3% sodium hypochlorite, and then thoroughly rinsed three times in sterile water. The explant was prepared by removing the bleached substance, seeding it in Murashige and Skoog's [21] basal media (MS-media) supplemented with 3% sucrose and 2.5 mg/L 6-Benzylaminopurine (BAP), and with 0.8% Difco Bacto-agar as a solidifying agent [22].

Explant Sterilization

Detergent solutions, water (running tap water, single or double-distilled, and sterilized), antifungal and antibacterial agents, calcium oxychloride, Sodium hypochlorite, ethyl alcohol, mercuric chloride, and antibiotics are the standard agents of surface sterilization of explants and restraining *in vitro* growth media contamination. Sodium hypochlorite, RBK (lab-made antimicrobial agent composed of bayleton, kocide and ridomile with the ratio of 1:1:1) detergent, ethyl alcohol, and mercuric chloride are the most generally used external sterilization agents in readying sterile explants for *in vitro* proliferation of numerous species [23].

The shoot tip explants were rinsed comprehensively with running water and soap solution to clear dirt trails. They were flushed with sterile purified water until all soap solution traces were cleared. The study was devised to experiment with the efficacy of 3 external sterilization agents, *viz*., 0.25%v/w RBK (3 treatments),

0.50%v/v Sodium hypochlorite (3 treatments) and 70%v/v ethanol (3 treatments) in combination with of 0.25% mercuric chloride. RBK (0.25%) surface sterilization treatment of the explants was done for 30, 45, and 60 minutes; surface sterilization utilizing 70% ethanol involved treatment for five, ten, and fifteen minutes. The sterilization approach using 0.5% Sodium hypochlorite involved the treatment of explants for ten, twenty, and thirty minutes. In all circumstances, the explants were sterilized externally with 0.25% mercuric chloride for ten minutes, additionally under the laminar airflow chamber. Eventually, the explants were flushed with sterile water to dispose of any traces of the mercuric chloride and were prepared for inoculation. Sodium hypochlorite (0.50%v/v) at twenty minutes of exposure rendered a remarkably highest mean number (8.00±1.73) of sterile explants after twenty-one days of gestation. Sodium hypochlorite at twenty and thirty minutes and 70% ethanol at five and ten minutes of exposure yielded significantly higher mean numbers of live explants (7.3±0.5 to 7.0±1.0). In contrast, 70% ethanol was less efficacious in disinfecting capability [5].

The soft yellow sprouting buds were surface sterilized using a mercuric chloride (0.2%) aqueous solution and then flushed several times in sterile purified water [24]. Pure rhizome portions with shoot tips were extensively cleansed five to six times in sterile water before being surface sterilized in 0.1% mercuric chloride solution for fifteen minutes. The ginger rhizomes' shoot tips were then sliced off, gathered in a Petri dish, and inoculated [25]. Disease-free rhizomes having sprouting rhizome buds are washed with water and then transferred in Tween 20 solution for 5-10 min and afterwards treated with Clorox solution. Finally, surface sterilization is done using 95% ethanol, 2.5% (v/v) sodium hypochlorite, followed by 0.1% (w/v) mercuric chloride. Afterwards, the rhizomes are washed with distilled water, and the shoots are carefully trimmed off to obtain rhizome explants of a final size of 1 cm; then, the explants are cultured in MS medium supplemented with sucrose and solidified with agar or Gelrite. The pH of the culture medium was adjusted to 5.5-5.8 before adding the solidifying agent. Then, it was autoclaved for 20 min at 121℃ and 104 kPa pressure. The culture vials or tubes were incubated at 25℃ to 27℃ for 16 hours under a photoperiod of 35-40 mol/m^2/s using a cool white fluorescent light [16, 26 - 29].

Culture Medium

The emerging buds from ginger rhizomes were inoculated in Difco Bacto-agar (0.8%) with 0.3% sucrose and 2.5 mg/L BAP in MS basal medium. The cultures were kept at 25°C for a 16-hour photoperiod. Excellent shoot multiplication was observed under these conditions. This medium also aided in developing strong roots, making it simple to deflask and establish plantlets in the glasshouse [22].

Emerging buds from ginger's rhizome were cultured on MS-medium with 3% sucrose, 0.5% agar, and various doses of Kinetin (Kn), 2,4-dichlorophenoxyacetic acid (2,4-D), and BAP. The medium's pH was adjusted to 5.8 before autoclaving. Shoots with readily available roots emerged in the same medium four weeks after inoculation [25]. Trimmed buds were inoculated into Gamborg B5 medium gelled with 2.8 grams per litre Gelrite and 30% sucrose as a carbon source. The medium was supplemented with two millilitres per litre of Plant Preservative Mixture (PPM) and 15 mg/L of tetracycline to restrict contamination. Before wet sterilization, the pH was adjusted to 5.8 with 1N potassium hydroxide or 0.1N hydrochloric acid [30].

Effect of Different Plant Growth Regulators and Other Parameters

Studies revealed that even when the cultures were kept longer than the typical monitoring period of four weeks, culture media lacked any growth regulators (control) and could not trigger the bud break action in the explants. The shoot length and the average number of nodes produced per shoot, MS medium with growth regulator supplements, gave better results concerning the explant response percentage. Moreover, the ideal multiplication of shoots was observed in MS medium supplemented with 2.0 mg/L BAP + 0.5 mg/L 1-Naphthaleneacetic acid (NAA). In contrast, MS with NAA (2.0 mg/L) was the standard growth regulator for profuse rooting in the explants for the Suprava and Suruchi varieties of *Z. officinale* [17].

It was reported that microshoots of *Z. officinale* var. *rubrum* on MS medium supplemented with BAP (3 mg/L) + NAA (0.5 mg/L) exhibited enhanced shoot multiplication (19.50 ± 2.30). However, it antagonistically reduced the shoot length. In contrast, BAP with three mg/L of NAA increased the root multiplication rate, thus increasing the number of roots (11.60 ± 2.10) of *Z. officinale* var. *rubrum*. The maximum number of plantlets (23.0 ± 2.50) and roots per explants (15.40 ± 2.40) were produced by microshoots kept on MS medium supplemented with 45 g/L sucrose, although the average length of roots (2.6 ± 0.2 cm) was reported to be decreased in size [26].

MS medium supplemented with Indole-3-butyric acid (IBA) at different pH, and cytokinin concentrations significantly affect the *in vitro* shoot and root propagation of *Z. officinale* fresh rhizome buds. The highest root and shoot development was achieved at pH 5.8 compared to the culture medium with pH 5.6 and 6.0. Moreover, explants cultured on MS basal medium supplemented with 2.0 mg/L Kn + 1.0 mg/L IBA gave the highest shoot multiplication rate, shoot length and root number. Further, it was concluded that Kn (2 mg/L) could be used as the best hormone for the ideal growth of the local Yemeni ginger variety [31]. Using cytophotometric determination of 4C nuclear DNA content and random amplified polymorphic DNA (RAPD), the genetic stability of micropropagated clones of *Z. officinale* obtained from axillary buds from unsprouted rhizomes was examined in MS culture at periodic intervals of 6 months up to 24 months. The genetic uniformity of *in vitro* grown clones of *Z. officinale* was revealed by cytophotometric analysis, which showed a monotonic distribution of the DNA content with high linearity to the 4C value (23.1 pg.), and RAPD analysis, which illustrated monomorphic bands indicating the absence of polymorphism in all 50 regenerants [32].

It was observed that the best substrate for tissue culture plantlets was found to be vermiculite + peat (1:1 v/v) with MS + 0.5 mg/L NAA + 0.1% Metalaxyl-M·Hymexazol + 80% relative humidity. Further, rootless ginger tissue culture plantlets were started in seedling trays grown in a chamber maintained under a 16 h/day photoperiod with 60 μ E m^{-2} s^{-1} of cool white light from fluorescent lamps at 25 ± 2 °C. *Ex-vitro* ginger seedlings were successfully developed after 40 days of culture, with a rooting rate of 100.0%. They could be transplanted directly into the fields [33].

It was reported that the *in vitro* vegetative growth of *Z. officinale* var. *rubrum* was considerably but variably impacted by the light emission of light-emitting diodes (LEDs). LED lights in green, red, and purple colours significantly aided microshoot production [34]. Purple LED light encouraged the production of shoots and leaves and rooting ability in *in vitro Z. officinale* var. *rubrum* culture, whereas red LED light source enhanced rooting and shoot length. As a result, purple LED will be the ideal light source for *Z. officinale* var. *rubrum* micropropagation.

Reports suggest that roughly 61% of the rhizomes of mother-derived plants rotted during storage; rhizomes obtained from tissue culture-derived plants did not exhibit any rotting of rhizomes caused by *F. oxysporum* f. sp. *zingiberi* infection after being stored in the sand for six months. Portions of micropropagated plant rhizomes were transferred to Potato dextrose agar (PDA) medium to check for *F. oxysporum* f. sp. *zingiberi* infection. After 8 to 10 days of incubation, mycelial growths were not seen on the medium, but 79% of the pieces of traditional rhizomes exhibited mycelial growth. As a result, the *in vitro* regeneration process can quickly produce many disease-free somaclones from the target cultivar [35].

Shoot and root propagules of the two ginger cultivars, var. "Jamaica" and var. "Varada," was established from aerial stem explants. The cultures with Thidiazuron (TDZ) and IBA showed excellent shoot and root regrowth. The development of roots and shoots was inadequate in cultures containing only TDZ.

According to histological research, the primordial root developed from the primary thickening meristem in the aerial stem. In contrast, the shoot initials came from the primary thickening meristem and apical meristem [36].

Reports suggest that in order to induce polyploid plants, the adventitious buds were placed in various concentrations of colchicine water solution for varying periods to induce polyploid plants, and 48 tetraploid lines of ginger were obtained. The induced buds were identified by determining the chromosomes at the root tip and observing the stomatal apparatus. The induction rate of the treated buds increased to 33.3%. All tetraploid plants displayed standard polyploidy traits. Compared to the diploid mother plant lines, 18 tetraploid ginger lines produced more rhizomes, volatile oils, and gingerol per plant [37].

Studies revealed that different amounts of sucrose, plant growth regulators, ammonium nitrate, and silver nitrate were examined to identify the ideal conditions for ginger microrhizome development. With the application of BAP, NAA, IBA, and a low ammonium nitrate concentration, microrhizome fresh weight and diameter rose to the maximum values, with 0.433 g at 9.03 mm, 0.437 g at 9.73 mm, 0.478 g at 10.80 mm, and 0.449 g at 9.53 mm, respectively. Furthermore, it was observed that Kn inhibits the growth of microrhizomes. It was reported that the highest microrhizomes were produced on MS media with 80 g/L sucrose, 1.9 mg/L silver nitrate ($AgNO_3$), 550 mg/L ammonium nitrate, 4 mg/L BAP, 6 mg/L NAA, and 4 mg/L IBA [38].

Callus Induction from Anther, Ovary, Sprouted Buds and Primordial Leaf Explants

Disease-free inflorescence was collected and washed thoroughly with running water; they were exposed to cold treatment for nine days. Afterwards, the flowers were separated, washed with mercuric chloride, and then washed with distilled water. Anthers were collected from the flower carefully and then transferred to a culture medium containing MS media and agar with some growth inducer such as 2, 4-D for callus initiation. Afterwards, the culture plates were incubated at 25 ℃ with a 14-hour photoperiod provided by cool white fluorescent lights giving 2000 to 3000 lux light intensity. Once the callus development was observed, the anthers were transferred to a basal medium containing MS media and agar with different concentrations of NAA and 2,4-D, BAP and Kn for regeneration. Further, to complete the regeneration process, the callus is transferred twice at monthly intervals to a subculture containing the same medium [39, 40]. The overview of procedures involved in micropropagation of *Zingiber officinale* is given in Fig. (**1**).

Fig. (1). Micropropagation of *Zingiber officinale***(A)** Explant for culture initiation; **(B)** Culture Establishment of Explants after surface sterilization; **(C)** *In vitro* shoot multiplication (after 10 days of inoculation on MS medium supplemented with 10 µM zeatin); **(D)** Multiple shoots produced in MS medium supplemented with 10 µM of zeatin after six weeks of inoculation; **(E)** *In vitro*-raised shoot used as explant for rooting; **(F)***In vitro*-rooted plantlet derived from MS medium supplemented with 7.5 µM NAA after four weeks of culture; **(G)** Acclimatized plantlets after three weeks of acclimatization; **(H)** *In vitro*-raised plant of *Zingiber officinale* after seven months of transplanting (Source: Zahid *et al.* [29]).

As per reports, the inflorescence was collected and washed with 5% teepol solution followed by distilled water; then, surface sterilization was done using mercuric chloride and washing the inflorescence with sterile distilled water for 4-5 times. The flowers were separated, then ovaries were removed from the flowers carefully by maintaining an aseptic condition; then, they were transferred to MS medium containing agar and supplemented with 2,4-D and BAP for inducing somatic embryogenic culture. Then, the embryoids were transferred in MS basal medium supplemented with NAA for callus development. Then, the cultures are incubated at 25 ℃ with a photoperiod of 14 hours [40]. Similarly, sterilized sprouting buds were grown on MS medium supplemented with various growth regulators. It was reported that MS-medium supplementation with BAP resulted in the highest percentage of shootlets multiplication [3, 40].

It was reported that *in vitro* callus development was induced from primordial leaves of eight-week-old plantlets cultures. For this, the disease-free basal primordial leaves were carefully isolated, cut into small segments, washed with distilled water, and transferred to the MS culture medium containing 3% sucrose, 0.7% agar, and different concentrations of BAP and 2,4-D. All culture plates were incubated at 25 ℃ with a 16-hour photoperiod provided by cool white fluorescent lights giving 2000 to 3000 lux light intensity [41].

Shoot Multiplication

It was reported that for shoot multiplication, the shoot explants having shoot buds were cultured in MS media supplemented with sucrose, mesoinositol and different concentrations of BA and NAA combined with Kn or other hormones [4, 16, 26, 28]. Reports suggest that the *in vitro* derived shoots were grown on various growth regulators supplied in MS medium for multiplication after four weeks. From a group of multiple shoots, single shoots generated *in vitro* were separated and sub-cultured on a medium with the same composition for multiplication and rooting [19]. As per reports, the appearance of new shoot primordia from the base of the primary shoot marks the start of the shoot proliferation. After 6.87 and 7.13 days of culture, the appearance of the new shoot primordia was observed in the zeatin and BAP treatments. Shoots began to grow much more slowly in the Kn and TDZ treatments than in the zeatin treatment. After 10.83 days of culture, the control treatment showed the slowest explant response to shoot initiation [29].

Root Induction

Studies indicated that after four to six weeks of culture, the few shoots grown in an *in vitro* culture medium were isolated and sub-cultured in the same growth regulator medium containing MS medium, agar, sucrose, and different concentrations of BAP, NAA or Indole-3-acetic acid (IAA), which stimulates the process of root development. The culture flask was incubated at 25 ℃ -27 ℃ with a 16-hour photoperiod provided by cool white 40 watt (W) fluorescent lights giving 2000 to 3000 lux light intensity. The relative humidity of 60 to 70% was maintained [4, 16, 26 - 28]. Reports indicate that auxins of various types and concentrations significantly altered ginger's number of roots and root length from the "Bentong" variety. The concentrations of various auxin types affected how well roots were induced. The number of roots per plantlet was not substantially different at a lower concentration of 2.5 μM IAA and NAA. However, the number of roots per plantlet varied dramatically at higher concentrations (5 or 7.5 μM). The number of roots per plantlet increased when the NAA concentration was raised; the most significant number of roots per plantlet (15.44 ± 0.8) was produced with 7.5 μM of NAA [29]. The report suggests that 4.5 mg/L BAP produced the most *in vitro* shoots and roots per explant for direct organogenesis, with 4 ± 0.35 shoots and 15 ± 0.46 roots [42].

Acclimatization

Well-developed micropropagated plantlets were deflasked, cleaned adequately under running water to remove media from the roots, and then planted in plastic pots (5 cm) with a potting combination of sand, soil, and farmyard manure (1: 1: 1 w/w). Different treatments were applied to the potting mixture: (1) steam sterilization in an autoclave for 2 hours at 121.6°C and 1.05 kg cm^{-2} pressure; (2) soaking in the 1 g/L fungicide Bavistin (methyl-2-benzimidazole carbamate, $C_9H_9N_3O_2$), and (3) leaving it untreated. A polythene bag was placed over the pots to keep the humidity high for five to six days to help the plants adjust to the outside environment. Ginger plantlets generated *in vitro* were transplanted to pots in both sterilized and non-sterile potting mixtures. Clones survived 97% of the time in sterilized and 93% in non-sterile soil. After five days of hardening, the plants were transferred to bigger pots (20 cm dia.). Even without sterilizing, the potting combination (1: 1: 1 w/w) of sand, soil, and farmyard manure was employed successfully for the high-frequency establishment of micropropagated plants [43]. As per studies, after four weeks of incubation, the shoot with a well-developed elongated root was carefully removed from the culture flask. Then, the root was washed with running tap water. They are planted in pots containing sterile soil mixed with decomposed coffee husk or cocopeat or organic manure and sand in equal ratios and kept under 30% shade in net condition. The relative humidity and temperature were maintained at 75-80% and 25 ℃ - 27 ℃, respectively. Watering was done at 2-day intervals to maintain humidity [4, 16, 18, 27, 28, 42, 44].

CONCLUSION

Micropropagation is the best method as a persistent source of supply of disease-free planting material for commercial use. Ginger is generally propagated through the rhizome. The rhizome is easily affected by bacteria and fungi, the newly grown leaves also get infested by pests easily. Therefore *in vitro* propagation or micropropagation of the crop will be fruitful for producing healthy planting material, which can help achieve the growing demand for ginger. Establishment of the explants plays a very important role in micropropagation. The explants usually used are the rhizome buds, leaf explants, internodes and roots. However, adventitious buds have been widely used for successful micropropagation. Careful determination of factors including the source, type, formative stage and size of explants, cleansing of explants, growth regulator, growth medium and culture conditions are the important elements for the successfulness of *in vitro* propagation of ginger. The optimization of these factors is a guide for successful micropropagation of ginger.

REFERENCES

[1] Randall RP. A Global Compendium of Weeds. 2nd. Western Australia: Department of Agriculture and Food 2012; p. 1124.

[2] Baliga MS, Latheef L, Haniadka R, Fazal F, Chacko J, Arora R. Ginger (*Zingiber officinale* Roscoe) in the treatment and prevention of arthritis. Bioactive food as dietary interventions for arthritis and related inflammatory diseases. San Diego: Academic Press 2013; pp. 529-44.

[3] Abbas MS, Taha HS, Aly UI, El-Shabrawi HM, Gaber ESI. *In vitro* propagation of ginger (*Zingiber officinale* Rosco). J Genet Eng Biotechnol 2011; 9(2): 165-72. [http://dx.doi.org/10.1016/j.jgeb.2011.11.002]

[4] Ayenew B, Tefera W, Kassahun B. *In vitro* propagation of Ethiopian ginger (*Zingiber officinale* Rosc.) cultivars: Evaluation of explant types and hormone combinations. Afr J Biotechnol 2012; 11(16): 3911-8.

[5] Tewelde S, Patharajan S, Teka Z, Sbhatu DB. Assessing the efficacy of broad-spectrum antibiotics in controlling bacterial contamination in the *In vitro* micropropagation of ginger (*Zingiber officinale* Rosc). ScientWorldJ 2020; 2020: 1-8. [http://dx.doi.org/10.1155/2020/6431301] [PMID: 32581658]

[6] Baruah J, Pandey SK, Begum T, Sarma N, Paw M, Lal M. Molecular diversity assessed amongst high dry rhizome recovery Ginger germplasm (*Zingiber officinale* Roscoe) from NE-India using RAPD and ISSR markers. Ind Crops Prod 2019; 129: 463-71. [http://dx.doi.org/10.1016/j.indcrop.2018.12.037]

[7] Syafitri DM, Levita J, Mutakin M, Diantini A. Review: Is ginger (*Zingiber officinale*. Roscoe) potential for future phytomedicine? Ind J Appl Sci 2018; 8(1): 8-13. [http://dx.doi.org/10.24198/ijas.v8i1.16466]

[8] Aleem M, Khan MI, Shakshaz FA, Akbari N, Anwar D. Botany, phytochemistry and antimicrobial activity of ginger (*Zingiber officinale*): A review. Int J Herb Med 2020; 8(6): 36-49. [http://dx.doi.org/10.22271/flora.2020.v8.i6a.705]

[9] Weidner MS, Sigwart K. The safety of a ginger extract in the rat. J Ethnopharmacol 2000; 73(3): 513-20. [http://dx.doi.org/10.1016/S0378-8741(00)00340-8] [PMID: 11091007]

[10] Zhang M, Zhao R, Wang D, *et al.* Ginger (*Zingiber officinale* Rosc.) and its bioactive components are potential resources for health beneficial agents. Phytother Res 2021; 35(2): 711-42. [http://dx.doi.org/10.1002/ptr.6858] [PMID: 32954562]

[11] Lim S, Moon M, Oh H, Kim HG, Kim SY, Oh MS. Ginger improves cognitive function *via* NGF-induced ERK/CREB activation in the hippocampus of the mouse. J Nutr Biochem 2014; 25(10): 1058-65. [http://dx.doi.org/10.1016/j.jnutbio.2014.05.009] [PMID: 25049196]

[12] Ryu MJ, Chung HS. [10]-Gingerol induces mitochondrial apoptosis through activation of MAPK pathway in HCT116 human colon cancer cells. *In vitro*. Cell Dev Biol Anim 2015; 51(1): 92-101. [http://dx.doi.org/10.1007/s11626-014-9806-6] [PMID: 25148824]

[13] Shidfar F, Rajab A, Rahideh T, Khandouzi N, Hosseini S, Shidfar S. The effect of ginger (*Zingiber officinale*) on glycemic markers in patients with type 2 diabetes. J Complement Integr Med 2015; 12(2): 165-70. [http://dx.doi.org/10.1515/jcim-2014-0021] [PMID: 25719344]

[14] Yadufashije C, Niyonkuru A, Munyeshyaka E, Madjidi S, Mucumbitsi J. Antibacterial activity of ginger extracts on bacteria isolated from digestive tract infection patients attended Muhoza Health Center. Asian J Med Sci 2020; 11(2): 35-41. [http://dx.doi.org/10.3126/ajms.v11i2.27449]

[15] Rai S. Management of ginger (*Zingiber officinale* Rosc.) rhizome rot in Darjeeling and Sikkim Himalayan Region. Programme Coordinator of Darjeeling Krishi Vigyan Kendra, Uttar Banga Krishi Viswavidyalaya, and Kalimpong India 2006.

[16] Miri SM. Micropropagation, callus induction and regeneration of ginger (*Zingiber officinale* Rosc.). Open Agric 2020; 5(1): 75-84. [http://dx.doi.org/10.1515/opag-2020-0008]

[17] Kambaska KB, Santilata S. Effect of plant growth regulator on micropropagation of ginger (*Zingiber officinale* Rosc.) cv-Suprava and Suruchi. Agric Technol Thail 2009; 5(2): 271-80.

[18] Seran TH. *In vitro* propagation of ginger (*Zingiber officinale* Rosc.) through direct organogenesis: A review. Pak J Biol Sci 2013; 16(24): 1826-35. [http://dx.doi.org/10.3923/pjbs.2013.1826.1835] [PMID: 24516998]

[19] Lincy AK, Remashree AB, Sasikumar B. Direct multiple shoot induction from aerial stem of ginger (*Zingiber officinale* Rose.). J Appl Hortic 2004; 6(2): 99-101. [http://dx.doi.org/10.37855/jah.2004.v06i02.21]

[20] Olivier JJ. The initiation and multiplication of ginger (*Zingiber officinale* Rosc.) in tissue culture. Ligntings Bull Inst Vin Trop Subtrop 1996; 291: 10-1.

[21] Murashige T, Skoog F. A revised medium for rapid growth and bio assays with tobacco tissue cultures. Physiol Plant 1962; 15(3): 473-97. [http://dx.doi.org/10.1111/j.1399-3054.1962.tb08052.x]

[22] Smith MK, Hamill SD. Field evaluation of micropropagated and conventionally propagated ginger in subtropical Queensland. Aust J Exp Agric 1996; 36(3): 347-54. [http://dx.doi.org/10.1071/EA9960347]

[23] Loyola-Vargas VM, Vazquez-Flota F. Methods in Molecular Biology: Plant Cell Culture Protocols. Totowa, NJ, USA: Humana Press 2006.

[24] Bhagyalakshmi B, Singh NS. Meristem culture and micropropagation of a variety of ginger (*Zingiber officinale* Rosc.) with a high yield of oleoresin. J Hortic Sci 1988; 63(2): 321-7. [http://dx.doi.org/10.1080/14620316.1988.11515865]

[25] Khatun A, Nasrin S, Hossain MT. Large scale multiplication of ginger (*Zingiber officinale* Rosc.) from shoot-tip culture. OnLine J Biol Sci 2003; (1): 59-64.

[26] Pavallekoodi G, Sreeramanan S. Micropropagation of ginger (*Zingiber officinale* var. *rubrum*) using buds from microshoots. Pak J Bot 2016; 48(3): 1153-8.

[27] Manisha T, Vishal S, Garima K. *In vitro* production of disease-free planting material of ginger (*Zingiber officinale* Rosc.) : A single step procedure. Res J Biotechnol 2018; 13: 3.

[28] Mehaboob VM, Faizal K, Shamsudheen KM, Raja P, Thiagu G, Shajahan A. Direct organogenesis and microrhizome production in ginger (*Zingiber officinale* Rosc.). J Pharmacogn Phytochem 2019; 8: 2880-3.

[29] Zahid NA, Jaafar HZE, Hakiman M. Micropropagation of ginger (*Zingiber officinale* Roscoe) 'Bentong'and evaluation of its secondary metabolites and antioxidant activities compared with the conventionally propagated plant. Plants 2021; 10(4): 630. [http://dx.doi.org/10.3390/plants10040630] [PMID: 33810290]

[30] Hamirah MN, Sani HB, Boyce PC, Sim SL. Micropropagation of red ginger (*Zingiber montanum* Koenig), a medicinal plant. Proc Asia Paci Conf Plant Tissue Agribiotech.

[31] Alqadasi AS, Al-madhagi I, Al-kershy A, Al-Samaei M. Effect of cytokinin type and pH level on regeneration of ginger *in vitro.* Int J Hortic Sci Technol 2022; 9(3): 265-74.

[32] Mohanty S, Panda MK, Subudhi E, Acharya L, Nayak S. Genetic stability of micropropagated ginger derived from axillary bud through cytophotometric and RAPD analysis. Z Naturforsch C J Biosci

2008; 63(9-10): 747-54.
[http://dx.doi.org/10.1515/znc-2008-9-1021] [PMID: 19040116]

[33] Zhou J, Guo F, Qi C, Fu J, Xiao Y, Wu J. Efficient *ex-vitro* rooting and acclimatization for tissue culture plantlets of ginger. Plant Cell Tissue Organ Cult 2022; 150(2): 451-8.
[http://dx.doi.org/10.1007/s11240-022-02296-3]

[34] Gnasekaran P, Rahman ZA, Chew BL, Appalasamy S, Mariappan V, Subramaniam S. Development of micropropagation system of *Zingiber officinale* var. *rubrum* Theilade using different spectrum light-emitting diode (LED) irradiation. Ind Crops Prod 2021; 170: 113748.
[http://dx.doi.org/10.1016/j.indcrop.2021.113748]

[35] Bhattacharya MA, Sen AR. Rapid *in vitro* multiplication of disease-free *Zingiber officinale* Rosc. Indian J Plant Physiol 2006; 11(4): 379.

[36] Lincy A, Sasikumar B. Enhanced adventitious shoot regeneration from aerial stem explants of ginger using TDZ and its histological studies. Turk J Bot 2010; 34(1): 21-9.
[http://dx.doi.org/10.3906/bot-0805-6]

[37] Shan-Lin G, Jian-Hua M, Kun-Hua W, He-Ping H. Generation of autotetraploid plant of ginger (*Zingiber officinale* Rosc.) and its quality evaluation. Pharmacogn Mag 2011; 7(27): 200-6.
[http://dx.doi.org/10.4103/0973-1296.84230] [PMID: 21969790]

[38] An NH, Chien TTM, Nhi HTH, *et al.* The effects of sucrose, silver nitrate, plant growth regulators, and ammonium nitrate on microrhizome induction in perennially-cultivated ginger (*Zingiber officinale* Roscoe) from Hue, Vietnam. Acta Agrobot 2020; 73(2): 7329.
[http://dx.doi.org/10.5586/aa.7329]

[39] Samsudeen K, Babu KN, Divakaran M, Ravindran PN. Plant regeneration from anther derived callus cultures of ginger (*Zingiber officinale* Rosc.). J Hortic Sci Biotechnol 2000; 75(4): 447-50.
[http://dx.doi.org/10.1080/14620316.2000.11511266]

[40] Babu KN, Samsudeen K, Minoo D, Geetha SP, Ravindran PN. Tissue Culture and Biotechnology of Ginger.In: Ginger. 1st. CRC Press 2016; pp. 201-30.

[41] Taha HS, Abbas MS, Aly UI, Gaber EI. New aspects for callus production, regeneration and molecular characterization of ginger (*Zingiber officinale* Rosc.). Med Aromat Plants 2013; 2(6): 2-8.
[http://dx.doi.org/10.4172/2167-0412.1000141]

[42] Ammar MAA, Mawahib EAME-N, Sakina MY. Callus induction, direct and indirect organogenesis of ginger (*Zingiber officinale* Rosc). Afr J Biotechnol 2016; 15(38): 2106-14.
[http://dx.doi.org/10.5897/AJB2016.15540]

[43] Sharma TR, Singh BM. High-frequency *in vitro* multiplication of disease-free *Zingiber officinale* Rosc. Plant Cell Rep 1997; 17(1): 68-72.
[http://dx.doi.org/10.1007/s002990050354] [PMID: 30732423]

[44] Nirmal Babu K, Samsudeen K, Divakaran M, *et al.* Protocols for *in vitro* propagation, conservation, synthetic seed production, embryo rescue, microrhizome production, molecular profiling, and genetic transformation in ginger (*Zingiber officinale* Roscoe.).Protocols for *In Vitro* Cultures and Secondary Metabolite Analysis of Aromatic and Medicinal Plants. 2nd . New York: Humana Press 2016; pp. 403-26.

CHAPTER 5

Micropropagation Protocol in *Atropa acuminata* Royle ex Lindl. and *Atropa belladonna* L.

S. Manjula[1] and **B. L. Manjula**[2,*]

[1] *Department of Botany, Government First Grade College, B. M. Road, Ramanagara-562159, Karnataka, India*

[2] *Department of Botany, Sri Jagadguru Renukacharya College of Science, Arts and Commerce, # 9 Race course road, Bengaluru 560 009, Karnataka, India*

Abstract: *Atropa*, a Solanaceae member, contains many active chemical compounds such as atropine, saponins, polyphenols, scopolamine and hyoscyamine. Because of the presence of these active principles, endangered species *Atropa acuminata* and *Atropa belladonna* have been indiscriminately exploited in traditional medicine for treating various disorders and thus *Atropa acuminata* has become an endangered species in some regions. Due to the threat of extinction, low seed germination and seedling survival rate, there is a need for conservation through efficient micropropagation protocols. In this regard, the current chapter is focused on micropropagation methods/protocols developed by various researchers using various explants of *Atropa acuminata* and *Atropa belladonna* and their responses to different media compositions with respect to direct and indirect organogenesis *in vitro*, as the technique of *in vitro* regeneration has played a pivotal role in the mass multiplication of many plant species.

Keywords: *Atropa acuminata*, *Atropa belladonna*, Callus, Explants, Micropropagation.

INTRODUCTION

Atropa, a genus belonging to Solanaceae, is a toxic perennial herbaceous plant that prefers temperate climates and alkaline soils, often growing in light shade in woodland environments associated with limestone hills and mountains. It contains tropane alkaloids such as atropine, hyoscyamine, scopolamine, and others as medicinal compounds [1]. *A. acuminata* and *A. belladonna* are the most well known members of the genus *Atropa*. *A.acuminata* is a subalpine tall perennial plant native to Asia and found throughout the north western Himalayas [2]. *A. bel-*

* **Corresponding author B. L. Manjula:** Department of Botany, Sri Jagadguru Renukacharya College of Science, Arts & Commerce, #9 Race course road, Bengaluru 560 009, Karnataka, India; Tel: 9945181890;
E-mail: manjulasrivats@gmail.com

T. Pullaiah (Ed.)

ladonna, also known as belladonna or deadly nightshade or poisonous plant [3], is native to Central and Southern Europe, North Africa, and Western Asia.

Tropane alkaloids and highly oxygenated triterpenes are active chemical compounds found in *A. acuminata*. Likewise, saponins, polyphenols, anthraquinones, phytosterols, tetrahydroxyoleane, oleanolic acid, and β -sitosterol are some of the other compounds found in *A. acuminata* [4]. Rhizomes of *A. acuminata* were used to treat arthritis, muscle and joint pain, and also muscle spasms [5, 6]. *A. belladonna*, on the other hand, contains belladonna alkaloids such as atropine, its isomer hyoscyamine and scopolamine, which are most important and used in medicine. Apoatropine, norhyoscyamine, belladonine, tropacocaine, noratropine, and meteloidine are some of the other alkaloids found in *A. belladonna* that are not widely used in medicine. *Atropa belladonna* has been used in folk medicine to improve wound healing, particularly in septic post traumatic wounds [7].

Atropine and hyoscyamine alkaloids from *Atropa* are used as an antidote to opium and as sympathetic nervous system stimulants. These are frequently used to control excessive salivation, sweating, and nasal secretions. Atropine has an effect on the respiratory and also on circulatory systems. Belladonna is used externally to treat neurologic pain and internally to treat asthma and whooping cough. Atropine is frequently used to treat muscle spasms (excessive muscular contractions) and in ophthalmologic examinations to dilate the eye pupil. Scopolamine, which depresses the parasympathetic nervous system, is used as a sedative or as an anti-insomnia medication. It is used in conjunction with morphine to induce twilight sleep. Tropane alkaloids have anticholinergic and spasmolytic properties, making them popular in eye surgery and as an anaesthetic and spasmolytic [8]. In traditional medicine, aerial parts of the *Atropa* plant are used to treat a variety of ailments including acute infections, anxiety, chicken pox, asthma [9, 10], acute inflammation, muscle and joint pain, peritonitis, pancreatitis, scarlet fever, neuroinflammatory disorders, Parkinson's disease [11, 12], conjunctivitis, and encephalitis fever [13, 14]. The root extracts were also used to treat sore throat, ulcerative colitis [15], as a sedative [16], and whooping cough [17].

Toporcer *et al.* [18] used *Atropa* for aseptic surgical skin wounds to differentiate between immunogenic and antibiotic effects, and they also demonstrated that *A. belladonna* has a positive effect on aseptic surgical wound healing. *A. acuminata* is a valuable medicinal plant due to the presence of tropane alkaloids such as atropine, hyoscyamine, and scopolamine, all of which have sedative, analgesic, mydriatic, antiasthmatic, anodyne, and antispasmodic properties [5, 19, 20]. The main tropane alkaloids found in this plant are atropine and scopolamine, both of which have anticholinergic properties and are used in pharmaceuticals [21].

A. belladonna also contains tropane alkaloids and other active agents, such as atropine and hyoscyamine, throughout its plant parts. Because of this reason, it has been used for centuries in traditional treatments for a variety of conditions such as headache, menstrual symptoms, peptic ulcer, histamine reactions, and motion sickness, among others. Hyoscyamine, is also found in anti-vertigo medications and other medications that help to prevent motion sickness [22]. Atropine can also be used to treat hypertension and lower blood pressure [23].

IMPORTANCE OF MICROPROPAGATION IN *ATROPA*

A. acuminata is considered as an endangered species because wild plants are being indiscriminately harvested for their medicinal value from natural resources with no regard for cultivation practices [24 - 27]. Climate change, unplanned development, habitat destruction, excessive tourist traffic, and legal or illegal harvesting of this plant for local use and pharmaceutical industry needs are significant threats to its extinction. As a result of this, the IUCN has designated *A. acuminata* as an endangered species in the Kashmir Himalayas [28]. Due to the extinction threat, it is critical to develop necessary conservation strategies and efficient propagation protocols [29]. The low seed germination and seedling survival rate is a major constraint in traditional *A. acuminata* propagation methods [30]. As a result, developing an appropriate micropropagation protocol is critical for the survival of this critically endangered medicinal herb. Tissue culture has emerged as a broad field of study with tremendous potential for the conservation of endangered and rare plant species. Furthermore, tissue culture methods significantly increase the *in vitro* production of several bioactive compounds (non-enzymatic antioxidants) such as tannins, flavonoids, phenols, and so on, which significantly contribute to plant bioenrichment. Hence, *in vitro* regeneration techniques are used for many species of the genus *Atropa* [1, 31].

MICROPROPAGATION IN *ATROPA*

Explant Sterilization

Ahuja and co-workers used shoot tips and nodal segments of *A. acuminata* as explants, treating them with Tween-20 (Sigma, St Louis, Missouri, USA) for 10 minutes before thoroughly washing them with tap water for 30 minutes. The explants were then surface sterilized with 0.1% (w/v) mercuric chloride for 4 minutes after being soaked in 0.1% Bavistin for 10 minutes. To remove chemical residues, the explants were thoroughly washed 5 times with sterile distilled water [24]. Al-Ashaal *et al.* used leaves of *A. belladonna* as explants, washed them with tap water first, then treated with 10% v/v sodium hypochlorite for 20 minutes and immersed in 70% ethyl alcohol for a few seconds before being rinsed twice with sterile distilled water in an aseptic laminar airflow cabinet [32].

Maqbool *et al.* [33] attempted callus production from *A. acuminata* leaf explants, which were thoroughly washed with running tap water to remove dirt and dust, followed by washing with Labolene detergent and surface sterilizer Tween-20. The detergent residues were removed by washing the explants with sterile double distilled water. Explants were then treated with 2% sodium hypochlorite as a chemical sterilant for 8-10 minutes before being washed with autoclaved double-distilled water and inoculated onto a sterilized nutrient medium [33].

Maqbool *et al.* developed a rapid micropropagation protocol for *A. acuminata* using petiole and nodal segments as explants, which were thoroughly washed with tap water to remove dust before being washed with detergent Labolene (1% v/v) and surfactant Tween-20 (1% v/v). The detergent residues were removed by washing the explants with double distilled water and then treating them with 2 or 4% sodium hypochlorite for 8-10 minutes. The explants were then washed with sterile double-distilled water before being inoculated onto a sterilized nutrient medium [34]. Khan *et al.* [35] and his colleagues used *A. acuminata* seeds as explants, surface sterilising them by soaking them in 70% ethanol for 3 minutes and then placing them for 1 minute in 0.1% mercuric chloride. This was followed by three washes with sterile distilled water to remove the chemicals [35]. Maqbool *et al.* [36] used *A. acuminata* leaf explants, which were surface sterilized for 30-45 minutes with Labolene detergent (Praxor Group) and surfactant Tween-20, then rinsed 3-4 times with distilled water. After surface sterilization, explants were treated for 8 minutes with 2% sodium hypochlorite before being washed with double-distilled water in a laminar air flow hood.

Rajput and Agrawal treated nodal explants of 3-month-old *A. acuminata* plants with 4% (v/v) Teepol for 10-15 minutes to remove all debris and dust particles, followed by the treatment with a systemic fungicide Bavistin (0.1% w/v) and 250μL of cefotaxime for 20 minutes with vigorous shaking to remove fungi and bacteria spores. The explants were treated with citric acid (1% w/v) to prevent browning. The explants were thoroughly washed with sterile double-distilled water after each treatment. The explants were surface sterilized for 1 minute with 0.1% w/v aqueous solution of mercuric chloride before being rinsed with sterile double-distilled water [37].

Nutrient Medium and Culture Conditions

Saito *et al.* reported the formation of *A. belladonna* adventitious shoots from herbicide-resistant hairy roots grown in B5 medium without the addition of phytohormones [38]. Similarly, Aoki *et al.* reported adventitious shoot formation in *A. belladonna* on 12 Macro MS medium supplemented with 1.5 percent sucrose, 0.5μM Naphthalene Acetic Acid (NAA), 5μM Benzyl adenine (BA), and 0.2 percent gelrite using nodal explants [39].

Ahuja *et al.* cultured *A. acuminata* shoot tip explants on Murashige and Skoog's (MS) medium supplemented with 3% (W/V) sucrose and 0.7% (W/V) agar for *in vitro* propagation and conservation. The MS medium was supplemented with 1mg/L NAA, Indole-3-Butyric Acid (IBA), and 2,4-Dichlorophenoxyacetic acid (2,4-D) alone and in combination with BAP and Kinetin (Kn). The cultures were transferred to a revised tobacco medium containing 1mg/L Indole-3-Acetic-Acid (IAA) to develop multiple shoots from the callus. Nodal segments were cultured on MS medium containing IBA+ 6-Benzylaminopurine (BAP) at a concentration of 1 mg/L in the same study. The *in vitro* developed shoots were subcultured on half and full strength revised tobacco and B5 medium containing various concentrations of IBA to initiate rooting. The pH of the medium was adjusted to 5.8 using 1N NaOH or HCl, and cultures were kept at 24±2 °C with 16 hours of light using cool white fluorescent tubes [24].

Mera and Takayama used hormone free liquid Gamborg's B5 medium containing 3% (W/V) sucrose to test the effects of medium components on the formation of adventitious buds from the hairy roots of *A. belladonna*. B5 and MS media containing 3% sucrose and 8% agarose were used in this study to assess the effects of Kn concentration on adventitious bud formation, with each medium supplemented with different concentrations of Kn (0, 0.5, 4.7, and 46.5 μM). In the same study, MS liquid media was prepared with a combination of Kn (0-0.5μM) and sucrose (0, 1, 3, 5 & 7%) to assess the effects of sucrose concentration on adventitious bud formation. The components of MS medium with 1% sucrose was divided into five different solution groups (solutions 0-fold to 2-fold), and all experiments were carried out at 25°C under continuous light [40].

To investigate the production of tropane alkaloids *in vitro* and to assess the anticonvulsant, antinociceptive, motor incoordination, and antioxidant activities of *A. belladonna*, Al-Ashaal *et al.* [32] used leaf explants that were aseptically cultured in glass jars containing Murashige and Skoog's (MS) media [41], with 3% sucrose, and augmented with different concentrations of growth regulators such as (Benzyl adenine-BA, Kinetin-Kn, Indole acetic acid-IAA & Naphthalene acetic acid - NAA). All cultures were kept at 26 ± 2°C and 16/8 h photoperiod, with the callus subcultured every month for shoot development and regenerated shoots were transferred to basal media to initiate root formation [32].

Asha Rani and Prasad used MS Media containing minor salt Iron EDTA, vitamins, 3% sucrose, and 6.5-8.0% agar, as well as phytohormones NAA, IBA and Kn, for organogenesis from *A. belladonna* leaf explants. Three hormone solutions of various concentrations NAA (0.2 mg/L), IBA (0.3 mg/L), KI (0.4 mg/L) were prepared and added to the best media culture: [MS (25.0g/l) + Sucrose (30g/l) + Agar (6.5g/l)] separately. *A. belladonna* grew well at 26±3°C with a photoperiod of 16 hours of light and 8 hours of darkness [42]. The authors used simple MS media supplemented with NAA and Kn in a comparative study on callus induction from fresh and reused *A. belladonna* explants, and the cultures were incubated at 25±3°C for 16 hrs light/8 hrs dark under aseptic conditions [43].

Maqbool *et al.* cultured *A. acuminata* leaf and petiole explants for callus induction and shoot regeneration on Murashige and Skoog's [41] medium, which contained 8% agar and various concentrations of auxins and cytokinins, both separately and in combination. 2, 4-D, IAA, NAA, IBA, BAP and Kn were used at concentrations ranging from 0.1 to 5 mg/L, and the cultures were kept at 22±4°C with regular photoperiod of 12 hr light and 12hr dark [34]. MS medium supplemented with different concentrations of cytokinins individually or in combination with auxin was used for the regeneration of shoots from leaf derived callus, where cytokinins such as BAP were used in concentrations of 2 mg/L, 3 mg/L, and 5 mg/L. MS medium containing different growth hormones (BAP, Kn, IAA, NAA, IBA & 2, 4-D) individually or in different combinations was used to develop callus from nodal explants (Tables **1-3**). *In vitro* cultured shoots were placed on full and half strength MS medium supplemented with auxins like IAA and IBA individually at various concentrations and in combinations with cytokinins like BAP for root initiation. The pH of the media was adjusted to 5.8 before autoclaving it at 121 °C and 15 lbs pressure. The cultures were incubated at 22±4 °C and kept at a 24-hour photoperiod [33].

Table 1. Composition of MS Media supplemented with different concentrations of hormones for callus production from leaf explants of *A. acuminata* [33].

MS medium	BAP mg/l	IAA mg/l	IBA mg/l	NAA mg/l
+	2	-	-	2
+	5	-	-	-
+	2	-	-	5
+	2	5	-	-
+	-	-	0.5	-

Table 2. Composition of MS Media supplemented with different concentrations of hormones for callus production from *in vitro* leaf explants of *A. acuminata* [33].

MS Medium	BAP mg/l	IAA mg/l
+	2	-
+	5	-
+	3	-
+	3	2

Table 3. Composition of MS media with different growth regulators for shoot formation from leaf derived callus of *A. acuminata* [33].

MS Medium	BAP mg/l	NAA mg/l	IAA mg/l
+	2	-	2
+	3	-	2
+	5	-	-
+	3	-	-

A. acuminata seeds were cultured for 45 days on half strength MS solid medium in a growth chamber with a 16/8-hour light/dark photoperiod under cool-white light. Leaves were collected from *in vitro* germinated plantlets and cultured on MS solid medium supplemented with various hormone concentrations such as Thidiazuron (TDZ), Kn, BA alone and in combination with NAA. For root initiation, regenerated shoots were placed on MS medium supplemented with IBA and NAA at various concentrations (0.5, 1.0, 2.0, 5.0, and 10.0 mg/L), with media containing 0.8% agar and 30 g/l sucrose. pH level of the media was adjusted to 5.7 [35]. For phytochemical screening and antioxidant activity, Maqbool *et al.* cultured surface sterilized leaf explants of *A. acuminata* on MS basal medium containing 3% (w/v) sucrose and 0.8% agar. The media was supplemented with various BAP and IAA combinations. The pH of the medium was adjusted to 5.8±0.1 using 1N NaOH and sterilized for 15 minutes at 121°C. The cultures were kept at 22±4 °C, 50-60% RH, and a 16-hour photoperiod using 40-W cool white fluorescent lamps [36].

MS media composition for callus production and shoot regeneration from leaf explants of *A. acuminata*

MS+BAP (1mg/l), MS+BAP (2 mg/l), MS+BAP (3 mg/l), MS+BAP (4 mg/l)

MS+BAP (5 mg/l)

MS media composition for root regeneration from shoots of *A. acuminata*

MS+IBA (0.2 mg/l), MS+IBA (0.3 mg/l), MS+IBA (0.5 mg/l), MS+IBA (1 mg/l)

MS+IBA (2 mg/l) [36]

Rajput and Agrawal used a semisolid MS medium containing 3% (w/v) sucrose and 0.65% agar for inoculation of *A. acuminata* nodal explants. The cultures were incubated in a growth chamber at 25 ± 2 °C and 55± 5% relative humidity. To develop callus from roots, 1cm long aseptic root explants were excised and inoculated onto the MS medium supplemented with different concentrations (0, 0.1, 1, 5, 10 µM) of IBA, NAA, 2,4-D, TDZ, or BA individually (Table **4**). Callus was further subcultured on freshly prepared medium (MS + 10 µM TDZ) for shoot regeneration, and shoots were transferred to hormone free MS medium for further elongation. MS medium supplemented with different concentrations (0-10 µM) of IAA, IBA, or NAA alone was used to induce roots from shoots (Table **5**) [37].

Table 4. Composition of media with different concentrations of plant growth regulators for callus induction using root explants of *A. acuminata* [37].

Plant Growth Regulators	Concentration Used in µM
IBA	0,0.1,1.0,5.0 and 10.0
2,4-D	0,0.1,1.0,5.0 and 10.0
NAA	0,0.1,1.0,5.0 and 10.0
BA	0,0.1,1.0,5.0 and 10.0
TDZ	0,0.1,1.0,5.0 and 10.0

Table 5. Composition of media with different concentrations of plant growth regulators for root induction from excised shoots of *A. acuminata* [37].

Plant Growth Regulators	Concentration Used in µM
IAA	Control 0, 0.1,1.0,5.0 and 10.0
IBA	Control 0, 0.1,1.0,5.0 and 10.0
NAA	Control 0, 0.1,1.0,5.0 and 10.0

Regeneration Response

Regeneration studies of *A. acuminata* has been conducted using a variety of explants including shoot tip, nodes [24, 34], petioles [34], leaves [32, 33, 35], and roots, since roots are the primary site for tropane alkaloids synthesis [37].

Ahuja *et al.* studied the regeneration response of *A. acuminata* and discovered that shoot tip explants had a better regeneration response than nodal explants. In this study, shoot tip explants responded well to the combination of BAP and IBA, with a maximum of 5-6 shoots production/culture, after a 40-day incubation period on MS medium containing 1mg/L BAP + 1mg/L IBA. After 6 weeks of continuous subculturing, callogenesis was observed at the lower portion of the shoots. In this case, the % shoot regeneration and number of shoots developed /node retained a high value of around 90%, whereas the induced shoots failed to elongate rapidly. On transfer to rooting media with various treatments, the differentiated shoots showed 80-100% rooting. In terms of length and number of shoots produced, full-strength RT medium supplemented with 1mg/L IBA proved to be the best rooting medium. It was also discovered that when IBA was replaced with NAA/IAA, there were fewer or no roots on half strength and full strength MS, Revised Tobacco medium (RT), and B5 media. According to the findings of this study, growth hormones are critical for the differentiation of shoots/roots or both [24].

Mera and Takayama studied the effects of Kn concentration, sucrose concentration, and inorganic salt strength on adventitious bud formation from *A. belladonna* hairy roots. In terms of Kn concentration, more adventitious buds were formed in the MS medium than in the B5 medium, and the weight of the adventitious buds was lower in the B5 medium than in the MS medium. This study concluded that adding Kn to media had no discernible effect on adventitious bud formation. Similarly, as sucrose concentration increased, the number and weight of adventitious buds formed decreased, whereas at 1% sucrose and 0μM Kn, both the number and weight of adventitious buds formed increased. On MS media, the number of adventitious buds formed increased with an increase in inorganic salt solution strength from 0-1 fold , whereas the number of adventitious buds formation at 2-fold of inorganic salt concentration from 0-1 fold slightly decreased, and there was no significant difference observed at concentration 0.125-fold to 2-fold. In general, the number and weight of adventitious buds were found to be greater in the MS Medium than in the B5 medium [40].

Al-Ashaal *et al.* recorded the induction of callus on MS media using leaf explants, where the media was supplemented with Kn (0.5mg/L), BA (0.5mg/L), and IAA (2mg/L) (culture I), and in MS media supplemented with BA (0.5mg/L) and NAA (1mg/L) (culture II). Successful differentiation was observed in the media containing 0.2 mg/L of BA and 2 mg/L of IAA [32]. According to Asharani *et al.'s* findings, 0.3 mg/L NAA produced the best root initiation when compared to higher concentrations. The lamellae explant on MS + 0.3mg/L NAA showed maximum root growth, but the other concentrations also produced good results. Taking callus dry weight into account, the authors concluded that the callus formation was greatest at 0.3mg/L IBA. In Kn media, indirect organogenesis was also observed. At a temperature of 26±2°C, the differentiation of roots from callus was successfully reported [42].

In a comparison study on callus induction of fresh and reused explants of *A. belladonna,* the results showed that the reused explants were slow in growth and callus yield was also less compared to fresh explants used. In reused explants, no root differentiation was observed, but callus development was observed. Indirect organogenesis was not observed in reused explants, but it was observed in fresh explants [43]. Callus was produced when leaf explants of *A.acuminata* were inoculated on MS medium supplemented with BAP + NAA (2 mg/l), BAP (5 mg/l), BAP (2 mg/l) + IAA (5 mg/l) and BAP (2 mg/l) + NAA (5 mg/l) in a period of 62, 29, 48 and 43 days of inoculation respectively. Callus was also produced when *in vitro* leaf explants were inoculated on MS medium fortified with BAP (2 mg/l, 3 mg/l, 5 mg/l) in a time period of 17, 34 and 26 days of inoculation. The authors also observed shoot regeneration when leaf derived callus was subcultured on MS medium supplemented with different concentrations of cytokinins individually or in combination with auxins. Cytokinins like BAP were used in a concentration of (2 mg/l), (3 mg/l) and (5 mg/l). BAP with IAA also gave good results in a combination of BAP (3 mg/l) + IAA (2 mg/l) and BAP (2 mg/l) + IAA (3 mg/l). Callus, after subculture, produced maximum shoot regeneration on MS medium supplemented with BAP (3 mg/l) after 19 days of culture. According to the findings of this study, MS medium supplemented with BAP (5 mg/L) resulted in the maximum callus development, whereas maximum number of shoots developed in MS medium supplemented with BAP (3 mg/L) [33].

For the conservation of this medicinally important plant, Maqbool *et al.* developed a rapid micropropagation protocol using petiole and nodal explants of *A. acuminata*. Callus was formed from petiole explants grown on MS medium supplemented with various growth hormones (BAP, Kn, IAA, NAA, IBA, and 2, 4-D) individually and in various combinations. After 18 days of inoculation, the maximum amount of callus was formed on MS medium supplemented with BAP at a concentration of 3 mg/L in 80% of cultures. In terms of percent culture response and number of days taken, the petiole explant produced more amount of callus on MS medium supplemented with BAP (3 mg/L) as compared to BAP (3 mg/L) in combination with IAA (2 mg/L). Furthermore, when the same callus was subcultured on MS medium supplemented with BAP (5 mg/L), shoot regeneration was observed, with a mean shoot length of 2.20±0.19 cm in 40% cultures within 48 days. Shoots obtained from MS medium supplemented with BAP (5 mg/L) were healthier and longer than the shoots obtained from BAP (3 mg/L) combined with IAA (2 mg/L) and in terms of percent culture response. Similarly, the maximum amount of callus was obtained from nodal explants in 80% of cultures in 17 days on MS medium supplemented with BAP (2 mg/L). When the callus was transferred to MS medium supplemented with BAP (2 mg/L), shoot regeneration occurred in 80% of the cultures within 14 days, with a mean shoot length of 2.0±0.20 cm. Shoots regenerated on MS medium supplemented with BAP (3 mg/L) grew longer and required fewer days to regenerate than the shoots regenerated on BAP (2 mg/L). Root differentiation with 100% response was achieved in *in vitro* grown shoots on MS medium supplemented with IBA (0.5 mg/L) after 18 days, with a mean number of 21.6±6.9 roots. Roots also regenerated in half strength MS medium with IBA (0.5 mg/L). In terms of percent culture response, roots regenerated from full strength MS medium enriched with IBA (0.5 mg/L) were more in number than IBA (1 mg/L) and ½MS+IBA (0.5 mg/L). However, the percentage of cultures showing rooting as well as the average number of roots/cultures decreased [34].

Khan *et al.* successfully regenerated *A. acuminata* from leaf explants derived from *in vitro* seed derived plants Figs. (**1A-F**). The researchers observed the formation of light green calli from the cut margins of leaf explants, which turned compact green after 28 days of incubation Fig. (**1A**). 1.0 mg/L TDZ alone produced the highest callogenic response (92%) of all the PGRs used. *A. acuminata* callus induction was inhibited by the addition of 1.0 mg/L NAA to other PGRs. After 35 days of subculture, data on various parameters of shoot regeneration were analyzed, and the highest shoot induction response of 89% was recorded for explants cultured on MS medium containing TDZ (1.0 mg/L) and NAA (1.0 mg/L) in combination and TDZ alone producing 88% at 1.0 mg/L Figs. (**1B-D**). A low concentration of TDZ alone or in combination with NAA stimulated shoot organogenesis, while a higher concentration inhibited it. In comparison to all other PGRs used, TDZ alone and BA in combination with NAA were more effective in inducing the maximum number of shoots/explants. The addition of NAA to the medium, along with either TDZ or BA, significantly increased the number of shoots/explants. According to the findings of this study, the MS medium supplemented with IBA (0.5 mg/L) was the best rooting medium for root organogenesis in *A. acuminata*, where 90% of regenerated plants showed rooting Fig. (**1E**). Moderate levels of IBA were found to be more beneficial for rooting, whereas higher levels inhibited the percent rooting response. *A. acuminata* rooting was not induced by NAA in MS medium [35].

Maqbool *et al.* used *in vitro* raised *A. acuminata* plants from leaf explants to study the phytochemical screening and antioxidant properties, where 100% callus was regenerated on MS medium with the highest frequency at 3 mg/L BAP concentration, among all phytohormones tested after one and a half week. The callus that formed was green and nodular in nature. When the callus was subcultured after 3 weeks of inoculation, 100% shoot regeneration was observed on MS medium, and the highest mean shoot number/explant recorded was 15.2 Similarly, *in vitro* rooting of microshoots on MS medium supplemented with 0.5mg/L IBA was successful. This study concluded that MS media with IBA and moderate IBA was the best rooting medium for root regeneration than higher concentrations of IBA [36].

Fig. (1). Stages of plant regeneration in *A. acuminata*. A. Callus induction B. Shoot regeneration C. Shoot multiplication D. Shoot elongation E. Rooting F. Regeneration of the plant (Khan *et al.*, 2017) (Permission taken).

Rajput and Agrawal developed a micropropagation protocol based on indirect organogenesis using root derived callus. The callus was formed on all types of media after 45 days of inoculation. But the quality and quantity of calli showed variable results. In this study, the highest percentage of explants developed callus on 10 µM TDZ, while callus formed on other media containing IBA, NAA, or 2,4-D were slow growing, brownish, and non-morphogenic, and callus formed on BA was green and compact but failed to grow further. After 28 days of subculturing on the same medium (10 µM TDZ), these callus pieces differentiated an average of 5.54± 0.65 shoots with a length of 2.38± 0.37 cm in 71.5% of the explants. Subculturing of callus on MS medium supplied with TDZ in the range of 0-10 µM was done in the same study to induce shoot development in the callus. In 71.5% of cultures on MS + 10 µM TDZ, the callus pieces differentiated an average of 5.54± 0.65 shoots with a length of 2.38± 0.37 cm. Shoot regeneration was not observed in the control or other TDZ concentrations. Shoot elongation was inhibited and stunted shoots were observed during subsequent subculturing of shoot buds on the same medium (10 µM TDZ). For elongation, these shoots were transferred to hormone free MS medium. MS + 1µM IBA produced the best rooting response, yielding an average of 8.08 ± 0.19 roots with a length of 7.28± 0.11 cm after 28 days of culture. However, poor rooting response was observed with all the concentrations of IAA or NAA tried. At low concentrations (0.1 and 1.0 µM), thin roots with minimal lateral branching were observed, whereas, at higher concentrations (5.0 and 10 µM), short and thin roots with profuse callus were seen. Therefore, 1.0 µM IBA proved to be the most effective PGR for rhizogenesis [37].

Acclimatization/Hardening

In a study of *in vitro* propagation of *A. acuminata* from the shoot tip explant for its conservation, the healthy plantlets which attained a height of 30-35 cm with 4-5 nodes along with roots were washed in running tap water to remove agar and then transferred to sterilized garden soil taken in small polypots. The potted plants were kept in a mist chamber with an RH of 80-90% and regular watering for three weeks. Following acclimatization, the potted plants were transferred to a greenhouse and hardened at a temperature of 18±2°C. Different morphological characteristics of the parental plant, such as branching pattern, leaf shape, petiole length, and area of leaf lamina, remained consistent [24].

For acclimatizing plants, Maqbool *et al.* used 18 day-old well developed plantlets that were washed with double distilled water to remove media remnants before being transferred to jiffy pots and earthen pots containing autoclaved soil and sand in the ratio 1:1 as well as compost. The transferred plantlets were watered on a regular basis and kept in a greenhouse under controlled conditions of 22±4°C and 60% RH. The plants were then hardened in 3 weeks, with an 80% survival rate in compost and a 60% survival rate in soil-sand mixture in 4 weeks. Under greenhouse conditions, *in vitro* raised plantlets were successfully acclimatized/hardened, and hardened plants were successfully transferred to field conditions [34]. Likewise, *A. acuminata in vitro* regenerated plants were transferred to potting soil mixture and placed in greenhouse covered with polythene bags to maintain 90% RH. After 25 days, polythene bags were removed and then exposed to greenhouse conditions. 80% of the plants survived [35].

In a phytochemical screening study of *in vitro* raised *A. acuminata*, three months old rooted plantlets were washed and then transferred to a mixture of vermicompost and garden soil in the ratio 3:1 (v/v), placed in styrofoam cups, and grown under controlled conditions at 27±1°C and 80-90% RH at 16 h/8 h photoperiod. Plants were hardened in a greenhouse for 10 days and watered with tap water on alternate days. Hardened plants were transferred to earthen pots containing soil and sand mixture in the ratio 3:1 (v/v) for three months, and then to field conditions after two months [36].

In a study of *A. acuminata* micropropagation *via* indirect organogenesis using root-derived callus, well rooted regenerated plants were removed from the culture medium and thoroughly washed with ddH_2O to remove adherent agar particles before transfer. These plantlets were then transferred to thermocol cups (with a hole in the bottom) containing 45 g of Soilrite™-Mix (75% Irish peat moss + 25% Perlite + Vermiculite), which were then incubated in a growth room under specific conditions. For the first week, the plants were sprinkled twice daily with 10 mL of one-fourth strength MS aqueous solution (without sucrose). After 4 weeks, thermocol cups containing plantlets were moved to a greenhouse for a week before being transplanted to larger pots filled with loamy soil with soil and sand in a ratio of 1:1 under natural conditions [37].

Studies carried out on *in vitro* propagation of *A. acuminata* and *A. belladonna* are summarized in Table **6**.

Table 6. Overall *in vitro* propagation work done till now in *A. acuminata* and *A. belladonna* with respect to explants used, washing surface sterilizers used, media, culture conditions, regeneration responses, acclimatization and hardening of *in vitro* raised plantlets.

S. No.	Explants Used	Washing & Surface Sterilization	Media Used & Culture Conditions	Regeneration Responses	Acclimatization and Hardening	References
1.	Shoot tips &	Tween-20 (10 m), tap water (30 m),	**For shoot tip explants**-MS medium with 3%(W/V) sucrose and 0.7% (W/V) agar containing 1mg/L each of NAA,	Development of callus, shoots & roots,	The potted plants were maintained inside an intermittent mist chamber at 80-90% RH for 3 weeks,	[24]

(Table 6) cont.....

S. No.	Explants Used	Washing & Surface Sterilization	Media Used & Culture Conditions	Regeneration Responses	Acclimatization and Hardening	References
-	nodal explants of *A. acuminata*	0.1% Bavistin (10 m), 0.1% (W/V) mercuric chloride (4 m) & sterile distilled water	IBA & 2,4, D alone and in combination with BAP & Kn Nodal explants- MS medium with IBA (1 mg/L) + BAP (1 mg/L) For rooting- Revised Tobacco Medium (RT) & B5 with different concentrations of IBA	where shoot tip explants showed better response than the nodal explants used	then potted plants were transplanted to green house and maintained at 18± 2 °C	
2.	Leaf explants of *A.belladonna*	-	**For Preculture-** B5 liquid media with 3% sucrose & no phyto hormones **To assess Kn conc**. B5 and MS media (3% sucrose and 8% agarose) with diff. con. of Kn (0,0.5,4.7 & 46.5μM) **To assess sucrose conc.** MS liquid media with 0-0.5 μM Kn & 0,1,3,5 & 7% Sucrose **To assess inorganic solution strength.** MS media (0- to 2-fold)	Formation of adventitious buds was more in MS media compared to B5 media & at 1% sucrose concentration & number of adventitious buds increased with an increase in inorganic salt strength	-	[40]

(Table 6) cont.....

S. No.	Explants Used	Washing & Surface Sterilization	Media Used & Culture Conditions	Regeneration Responses	Acclimatization and Hardening	References
3.	Lamellae, midrib & petiole explants of *A. belladonna*	-	MS Media with minor salts, Iron EDTA, vitamins, Sucrose 3%, Agar-6.5-8.0% along with growth regulators like NAA (0.2 mg/L) IBA (0.3 mg/L) & Kn (0.4 mg/L)	Direct Organogenesis on MS+ NAA, indirect organogenesis on Kn media and callogenesis from roots	-	[42]
4.	Leaf explants of *A. belladonna*	Sodium hypochlorite (10% v/v) for 20 m, 70% ethyl alcohol for a few seconds	MS media (3% sucrose) supplemented with different concentrations of BA, Kn, IAA, and NAA. Cultures were maintained at 26 ± 2°C and 16/8 h photoperiod.	Callus formation & shoot differentiation	-	[32]
5.	Fresh & reused leaf explants of *A. belladonna*	-	MS media supplemented with NAA and Kn cultured at 25±3°C for 16hrs light/8h dark at sterile conditions	Both callogenesis & indirect organogenesis. Fresh explants with NAA showed organogenesis but reused explants showed only callogenesis whereas fresh explants with Kn showed callogenesis as well as organogenesis but reused explants showed only callogenesis	-	[41]

(Table 6) cont.....

S. No.	Explants Used	Washing & Surface Sterilization	Media Used & Culture Conditions	Regeneration Responses	Acclimatization and Hardening	References
6.	Leaf explants of *A. acuminata*	Detergent Labolene, surfactant Tween-20 & 2% sodium hypochlorite (8-10 m)	MS medium (8% agar) supplemented with BAP (2 mg/L) + NAA (2 mg/L), BAP (5 mg/L), BAP (2 mg/L) + IAA (5mg/L) & BAP (2 mg/L) + NAA (5 mg/L) Cultures were incubated at 22±4 °C, exposed 24 hrs of photoperiod	Callus formation was more on MS media + BAP (5mg/L) & shoot regeneration was more on MS media + BAP (3mg/L)	-	[33]
7.	Petiole & nodal explants of *A. acuminata*	Washing with detergent labolene (1% v/v) and surfactant Tween-20 (1%v/v), double distilled water, 4% sodium hypochlorite (8-10 m) & double distilled water	**For callus & shoot regeneration:** MS medium (8% agar) supplemented with BAP, Kn, IAA, NAA, IBA, 2,4, D (0.1- 5mg/L) **For rooting:** Full & half strength MS media with IAA, IBA & in combination with BAP of various concentrations used. The cultures were incubated at 22±4 °C under 12hr light & 12 hr dark photoperiod	Callus formation, shoot regeneration & root differentiation	Plantlets were sown in pots containing soil and sand in the ratio 1:1 along with compost and maintained at (22±4°C) and (60%) RH under greenhouse. The hardening was achieved within 3 weeks with 80% survival rate in compost & 60% survival rate in soil-sand mixture within 4 weeks.	[34]

(Table 6) cont.....

S. No.	Explants Used	Washing & Surface Sterilization	Media Used & Culture Conditions	Regeneration Responses	Acclimatization and Hardening	References
8.	Seeds of *A. acuminata*	70% ethanol (3 m) and 0.1% mercuric chloride (1 m) followed by washing three times with sterile distilled water	**Seeds**: Half strength MS solid medium (0.8% agar & 30g/l sucrose) **Leaves**: MS solid medium with different hormones like TDZ, Kn, BA alone & in combination with NAA of various concentrations **For Rooting:** MS + IBA (0.5,1.0, 2.0, 5.0 & 10 mg/L) Plants were cultured at 16hrs light/8hrs dark photoperiod	Callus formation, shoot regeneration & root differentiation	Plantlets were transferred to pots containing soil mixture & covered with polythene bags and placed under greenhouse conditions	[35]
9.	Leaf explants of *A. acuminata*	Labolene &Tween-20 (30-45 m) and rinsed 3-4 times with distilled water, treated with 2% sodium hypochlorite (8 m)	MS basal medium, (3% (w/v) sucrose & 0.8% Agar) with different combinations of BAP and IAA of various concentrations. The cultures were kept under 16 hr photoperiod using 40-W white fluorescent lamps. Cultures were maintained at 22±4 °C and 50–60% RH (relative humidity)	Callus formation, shoot regeneration & root differentiation	Plants were transferred to cups containing a mixture of soil & vermicompost in the ratio 3:1 v/v and grown under a controlled environment chamber at 27±1°C under 16 h/8 h photoperiod and 80–90% RH and hardened for 3 months under greenhouse conditions	[36]

(Table 6) cont.....

S. No.	Explants Used	Washing & Surface Sterilization	Media Used & Culture Conditions	Regeneration Responses	Acclimatization and Hardening	References
10.	Nodal explants of *A. acuminata*	4% (v/v) Teepol (10–15 m), Bavistin (0.1% w/v) and 250 μL of cefotaxime as antibiotic (20 m), (1% w/v) citric acid & mercuric chloride (0.1% w/v) for 1 min followed by rinsing with sterile ddH2O	Semi solid MS medium 3% (w/v) sucrose & 0.65%, agar. The cultures were maintained at 25 ± 2 °C temperature and 55 ± 5% relative humidity. **For Callus induction:** MS + different growth regulators at different concentrations **For shoot regeneration:** MS+ 10 μM TDZ of various concentrations **For rooting:** MS+ 0-10 μM IAA, IBA or NAA alone	Callus formation, shoot regeneration & root differentiation	Plantlets were transferred to thermocol cups containing 45 g of Soilrite™-Mix (75% Irish peat moss + 25% Perlite + Vermiculite) & incubated in a growth room under specific conditions. For hardening, thermocol cups containing plantlets were shifted to the greenhouse for a week	[37]

CONCLUSION

The current chapter sheds light on various micropropagation protocols developed by many researchers using different explants of *A. belladonna* and an endangered *A. acuminata* of medicinal importance. The studies showed that the shoot tip and petiole explants of *A. acuminata* produced more calli than the nodal segments. The addition of NAA to the MS medium containing either TDZ or BA increased the number of shoots/explants significantly. MS medium supplemented with IBA was the best rooting medium for root organogenesis. The Leaf explants of *A. acuminata* inoculated on MS media supplemented with NAA showed direct organogenesis, whereas leaf explants inoculated on MS media supplemented with Kn showed indirect organogenesis. When it came to growth, the fresh explants outperformed the reused explants. In *A. belladonna*, herbicide resistant hairy root explants do not require the addition of phytohormones to B5 media for adventitious root development. Whereas the nodal segments of *A. belladonna* required the addition of NAA, BA, Kn as well as Gelrite to MS media for adventitious root production. The studies showed that the fresh explants of *A.*

belladonna have more potential in producing callus and direct organogenesis compared to the reused explants, which produced less callus and indirect organogenesis. Finally, one can conclude that the micropropagation protocols developed by the researchers could be used for large-scale production of *A. belladonna* and an endangered species *A. acuminata,* which are of high therapeutic value.

REFERENCES

[1] Bajaj YPS, Simola LK. *Atropa belladonna* L. : *In vitro* culture, regeneration of plants, cryopreservation and the production of tropane alkaloids. In: Biotechnology in Agriculture and Forestry. Berlin Heidelberg: Springer 1991; 15: pp. 1-23.

[2] Dhar U, Kachroo P. Alpine flora of Kashmir Himalayas. Jodhpur: Scientific Publishers 1983. [http://dx.doi.org/10.2307/2805987]

[3] Largo M. Big, Bad Botany: Deadly Nightshade (*Atropa belladonna*), the Poisonous A-Lister. 2014. Available from: https://slate.com/technology/2014/08/poisonous-plants-belladonna-nightshade-i--the-celebrity-of-deadly-flora.html

[4] Mehmood A, Malik A, Anis I, *et al.* Highly oxygenated triterpenes from the roots of *Atropa acuminata.* Nat Prod Lett 2002; 16(6): 371-6. [http://dx.doi.org/10.1080/10575630290033097] [PMID: 12462340]

[5] 1948 Wealth of India The dictionary of Indian raw materials and Industrial products. New Delhi, India: Council of Scientific and Industrial Research 1952; p. 227.

[6] Kaul MK. Medicinal plants of Kashmir and Ladakh: Temperate and Cold Arid Himalaya. New Delhi: Indus Pub. Co 1997.

[7] Grieve M. A modern Herbal. Available from: https://www.botanical.com/botanical/mgmh/mgmh.html (accessed on 14-09-2022).

[8] Tyler VE, Brady LR, Robbers JE. Pharmacognosy. 9th., Philadelphia: Lea and Febiger 1988.

[9] Kumar S, Shukla YN, Lavania UC, Sharma A, Singh AK. Medicinal and aromatic plants: Prospects for India. J Med Arom Plant Sci 1997; 19: 361-5.

[10] Bettermann H, Cysarz D, Portsteffen A, Kümmell HC. Bimodal dose-dependent effect on autonomic, cardiac control after oral administration of *Atropa belladonna.* Auton Neurosci 2001; 90(1-2): 132-7. [http://dx.doi.org/10.1016/S1566-0702(01)00279-X] [PMID: 11485281]

[11] Kahn A, Rebuffat E, Sottiaux M, Muller MF, Bochner A, Grosswasser J. Prevention of airway obstructions during sleep in infants with breath-holding spells by means of oral belladonna: A prospective double-blind crossover evaluation. Sleep 1991; 14(5): 432-8. [http://dx.doi.org/10.1093/sleep/14.5.432] [PMID: 1759096]

[12] King JC. Anisotropine methylbromide for relief of gastrointestinal spasm: Double-blind crossover comparison study with belladonna alkaloids and phenobarbital. Curr Ther Res Clin Exp 1966; 8(11): 535-41. [PMID: 4962781]

[13] Ceha LJ, Presperin C, Young E, Allswede M, Erickson T. Anticholinergic toxicity from nightshade berry poisoning responsive to physostigmine. J Emerg Med 1997; 15(1): 65-9. [http://dx.doi.org/10.1016/S0736-4679(96)00244-2] [PMID: 9017490]

[14] Duncan G, Collison DJ. Role of the non-neuronal cholinergic system in the eye. Life Sci 2003; 72(18-19): 2013-9. [http://dx.doi.org/10.1016/S0024-3205(03)00064-X] [PMID: 12628451]

[15] Shanafelt TD, Barton DL, Adjei AA, Loprinzi CL. Pathophysiology and treatment of hot flashes. In Mayo Clinic proceedings Mayo Clinic, 2002; 77: 1207– 1218. [http://dx.doi.org/10.4065/77.11.1207]

[16] Rhodes JB, Abrams JH, Manning RT. Controlled clinical trial of sedative-anticholinergic drugs in patients with the irritable bowel syndrome. J Clin Pharmacol 1978; 18(7): 340-5. [http://dx.doi.org/10.1002/j.1552-4604.1978.tb01603.x] [PMID: 353089]

[17] Walach H, Köster H, Hennig T, Haag G. The effects of homeopathic belladonna 30CH in healthy volunteers : A randomized, double-blind experiment. J Psychosom Res 2001; 50(3): 155-60. [http://dx.doi.org/10.1016/S0022-3999(00)00224-5] [PMID: 11316508]

[18] Toporcer T, Grendel T, Vidinský B, Gal P, Sabo J, Hudák R. Mechanical properties of skin wounds after *Atropa belladonna* application in rats. J MetalsMaterials and Minerals 2006; 16(1): 25-9.

[19] Nasir E, Ali SI. Flora of West Pakistan. Pakistan Agriculture Research Council Islamabad 1982; 39: 168.

[20] Singh J, Chand R. Medicinal and aromatic plants. Adv Hortic Sci 1995; 1: 283-96.

[21] Jafe RJ, Novakovic V, Peselow ED. Scopolamine as an anti-depressant: A systematic review. Clin Neuropharmacol 2013; 1: 24-6. [http://dx.doi.org/10.1097/WNF.0b013e318278b703]

[22] Abraham S, Cantor EH, Spector S. Studies on the hypotensive response to atropine in hypertensive rats. J Pharmacol Exp Ther 1981; 218(3): 662-8. [PMID: 6115051]

[23] Butcher RW. *Atropa belladonna* L. J Ecol 1947; 34(2): 345-53. [http://dx.doi.org/10.2307/2256722]

[24] Ahuja A, Sambyal M, Koul S. *In vitro* propagation and conservation of *Atropa acuminata* Royle ex Lindl : An indigenous threatened medicinal plant. J Plant Biochem Biotechnol 2002; 11(2): 121-4. [http://dx.doi.org/10.1007/BF03263148]

[25] Rauf A, Baloch MK, Abbasi FM, Chatta RM, Mahmood TZ. Status, utilization and trade of Hazara areas healing plants of Pakistan. J Food Agriculture and Environment 2007; l5: 236-42.

[26] Banerjee S, Madhusudanan KP, Chattopadhyay SK, Rahman LU, Khanuja SPS. Expression of tropane alkaloids in the hairy root culture of *Atropa acuminata* substantiated by DART mass spectrometric technique. Biomed Chromatogr 2008; 22(8): 830-4. [http://dx.doi.org/10.1002/bmc.998] [PMID: 18386250]

[27] Shinwari ZK, Qaisar M. Efforts on conservation and sustainable use of medicinal plants of Pakistan. Pak J Bot 2011; 43(SI): 5-10.

[28] Tali BA, Ganie AH, Nawchoo IA, Wani AA, Reshi ZA. Assessment of threat status of selected endemic medicinal plants using IUCN regional guidelines: A case study from Kashmir Himalaya. J Nat Conserv 2015; 23: 80-9. [http://dx.doi.org/10.1016/j.jnc.2014.06.004]

[29] Shinwari ZK, Gilani SA, Khan AL. Biodiversity loss, emerging infectious diseases and impact on human and crops. Pak J Bot 2012; 44(SI): 137-42.

[30] Wani PA, Nawchoo IA, Wafai BA. Improvement of sexual destination in *Atropa acuminata* Royle (Solanaceae) : A critically endangered medicinal plant of Northwestern Himalaya. Pak J Biol Sci 2007; 10(5): 778-82. [http://dx.doi.org/10.3923/pjbs.2007.778.782] [PMID: 19069863]

[31] Zarate R, Cantos M, Troncoso A. Induction and development of adventitious shoots of *Atropa baetica* as a means of propagation. Euphytica 1997; 94(3): 361-6. [http://dx.doi.org/10.1023/A:1002953929890]

[32] Al-Ashaal H, Aboutabl M, Maklad Y, El-Beih A. Tropane alkaloids of *Atropa belladonna* L.: *in vitro* production and pharmacological profile. Egypt Pharmac J 2013; 12(2): 130-5. [http://dx.doi.org/10.4103/1687-4315.124012]

[33] Maqbool F, Singh S, Kaloo ZA, Jan M, Meraj M, Meraj M. Callus induction and shoot regeneration of *Atropa acuminata* Royle-a critically endangered medicinal plant species growing in Kashmir Himalaya. J Scient Innov Res 2014; 3(3): 332-6. [http://dx.doi.org/10.31254/jsir.2014.3310]

[34] Maqbool F, Singh S, Zahoor A, Kaloo A, Jan M. A rapid micropropagation protocol of *Atropa acuminata* Royle ex Lindl. : A threatened medicinal plant species of Kashmir Himalaya. Indian J Biotechnol 2016; 15: 576-80.

[35] Khan FA, Abbasi BH, Shinwari ZK, Shah SH. Antioxidant potential in regenerated tissues of medicinally important *Atropa acuminata.* Pak J Bot 2017; 49(4): 1423-7.

[36] Maqbool F, Singh S, Wani TA, Ahmad M, Ganai BA, Jan M. Phytochemical screening and antioxidant activity of methanolic extract of *in vitro* raised plants of *Atropa acuminata* Royle. Int J Sci 2018; 7: 1-11.

[37] Rajput S, Agrawal V. Micropropagation of *Atropa acuminata* Royle ex Lindl. (a critically endangered medicinal herb) through root callus and evaluation of genetic fidelity, enzymatic and non-enzymatic antioxidant activity of regenerants. Acta Physiol Plant 2020; 42(11): 160. [http://dx.doi.org/10.1007/s11738-020-03145-6]

[38] Saito K, Yamaai M, Anai H, Yoneyama K, Murakoshi I. Transgenic herbicide resistant *Atropa belladonna* using an Ri binary vector and inheritance of the transgenic trait. Plant Cell Replication 1992; 11: 219-24.

[39] Aoki T, Matsumoto H, Asako Y, Matsunaga Y, Shimomura K. Variation of alkaloid productivity among several clones of hairy roots and regenerated plants of *Atropa belladonna* transformed with *Agrobacterium rhizogenes* 15834. Plant Cell Rep 1997; 16(5): 282-6. [http://dx.doi.org/10.1007/BF01088281] [PMID: 30727663]

[40] Mera N, Takayama S. Effects of medium components and shear conditions on the formation and growth of adventitious bud derived from hairy roots of *Atropa belladonna* L. Environ Control Biol 2012; 50(4): 393-406. [http://dx.doi.org/10.2525/ecb.50.393]

[41] Murashige T, Skoog F. A revised medium for rapid growth and bioassays with tobacco tissue cultures. Physiol Plant 1962; 15(3): 473-97. [http://dx.doi.org/10.1111/j.1399-3054.1962.tb08052.x]

[42] Asha Rani NS, Prasad MP. Studies on the organogenesis of *Atropa belladonna* in *in-vitro* conditions. Intern J Biotechnol Bioeng Res 2013; 4(5): 457-64.

[43] Asha Rani NS, Prasad MP. Comparison studies on callus induction of fresh and reused explants of *Atropa belladonna.* Asian J Plant Sci Res 2014; 4(2): 50-3.

CHAPTER 6

Advances in Micropropagation Techniques of *Aegle marmelos* (L.) Corr.: A Review

Kalpana Agarwal[1] and **Richa Bhardwaj**[1,*]

[1] *Department of Botany, IIS (Deemed to be University), Jaipur, Rajasthan, India*

Abstract: *Aegle marmelos* (L.) Corr. is a plant of religious and medicinal importance in India. All of its plant parts have been reported to possess medicinal uses due to the presence of various phytoconstituents. Looking at its perspectives, *Aegle* is successfully propagated *in vitro*, primarily through organogenesis, using numerous explants. Efficient micropropagation is ensured by proper sterilization, preparation of explants, and use of antioxidants to avoid media browning. Various factors that affect the regeneration rate include season of explant collection, explant origin, phenological growth stage, concentration and combination of Plant Growth Regulators (PGR), culture media composition, and addition of additives to the media to enhance the micropropagation rate. The present review chapter compiles numerous reports of the effective micropropagation of *A. marmelos* and factors that affect the rate of micropropagation.

Keywords: *Aegle marmelos*, Bael, Micropropagation, Organogenesis, PGR.

INTRODUCTION

Aegle marmelos (L.) Corr, commonly called wood apple or stone apple, is indigenous to India. It is associated with spiritual and religious values. It belongs to the family Rutaceae. It is a slow-growing, spinous, tough, subtropical medium-sized tree. The fruits are used for both dietary and medicinal purposes.

A. marmelos contain several phytoconstituents namely marmenol, marmin, marmelosin, marmelide, psoralen, alloimperatorin, rutaretin, scopoletin, aegelin, marmelin, fagarine, anhydromarmelin, limonene, α-phellandrene, betulinic acid, marmesin, imperatorin, marmelosin, luvangentin and auroptene [1]. The Ayurvedic system considers it to be a healing tree that gives strength to the body, and this accreditation is due to its diverse medicinal properties considering all parts of this tree, namely, roots, leaves, trunk, fruit, and seeds, are used to cure va-

* **Corresponding author Richa Bhardwaj:** Department of Botany, IIS (Deemed to be University), Jaipur, Rajasthan, India; Tel: 9116627902; E-mails: richa.bhardwaj@iisuniv.ac.in and niki86ster@gmail.com

T. Pullaiah (Ed.)

rious human ailments and diseases [2 - 4]. Although all parts are useful, root and root bark are mainly used in medicinal preparations, which is why Bael suffers from destructive harvesting, as the tree has to be uprooted wholly to procure roots. Indiscriminate harvesting of the tree has posed so much threat that 'Red Data List of Indian Plants' has placed it in vulnerable threat status

MICROPROPAGATION

Aegle marmelos, a tree with immense medicinal importance, is traditionally cultivated through seed germination. Genetic variability, short viability, and vulnerability to insect and termite attacks make them unsuitable for propagation. Bael is also propagated vegetatively by root suckers, root cuttings, and layers, but the process is tedious and time-consuming [5]. The conventional plant propagation methods for woody trees are often difficult and slow because of high levels of heterozygosity and the long generation time between successive crosses.

Micropropagation is the only remedy that can help in overcoming these problems. In view of this scenario, a plethora of research work has been conducted on the morphogenic capacity of *Aegle*, and still, much more effort is required in this line [6, 7]. Trees have phenolic compounds; their life cycles are complex and long, are large sized, and therefore difficult to propagate. The intervention of modern propagation and improvement techniques for trees has become a necessity. Micropropagation using explants like shoot tip, root tip, shoot segments, nodal explants, cotyledonary nodes, zygotic embryos, *etc.*, has been a boon in this direction. Plant tissue culture not only helps in the mass multiplication of superior genotypes but also is a basic requirement of transgenesis.

A large number of factors are responsible for the success of a plant tissue culture protocol like explants- type, age, physiological status of donor plant, type and concentration of growth regulators, carbon and nitrogen source, gelling agents and their concentration, and even method of explant sterilization. Suitable explant choice at a particular receptive stage, alteration in nutrient media composition, and growth additives could help lessen recalcitrance and give satisfactory micropropagation results. Different types of basal media like MS (Murashige and Skoog) [8], B5 medium (Gamborg Medium), WPM medium (Woody Plant Medium) were tested by Arumugam *et al.* [9]. Pati and Muthukumar [10] attained somatic embryogenesis and genetic transformation in *Aegle marmelos* using a half-strength (½) MS medium. Above all, regulated physical conditions like light, temperature, magnetic or electromagnetic fields, photoperiod [11], humidity, and other factors also play an important role in Bael tissue culture.

Proper sterilization and preparation of explants is a prerequisite step for the establishment of a successful *in vitro* plant regeneration protocol. The problem of contamination is prominent in field-grown explants as compared to *in vitro* grown. There are several recommendations for decontaminating the explants before culture. For example, the use of teepol, a liquid detergent [12], NaOCl (Sodium Hypochlorite) [13], Clorox, Savlon [14, 15], or 5–10% Labolene [7, 16]. Many scientists have preferred using fungicides for disinfecting explants like Bavistin [12], Thiram, or 0.1% mercuric chloride ($HgCl_2$) [9, 15 - 22]. Princy *et al.* [11] were of the opinion that not only the concentration of $HgCl_2$ but the time duration of the treatment is also crucial for disinfection. Explant browning, necrosis, and decline in viability were noticed on increasing the time of sterilization from 5 minutes. A long period of surface disinfection is not suitable for juvenile explants such as shoot tips. Yadav and Singh [20] observed necrosis of shoot tips on being disinfected with 0.1% $HgCl_2$ for 8 minutes. They reported an optimal sterilization protocol for Bael tissue culture through experiments with elite varieties of Bael. According to them, soaking explants in 2.5% fungicide solution for two hours is the best for sterilization. Sometimes, a combination of antibiotics like rifampicin in combination with ethyl alcohol was also recommended [22].

Browning of cultured tissues due to the exudation of phenolic compounds from cut ends is another problem of concern in Bael micropropagation. To deal with this, antioxidants like citric acid, ascorbic acid, activated charcoal, and polyvinyl pyrrolidine (PVP) were added to the medium to avoid the browning of *A. marmelos* cultures [23]. Raj *et al.* [24] stated that the addition of activated charcoal in the medium for controlling browning intensity in cultures of Bael is unsurpassed.

Micropropagation in *Aegle* mainly relies on organogenesis using various explants, but somatic embryogenesis, androgenesis, and protoplast culture have also been explored, although not as successful as organogenesis [25].

ORGANOGENESIS

Correct choice of explants, explant origin, physiology of donor plant growth hormones, seasonal variations, *etc.,* affect organogenesis.

Explants

Different explants reported to demonstrate morphogenic potential in *Aegle* species are twigs [12], nodal segments [26], immature and mature leaf segments, tendrils, cotyledonary nodes [27 - 29], hypocotyls [22, 30], epicotyls [31], shoot tips [14], leaves from seedlings [9], root segments [9, 22], stem segments, *i.e.,* nodes and internodes [18, 32], nucellar tissue [33, 34], zygotic embryos [35], nodes from field-grown trees [7, 11, 12, 16, 21], nodes from root suckers [16], internodes or twigs [20, 36], leaves [36], shoot tips [20], zygotic embryos from immature fruits [10], cotyledons and shoot tips from *in vitro* seedlings [37 - 40], *etc.* Explants taken from *in vitro* grown seedlings like epicotyls, hypocotyls, cotyledonary nodes, shoot tips, and leaves are desirable over those taken from fully grown trees of Bael. *In vitro* explants are juvenile and meristematic, by virtue of which they lack recalcitrancy, increasing the possibility of morphogenesis, and are expected to be free of contamination, which otherwise is a great threat to Bael Tissue Culture. Moreover, the dependence on the season for the collection of explants is also waived. The season of explant collection also plays a vital role in micropropagation [18]. The time taken for the explant to respond also varies according to the explant. Cotyledonary nodes were shown to give slow shoot induction from that of axillary buds [35].

According to Raghu *et al.* [41], Kishore *et al.* [42] and Vasava [43], the phenological stage of the plant has a strong influence on the regeneration potential of Bael. They postulated that *in vivo* explants for *Aegle* tissue culture should be collected during April – June when the tree is not blooming but is at the bursting vegetative stage. In *Aegle marmelos,* hormones produced in the plant during flowering and fruiting provide hinderances to the success of regeneration.

Indhumathi and Rajamani [44] conducted experiments to study the germination of seeds of *A. marmelos* and concluded that germination was better with seeds without seed coats (5.57 shoots/ seed). Seeds with seed coats germinated to produce a single seedling.

Nodal segments are reported to be the most amenable explants for *in vitro* shoot multiplication of *Aegle marmelos*. Most of the studies have reported a mass multiplication of shoots using a single node segment [7, 11, 45]. Princy *et al.* [11] observed the multiplication of shoots at 0.5 mg/L of BA (6-Benzyl Adenine). They even confirmed the clonal fidelity of the shoots using ISSR markers, proving the reliability of the protocol to micropropagate true-to-type plantlets. The highest success in terms of shoot number and shoot length, using BAP (6-Benzyl Aminopurine) at 1 mg/L concentration, was also reported by many others. Further increase in this concentration to 4mg/L resulted in vitrification of multiplied shoots. Various other cytokinins, alone or in different combinations, were also tried by them, but none of them gave promising results. Indirect regeneration with the intervention of organogenic callus on MS medium with BAP and NAA (1-Naphthalene Acetic Acid) was also seen in the plant [46].

Proliferative rooting of shoots was observed on being transferred to the rooting medium. As expected, the results were not positive on plain MS medium, indicative of its hormonal requirements. This contrasts with many studies which reported zero hormonal requirements for rooting. The level of sucrose in the culture medium was varied to observe its effect on rooting and concluded that the normal MS level of sucrose, *i.e.,* 3%, supplemented with NAA, gave the best results in terms of rooting and plant growth. On the contrary, Princy *et al.* [11] reported rooting at half MS medium with 1.5% sucrose with NAA. Though not recent yet similar results were also reported by many other scientists like Pradeepa *et al.* [46]. Different scientists have found different hormonal requirements for the rooting of shoots. Gupta *et al.* [47] were of the opinion that IAA (Indole-3- Acetic Acid) and IBA (Indole-3- Butyric Acid) may be the appropriate hormones for rooting of shoots.

Micropropagation through nodal explants was remarkably superior to that from cotyledonary nodes. This may be an outcome of their heterozygous nature [12, 27, 28, 47 - 49]. Gupta *et al.* [29] germinated immature seeds under *in vitro* conditions, and cotyledonary nodes from 1-2 cm long seedlings were cultured on an MS medium supplemented with various combinations of Kn (Kinetin), BAP and NAA. High cytokinin (adenine sulphate) to low auxin (NAA) ratio was recommended for subculture and multiplication of shoots initiated from the cotyledonary node. They reported the development of healthy shoots, which were transferred to the rooting medium for root initiation and development.

Substantial mass multiplication of shoots of *Aegle marmelos* through axillary buds and shoot tip cultures has been reported. Kumari *et al.* [35] noticed direct organogenesis from axillary buds on WPM medium supplemented with BAP, NAA, and Kn. Rooting of these shoots was witnessed on the same media but at half mineral strength with BAP and NAA [50]. Gandhi *et al.* [51] in their study developed a protocol for micropropagation of Bael using axillary buds and shoot tip explants from *in vitro* grown seedlings. Maximum regeneration was observed on a medium supplemented with 11.11 μM/l of BAP and 1.07 μM/l of NAA. Similar were the results of Puhan and Rath [7] who reported a rapid bud proliferation using BAP followed by shoot development on transfer to MS medium with BAP in combination with Kn or GA_3 (Gibberellic Acid).

Epicotyl of *in vitro* grown seedling of *Aegle marmelos* also exhibited the potential of shoot induction, growth, and multiplication on being cultured with suitable growth regulators. According to the previous reports, in the case of nodal segments and cotyledonary nodes, BAP and Kn were potent to give promising results with epicotyl cultures as well [31]. Behera *et al.* [22] studied different explants, but epicotyl was reported to be the best of them. The hormone combination reported for shoot multiplication through epicotyl cultures was pertinent to previous studies, *i.e.,* BAP, Kn, and IAA alone or in different combinations, but the rooting hormone which emerged out was IBA instead of BAP. Similar were the results achieved by Parihar and Kumar [31].

Besides these, other explants have also been studied, with inadequate or zero success. There are many such unsuccessful reports, namely, nucellar calli by Hossain *et al.* [34], leaf explants [37], leaf and internode [52]. According to Mishra *et al.* [52], unsatisfactory adventitious shoot formation was obtained from internode explants of *in vitro* grown seedlings of *A. marmelos* on the medium supplemented with BA. In the presence of 2,4-D(2,4-Dichlorophenoxyaceticacid) along with BA, callogenesis occurred. Leaf explants were identified to induce only callogenesis in the same study. Futile efforts to raise cultures from seedling parts, mainly cotyledons, and hypocotyls, were also done by Arumugam and Rao [53]. Mature and immature leaves of *A. marmelos* were also tried by Fonseka and Aluthgamage [40] and were discovered to be inferior due to the low frequency of micropropagation and high frequency of contamination.

Besides PGR and explants, Shahina [54] studied the role of explant origin in *A. marmelos.* According to them, shoot multiplication from root explants was better in comparison to that from nodal explants procured from aerial shoots. Even the organogenic callus produced from root explants was much more in amount, resulting in better regeneration. 10 μM TDZ (Thidiazuron) was reported to give the best results.

Plant Growth Regulators

The critical role of plant growth regulators in plant growth and development, either in nature or in culture conditions, is unquestionable. Medium without growth regulators displayed a sluggish rate of bud break in *Aegle marmelos*. An exogenous supply of growth regulators to cultures is, therefore, a must for inducing bud break. Of all PGRs, cytokinins and auxins are known to play a pivotal role. Appropriate combination and optimal ratio of both hormones have a synergistic effect on cultures. *In vitro* propagation in *Aegle* species is also under the extreme influence of plant hormones and other growth regulators. A high cytokinin to low auxin concentration aids in shoot multiplication. The effectiveness of cytokinin in promoting *in vitro* axillary bud development is well documented.

MS medium supplemented with cytokinin alone or in combination with an auxin is mainly used for axillary shoot multiplication in *Aegle* species, but the type, concentration, and combination of PGR varies with the genotype and physiological status of the explant.

Behera *et al.* [22] extensively studied shoot multiplication and plantlet regeneration in *Aegle marmelos* using different explants from axenically grown seedlings. Maximum shoots were obtained on MS medium with 2.2 µM BAP with 1.425 µM IAA. Although BAP alone was found to be sufficient for the induction and multiplication of shoots, the addition of auxin enhanced the regeneration potential. Similarly, Bindhu [12] reported synergism between BAP and IAA for organogenesis through nodal explants. Gupta *et al.* [55] were also of the view that the addition of IAA, even at a minimal concentration of 0.2 mg/L, improves shoot induction and multiplication. Kn (0.5mg/L) and 0.1 mg/L GA_3 were used for shoot elongation. Shoot induction from cotyledonary explants on WPM medium with various PGRs (Plant Growth Regulators) was slow in comparison to that of axillary bud on medium with 0.5 mg/L Kn + 0.5 mg/L BAP + 0.5 mg/L IAA [35]. Gupta *et al.* [29] achieved maximum frequency of regeneration, number and length of shoots on BAP with NAA instead of IAA. The results were even more pronounced on augmenting medium with ammonium sulphate as an additive. But for shoot elongation, the subculture medium should be devoid of cytokinin. Similar observations were made using cotyledonary explants with similar hormonal combinations [55]. The studies show that the proximal part of cotyledon had the highest regeneration potential. Gandhi *et al.* [51] also studied the influence of PGR on organogenesis and shoot multiplication using axillary bud and shoot tip explants from *in vitro* seedlings.

The induction medium used for the study was supplemented with BAP (2.22, 4,43, 6.65, 8.87, and 11.11 µM) alone or in combination with NAA (0.54 & 1.07 µM). Initiation of shoots was observed in all treatments but with different efficiency. Minimum shoot induction was observed on 2.22 µM BAP alone (15.17%), whereas maximum on medium with 11.11 µM BAP + 1.07 µM NAA (81.40%). Mean shoot length, number of shoot buds per explant, and shoot multiplication of above-induced shoots were also best on this medium. The results were in accordance with that of Raghu *et al.* [41]. Gupta *et al.* [29] postulated that BAP with IAA (instead of NAA) gave superlative outcomes. Contrarily, Princy *et al.* [11] found that 0.5 mg/L BAP alone was best in terms of shoot induction and multiplication, whereas shoot elongation and growth were optimal at 1.0 mg/L BAP alone. In comparison to BAP, Kn and 2iP (Dimethylallylaminopurine) did not show any favorable response either individually or with BAP. Instead of being positive, the medium with BAP, when augmented with 2iP, proved negative for shoot induction. Ajithkumar and Seeni [56] also achieved shoot induction and multiplication on the medium fortified with BAP alone. Princy *et al.* [11] also got highest success rate, both in terms of shoot length and shoot number on medium with 1.0 mg/L BAP alone. On increasing the concentration to 4.0 mg/L, shoots suffered from hyperhydricity.

Hormonal regime of the medium, nitrogen content, and carbon percentage had dramatic effects on the rooting of the shoots as well. However, there is no consensus as to the hormonal requirement for rooting. Princy *et al.* [11] received a good rooting response on full MS medium with 6% sucrose and 1 mg/L NAA. Gupta *et al.* [29, 55] who described IAA to be the best for rooting, in their later studies, witnessed a better role of IBA and NAA. Gandhi *et al.* [51] tried full-strength and half-strength MS medium with various concentrations of IBA. Full strength MS with 4.92 µM IBA gave the poorest rooting response of only 23.80% with a mean number of 1.35 roots /shoot. On the other hand, half-strength MS medium with 19.68 µM IBA gave the maximum percentage of root induction (84.0%) and the highest mean number of shoots. Other hormonal combinations reported for successful rooting of shoots of *A. marmelos* are 2.85 µM IAA [22], 0.1 mg/L IAA [12], IBA + IAA [46], and IAA [56].

Phenological Stage

According to Raghu *et al.* [41], Kishore *et al.* [42] and Vasava [43], the phenological stage of the plant has a strong influence on the regeneration potential of Bael. They postulated that *in vivo* explants for *Aegle* tissue culture should be collected during April – June when the tree is at full vegetative growth. Hormones produced in the plant during flowering and fruiting provide hindrances in the success of regeneration in *Aegle marmelos.*

Indhumathi and Rajamani [44] conducted experiments to study the germination of seeds and concluded that germination was better on seeds without seed coats (5.57 shoots/seed). Seeds with seed coats germinated to produce a single seedling.

Somatic Embryogenesis

Somatic embryogenesis, the development of embryos from a single somatic cell, offers great advantages in the field of Plant Tissue Culture. It has a minimal chance of having genetic variants or chimeras, making it a suitable choice for genetic engineering. Other potential benefits of Somatic Embryogenesis are 1) easy and inexpensive mass multiplication. 2) bipolar nature having shoot and root end, which alleviates them from the hassle of being transferred to separate rooting medium. 3) competent in prolonged germplasm conservation 4) synthetic seed formation 5) cryopreservation.

Selection of proper explant, medium composition, growth hormones, and other adjuvants can enhance the feasibility of culture, despite all problems in somatic embryogenesis like slow side-shoot production, limited number, not so easy and tedious process.

Murashige [57], for the first time, reported somatic embryogenesis in *A. marmelos* using zygotic embryos as explants in the presence of 2,4-D and BA. Development of mature somatic embryos took approximately 42 days. After about a gap period of 20 years, successful somatic embryogenesis and subsequent plant regeneration in *Aegle marmelos* were given by Islam *et al.* [58] from cotyledons. Callus formation occurred on MS +2,4-D + Kn, whereas for regeneration from this embryogenic callus, MS media was augmented with BA and IAA. The report of Islam *et al.* [58] was followed by that of Islam *et al.* [59] from the zygotic embryo axis of matured seeds and Arumugam [60] from immature leaflets. Arumugam and Rao [53] established embryogenic cultures using cotyledons and hypocotyl explants from *in vitro* grown seedlings on MS +IAA +BA and MS +2,4-D +BA+ Glutamine. With 0.3 mg/L BA and 2.0 mg/L 2,4-D, a profuse callus was obtained from nodal explants of *Aegle marmelos* [26].

All these studies reported indirect regeneration from somatic embryos. Hazeena and Sulekha [61] witnessed callus induction and plant regeneration from cotyledon explants of Bael on 2.2 μM BA+ 2.26 μM 2,4-D. Superior callus formation from leaf explants was obtained on B5 medium with 2,4-D + BAP. Thangavel *et al.* [62] noticed proliferative embryogenic callus from a leaf disc with 2 μM Kn and 6.0 μM 2,4-D. When NAA was added in place of 2,4-D, slow growing callus formation occurred. A combination of BAP (2.2 μM) and 2,4-D (2.26 μM) was optimal for callus formation and its maintenance, which required BA (8.8 μM) and IAA (2.8 μM) for regeneration. Das *et al.* [15] mainly used 2,4-D alone for obtaining embryogenic callus from seed explants. The callus was, however whitish and friable in nature. Warrier *et al.* [45] in their studies found that a WPM medium with BAP at a concentration of 0.10 ppm- 1.0 ppm was required for the initiation of callus formation, but for regeneration, the BAP level has to be elevated. Akin to previous studies, Ramanathan *et al.* [63], reported indirect somatic embryogenesis from leaf primordia on MS with various combinations of NAA, Kn, and 2,4-D. Fonseka and Aluthgamage [40] developed a successful protocol for embryogenic callus development from seeds of *A. marmelos* on 0.5 BAP mg/L +1.0 2,4-D mg/L They communicated successful embryogenic callus formation in *A. marmelos* using cotyledon explants. Their main purpose was to isolate secondary metabolites from the callus and so did not proceed further to regeneration from somatic embryos.

FACTORS AFFECTING MICROPROPAGATION OF *AEGLE MARMELOS*

Season of Collection of Explants

Bael shows an effective reproduction mode by micropropagation techniques. With advances in research in the tissue culture of *Aegle*, explants showed marked seasonal variation in their response under *in vitro* conditions. Explants collected after the fruit setting period, that is, November, showed the most efficient growth response. Growth gradually declined during the leaf fall time, while it again peaked during the new leaf emergence in May [41]. Another study conducted by Pati *et al.* [64] showed maximum bud burst explants excised during September-October. Explants excised during these months showed the least inborn contamination in the explant, whereas bud burst decreased during other months.

Origin of Explant

Since *Aegle marmelos* displays heterozygosity and variation in the quality and size of fruits, it is difficult to propagate new plants *via* seeds. Propagation using stem and root modifications is also a cumbersome job. With the need to acquire *in vitro* cultures as propagation techniques, various explants were then checked for their efficacy to yield healthy plants: cotyledons, hypocotyls, root segments, nucellar tissues, single root segments, and cotyledonary nodes. Of all explants tested, the epicotyl segment resulted as the best source of explants for multiple shooting in all media compositions. To obtain maximum regeneration, the concentration of cytokinins also varied with the explants used [22]. The nodal stem segment of the mature fruit bearing tree resulted in rapid clonal micropropagation, the media was supplemented with optimum concentrations of auxins and cytokinins to yield maximum shoot proliferation [64]. Leaf explants of *in vitro* grown seedlings were also used as explants to initiate adventitious buds by supplementing media with adequate concentrations and combinations of auxins and cytokinins [24]. Cotyledon explants also formed alternatives for callus induction and plantlet regeneration in well-supplemented MS media [61]. Root sucker-derived nodal segments also resulted in increased shoot proliferation and regeneration [54].

Culture Media

With the progress in plant tissue culture, growing plants in any season seemed to solve the problem of plant propagation. An adequately supplemented media with optimum phytohormones and micronutrients could help in the regeneration of almost any plant.

Antioxidants

The influence of antioxidants were reported by Raj *et al.* [24]. They used antioxidants like activated charcoal, ascorbic acid, citric acid, and polyvinyl pyrrolidone, along with the best concentration of plant growth regulators in MS media. When tested on micropropagation of *Aegle marmelos*, these were reported to control the accumulation of phenolic compounds in the culture medium and augment the rate of micropropagation [65]. Results of the study showed that activated charcoal was most effective in maximizing shoot bud proliferation in nodal segment explants, while polyvinylpyrrolidone showed more callus induction in leaf explants.

Phenological Growth Stage

Phenological stages are referred to as any principal stages like bud development, leaf development, shoot/branch development, inflorescence emergence, flowering, fruit development, fruit maturity, senescence, and the beginning of dormancy. Therefore, determining the best phenological stage to enhance the micropropagation of *Aegle* is a must [36]. Bael shows phenological growth differences depending on rainfall, temperature, and fruiting. With the increase in micropropagation studies on Bael, it was shown that October to December is the finest period to collect explants in Kerela, India, as the plant is at its best phenological stage and results in the best micropropagation. It was seen that in the flowering and fruiting stages of the plants, the hormonal profiles of explant stages frequently delay effective regeneration. The study reported that leaves and twigs are the best explants for successful micropropagation studies. Another report by Pathirana *et al.* [65] suggested that successful micropropagation is possible if the explants are harvested from April to June, immediately after the fruiting season of the plant in Sri Lanka.

Additional Additives to the Media

Experimental data from various studies suggest that the addition of various additives to media proves to be effective in shoot multiplication effect, the addition of BAP, Kn and TDZ, 2iP, zeatin, and metatopolin in different concentrations and combinations like BAP and Kn, BAP and TDZ are effective in axillary shoots proliferation of *Aegle marmelos* [23]. Other additives like $AdSO_4$, glutamine, tryptophan, proline, and arginine in combination with BAP were also checked. The results showed that all three, tryptophan, arginine, and proline, showed the best results in combination with BAP with an increase in the average number of shoots. They were also used for synthesizing the indole ring of auxin, protein synthesis, and pollen fertility and viability. $AdSO_4$ could greatly reduce bud breakage, it also showed improved growth and increased number of shoots per explant [23].

The current review suggests that the protocol for micropropagation of *Aegle* depends on many factors, which include the use of various explants, of which epicotyl segments have reported efficient in *in vitro* propagation. Season of explant collection also plays an important role. Studies suggest fruit setting time displays the best results for micropropagation, antioxidants work on augmenting shoot proliferation apart from preventing browning of media. All protocols and their modifications reported in various studies present an effective mode of micropropagation to produce a mature plant without any genetic instability.

Different protocols for micropropagation of *Aegle marmelos* are given in Table **1**.

Table 1. Micropropagation in *Aegle marmelos*.

Explant	PGR/s	References
Organogenesis and Shoot Multiplication		
Cotyledons from *in vitro* seedlings	2.0 mg/L BA + 0.2 mg/L NAA	[37]
	BAP (2.5 mg/L) with NAA (0.5 mg/L)	[39]
	WPM medium supplemented with BAP (0.5 mg/L), IAA (0.5 mg/L) and Kn (0.5 mg/L)	[35]
Axillary buds from *in vitro* seedlings	WPM medium supplemented with BAP (0.5 mg/L), IAA (0.5 mg/L) and Kn (0.5 mg/L)	[35]
Axillary bud and shoot tips from the *in vitro* germinated seedlings	11.11µM of BAP and 1.07µM of NAA	[51]
Auxiliary bud from axenic shoot	2.5 mg/L BAP + 1.0 mg/L IAA	[56]
Shoot tips from *in vitro* seedlings	2.0 mg/L (BAP) + IAA (0.5 mg/L).	[38]
	WPM medium supplemented with BAP (0.5 mg/L), IAA (0.5 mg/L) and Kn (0.5 mg/L)	[35]
	2.0 mg/L BAP+0.3 mg/L NAA	[14].

(Table 1) cont.....

Explant	PGR/s	References
Cotyledonary nodes from *in vitro* seedlings	2.22 µM BA and 54.2 µM adenine sulphate	[43]
	BAP (1.0 mg/L) and NAA (0.5 mg/L)	[29]
	BA 2.0 mg/L + IBA 0.5 mg/L	[28]
	2.5 mg/L BAP and 0.5 mg/L Kn	[27]
	MS + 6.6 µM BA + 1.14 µM IAA	[48].
	2.0 mg/L BAP + 0.1 mg/L NAA	[15].
	BAP (0.1 mg/L)	[56]
	0.5 mg/L BA	[7]
	BAP (2.0 mg/L)	[47]
Hypocotyls from *in vitro* grown plantlets	2.2 µM BAP + 1.425 µM IAA	[22]
Root segments *in vitro* seedlings	BAP (2.2 µM)	[22]
Epicotyl from *in vitro* grown plantlets	BAP (2.2 µM) + IAA (1.425 µM)	[22]
	BAP (1.5 mg/L) + Kn (1.5 mg/L)	[31]
Nodes from field-grown trees	MS+BA 5.0 µM+NAA 0.5 µM	[16]
	2.0 mg/L BAP	[12]
	0.5 mg/L BA	[11]
	BAP (2.5 mg/L) and NAA (0.5 mg/L)	[46]
	BA 0.5 + GA_3 0.5+ KN 0.1	[7]
	BA + Kn +2,4-D	[49]
	2.0 mg/L BAP	[20]
	11.1 µM BA	[45]
	BA (3.0mg/L) + Kn (0.5mg/L) +NAA (0.01mg/L)	[9].
	2.5 mg/L BAP and 1.0 mg/L IAA	[56]
	BAP 8.84 µM + IAA 5.7 µM	[64]
	MS+BA 5.0 µM+NAA 0.5 µM	[54]
Nodes from root suckers	MS+BA 5.0 µM+NAA 0.5 µM	[16]
Internodes from 10 year old tree	BAP (2.0 mg/L) + IAA (0.5 mg/L)	[20]
Leaves from field grown tree	2,4-D (0.5 mg/L) and BA (0.2 mg/L).	[11]
	1 mg/L of BAP	[12]
Axillary bud from mature tree	DA (3.0mg/L) + Kn (0.5mg/L) +NAA (0.01mg/L)	[9]
Cotyledon from mature seeds	2.0 mg/L BAP+0.2 mg/L NAA	[14]
Nucellar tissue from mature fruit	4.4 µm BA and 2.7 µM NAA	[33]
	-	[34]

(Table 1) cont.....

Explant	PGR/s	References
Nodal segments from root suckers	MS+BA 5.0 µM+NAA 0.5 µM	[54]
Somatic Embryogenesis		
Cotyledon	2.2 µM BA+ 2.26 µM 2,4-D	[61]
	MS+2,4-D (0.2 mg/L) + BA (0.2 mg/L) + glutamine (10.0 mg/L)	[18]
	2 mg/1 2,4-D + 1 mg/L BAP	[58]
Leaf primordia	MS + 1.5 mg/L Kn + 0.5 mg/L 2,4-D + 1.0 mg/LNAA	[63]
Nodal segments from 10 year old tree	2.0 mgl-1 BAP + 0.5 mg/L 2,4-D	[20]
Cotyledon and hypocotyl explants,	2% sucrose and 1 µM BA	[59]
	MS+NAA (0.2 mg/L) + BA (0.1 mg/L)	[60]
Embryo axis of cotyledon	MS+2,4-D (0.25 mg/L) +BA (0.25 mg/L) + CW (20%	[59]
Globular embryo	1/2MS+Sucrose (2%)+13A (0.25 mg/L)	[59]
Immature leaflet	MS+2,4-D (0.2 mg/L) + BA (0.2 mg/L) + glutamine (10.0 mg/L)	[60]
Hypocotyl	Liquid MS+2A-D (0.1 mg/L) + BA (0.1 mg/L) + glutamine (10.0 mg/L)	[60]

CONCLUSION

The present chapter gives an overview of micropropagation of *A. marmelos* which will be useful for clonal propagation. Efforts have to be made using biotechnological tools to develop clones with higher quantities of secondary metabolites and *in vitro* production of secondary metabolites of medicinal value.

REFERENCES

[1] Baliga MS, Thilakchand KR, Rai MP, Rao S, Venkatesh P. *Aegle marmelos* (L.) Correa (Bael) and its phytochemicals in the treatment and prevention of cancer. Integr Cancer Ther 2013; 12(3): 187-96. [http://dx.doi.org/10.1177/1534735412451320] [PMID: 23089553]

[2] Atul NP, Nilesh VD, Akkatai AR, Kamlakar SK. A review on *Aegle marmelos*: A potential medicinal tree. Intern Res J Pharm 2012; 3: 86-91.

[3] Rahman S, Parvin R. Therapeutic potential of *Aegle marmelos* (L.) : An overview. Asian Pac J Trop Dis 2014; 4(1): 71-7. [http://dx.doi.org/10.1016/S2222-1808(14)60318-2]

[4] Pathirana CK, Madhujith T, Eeswara J. Bael (*Aegle marmelos* L. Corrêa), a medicinal tree with immense economic potentials. Adv Agric 2020; 2020: 1-13. [http://dx.doi.org/10.1155/2020/8814018]

[5] Hiwale S, Ed. Bael (*Aegle marmelos* Correa). Sustainable horticulture in semiarid dry lands. New Delhi: Springer 2015; pp. 177-96. [http://dx.doi.org/10.1007/978-81-322-2244-6_12]

[6] Singh AK, Singh S, Saroj PL, Krishna H, Singh RS, Singh RK. Research status of bael (*Aegle marmelos*) in India: A review. Indian J Agric Sci 2019; 89(10): 1563-71. [http://dx.doi.org/10.56093/ijas.v89i10.94576]

[7] Puhan P, Rath SP. *In vitro* propagation of *Aegle marmelos* (L.) Corr., a medicinal plant through axillary bud multiplication. Adv Biosci Biotechnol 2012; 3(2): 121-5. [http://dx.doi.org/10.4236/abb.2012.32018]

[8] Murashige T, Skoog F. A revised medium for rapid growth and bio assays with tobacco tissue cultures. Physiol Plant 1962; 15(3): 473-97. [http://dx.doi.org/10.1111/j.1399-3054.1962.tb08052.x]

[9] Arumugam S, Rao AS, Rao MV. *In vitro* propagation of *Aegle marmelos* (L.) Corr., a medicinal tree. Micropropagation of woody trees and fruits. Dordrecht: Springer 2003; pp. 269-315. [http://dx.doi.org/10.1007/978-94-010-0125-0_10]

[10] Pati R, Muthukumar M. Genetic transformation of Bael (*Aegle marmelos* Corr.). Biotechnology of neglected and underutilized crops. Dordrecht: Springer 2013; pp. 343-65. [http://dx.doi.org/10.1007/978-94-007-5500-0_14]

[11] Princy PS, Gangaprasad A, Sabu KK. *In vitro* propagation and evaluation of genetic fidelity of *Aegle marmelos* (L.) Correa, a highly sought-after sacred medicinal tree. Phytomorphology 2015; 65(1/2): 1-10.

[12] Bindhu KB. *In vitro* propagation of *Aegle marmelos* through nodal explants. Int J Sci Res 2015; 4: 2319-7064.

[13] Singh O, Misra KK. Standardization of micropropagation technique in bael (*Aegle marmelos* Correa). Progress Hortic 2008; 40(2): 203-8. [http://dx.doi.org/10.5958/2249-5258.2020.00030.5]

[14] Das R, Hasan MF, Rahman MS, Rashid MH, Rahman M. Study on *In vitro* propagation through multiple shoot proliferation in wood apple (*Aegle marmelos* L.). IntJ Sustain Crop Prod 2008; 3(6): 16-20.

[15] Das R, Hasan MF, Rashid H, Rahman M. Plant regeneration through nodal explant derived callus in wood apple (*Aegle marmelos* L.). Bangladesh J Sci Ind Res 1970; 44(4): 415-20. [http://dx.doi.org/10.3329/bjsir.v44i4.4590]

[16] Parveen S, Shahzad A, Nazim F. Effect of explant origin on clonal propagation of *Aegle marmelos* (L.) Corr. : A multipurpose tree species. Indian J Biotechnol 2015; 14(4): 566-73.

[17] Rawal K, Keharia H. Prevention of fungal contamination in plant tissue culture using cyclic lipopeptides secreted by *Bacillus amyloliquefaciens* AB30a. Plant Tissue Cult Biotechnol 2019; 29(1): 111-9. [http://dx.doi.org/10.3329/ptcb.v29i1.41983]

[18] Pati R, Chandra R, Chauhan UK, *et al. In vitro* plant regeneration from mature explant of *Aegle marmelos* Corr. cv. CISH-B2. Sci Cult 2008; 74: 359-67.

[19] Puhan P, Thirunavoukkarasu M. Direct organogenesis of *Aegle marmelos* (L.) Corr. from cotyledon explants. Afr J Biotechnol 2011; 10(82): 18986-90.

[20] Yadav K, Singh N. *In vitro* propagation and biochemical analysis of field established wood apple (*Aegle marmelos* L.). An Univ Oradea Fasc Biol 2011; 18(1): 23-8.

[21] Raj P, Jakhar ML, Ahmad S, Mtilimbanya KY, Chahar S, Jat HR. Effect of culture media on shoot proliferation and callus induction of Bael (*Aegle marmelos* L.). Int J Curr Microbiol Appl Sci 2020; 9(3): 243-8. [http://dx.doi.org/10.20546/ijcmas.2020.903.029]

[22] Behera P, Manikkannan T, Chand PK. Adventitious plantlet regeneration from different explants of *Aegle marmelos* (L.) Corr. J Med Plants Res 2013; 7(37): 2761-8.

[23] Mandal S, Parsai A, Tiwari PK, Nataraj M. The effect of additional additives on the axillary shoot micropropagation of medicinal plant *Aegle marmelos* (L.). Corrêa World News of Natural Sciences. 2021; p. 34.

[24] Raj P, Jakhar ML, Ahmad S, Chahar S, Mtilimbanya KY, Jat HR. A study on effects of antioxidants in micropropagation of Bael (*Aegle marmelos* L.). J Pharmacogn Phytochem 2020; 9(1): 1687-90.

[25] Parmar N, Singh S, Patel B. Various pathya kalpana of Bilva [*Aegle marmelos* (L.) Correa ex Roxb.] : A review. Glob J Res Med Plants Indig Med 2016; 5(2): 57.

[26] Islam MR, Zaman S, Nasirujjaman K. Regeneration of plantlets from node derived callus in *Aegle marmelos* Corr. Biotechnology 2007; 6: 72-5.

[27] Akter S, Banu TA, Habib MA, *et al. In vitro* clonal multiplication of *Aegle marmelos* (L.) Corr. through cotylodonary node culture. Bangladesh J Sci Ind Res 2013, 48(1): 13-8. [http://dx.doi.org/10.3329/bjsir.v48i1.15408]

[28] Deepa E, Nair DS, Alex S, Soni KB, Viji MM. Effect of plant stimulants on shoot proliferation from nodal segments of *Aegle marmelos* L. (Corr.). Int J Phytomed RelaInd 2018; 10(1): 58-64. [http://dx.doi.org/10.5958/0975-6892.2018.00009.6]

[29] Gupta A, Thomas T, Khan S. Regeneration of *Aegle marmelos* (L.) through enhanced axillary branching from cotyledenory node. Pharmaceut Biosci J 2018; 6(2): 24-8.

[30] Hossain M, Islam R, Islam A, Joarder OI. Direct organogenesis in cultured hypocotyl explants of *Aegle marmelos* Corr. Plant Tissue Cult 1995; 5(1): 21-5.

[31] Parihar N, Kumar S. *In vitro* seed germination and clonal propagation through epicotyls explants of *Aegle marmelos.* Int J Res Stud Biosci 2015; 3: 67-70.

[32] Bhardwaj L, Merillon JM, Ramawat KG. Changes in the composition of membrane lipids in relation to differentiation in *Aegle marmelos* callus cultures. Plant Cell Tissue Organ Cult 1995; 42(1): 33-7. [http://dx.doi.org/10.1007/BF00037679]

[33] Hossain M, Karim MR, Islam R, Joarder OI. Plant regeneration from nucllar tissues of *Aegle marmelos* through organogenesis. Plant Cell Tissue Organ Cult 1993; 34(2): 199-203. [http://dx.doi.org/10.1007/BF00036102]

[34] Hossain M, Islam R, Islam A, Joarder OI. Direct organogenesis in cultured hypocotyl explants of *Aegle marmelos* Corr. Plant Tissue Cult 1994; 5(1): 21-5.

[35] Kumari RU, Lakshmi SM, Thamodharan G. Effect of growth hormones for direct organogenesis in bael (*Aegle marmelos* (L.) corr). J Appl Nat Sci 2015; 7(1): 98-101. [http://dx.doi.org/10.31018/jans.v7i1.570]

[36] Pathirana CK, Attanayake AMURK, Dissanayake DMUSK, *et al.* Effect of phenological growth stage on establishment of *in-vitro* cultures of bael (*Aegle marmelos* (L.) Corr.). Trop Agric Res; Proc Annu Congress Postgrad Inst Agric, Peradeniya Univ Peradeniya Postgrad Inst Agric Congress 2018; 29(3): 268-75. [http://dx.doi.org/10.4038/tar.v29i3.8266]

[37] Islam MR, Nasirujjaman K, Zaman S. Production of plantlets from *in vitro* cultured cotyledons of *Aegle marmelos* Corr. Life Earth Sci 2006; 1(2): 31-3.

[38] Devi CP, Gopal RM, Settu A. Direct regeneration of *Aegle marmelos* (L.) Corr. from shoot tip explants through *in vitro* studies. Asian J Plant Sci Res 2014; 4(3): 22-5.

[39] Devi CP, Gopal RM, Settu A. Plant regeneration of *Aegle marmelos* (L.) Corr. from cotyledon explants through *In vitro* studies. J Nat Prod Plant Resou 2014; 4(2): 52-5.

[40] Fonseka DL, Aluthgamage HN. Callus production protocol for *Aegle marmelos* (L.) Corr.: As a tool for extraction of secondary metabolites. Agricul Sci DigA Res J 2002; 41(3): 455-9.

[41] Raghu AV, Geetha SP, Martin G, Balachandr I, Ravindran PN, Mohanan KV. An improved

micropropagation protocol for Bael : A vulnerable medicinal tree. J Bot 2007; 2(4): 186-94. [http://dx.doi.org/10.3923/rjb.2007.186.194]

[42] Kishore K, Mahanti KK, Samant D. Phenological growth stages of bael (*Aegle marmelos*) according to the extended Biologische Bundesantalt, Bundessortenamt und Chemische Industrie scale. Ann Appl Biol 2017; 170(3): 425-33. [http://dx.doi.org/10.1111/aab.12347]

[43] Vasava D, Kher MM, Nataraj M, Teixeira da Silva JA. Bael tree (*Aegle marmelos* (L.) Corrêa): Importance, biology, propagation, and future perspectives. Trees 2018; 32(5): 1165-98. [http://dx.doi.org/10.1007/s00468-018-1754-4]

[44] Indhumathi K, Rajamani K. *In vitro* germination of Bael (*Aegle marmelos* (L.) Corr.) seeds for clonal propagation. Ann Plant Soil Res 2020; 22(4): 449-53.

[45] Warrier R, Viji J, Priyadharshini P. *In vitro* propagation of *Aegle marmelos* L. (Corr.) from mature trees through enhanced axillary branching. Asian J Exp Biol Sci 2010; 1(3): 669-76.

[46] Pradeepa CB, Gopal RM, Settu A. Plant regeneration of *Aegle marmelos* (L.) Corr. from cotyledon explants through *In vitro* studies. J Nat Prod Plant Resour 2014; 4(2): 52-5.

[47] Gupta S, Chauhan D, Bala M. Micropropagation of Bael [*Aegle marmelos* (L.) Corr.] : An indigenous medicinal fruit tree of India. Indian J Plant Genet Resour 2008; 21(3): 213-6.

[48] Nayak P, Behera PR, Manikkannan T. High frequency plantlet regeneration from cotyledonary node cultures of *Aegle marmelos* (L.) Corr. *In Vitro* Cell Dev Biol Plant 2007; 43(3): 231-6. [http://dx.doi.org/10.1007/s11627-006-9013-6]

[49] Raj JV, Basavaraju R. *In vitro* nodal explants propagation of *Aegle marmelos* (L.) Correa. Ind J Innov Develop 2012; 1(8): 575-87.

[50] Prematilake DP, Nilmini HAS. Micropropagation of *Aegle marmelos* L. : A medicinal tree through culture of axillary bud and shoot tip explants from *in vitro* germinated seedlings. Int J Sci Res Biol Sci 2006; 5: 3.

[51] Gandhi K, Rajesh E, Saravanan S, Elamvaluthi M. Micropropagation of *Aegle marmelos* L.: A medicinal tree through culture of axillary bud and shoot tip explants from *in vitro* germinated seedlings. Int J Scient Res Biol Sci 2018; 5(3): 65-71. [http://dx.doi.org/10.26438/ijsrbs/v5i3.6571]

[52] Mishra Y, Bhadrawale D, Mishra JP, Singh V, Anon D, Shirin F. Hormonal regulation for de novo organogenesis in *in vitro* derived leaf and internode of *Aegle marmelos* L. Indian For 2021; 147(1): 64-70. [http://dx.doi.org/10.36808/if/2021/v147i1/154191]

[53] Arumugam S, Rao MV. Somatic embryogenesis in *Aegle marmelos* (L) Corr., a medicinal tree. Somatic embryogenesis in woody plants. Dordrecht: Springer 2000; pp. 605-55. [http://dx.doi.org/10.1007/978-94-017-3030-3_22]

[54] Shahina P, Anwar S, Farah N. Effect of explants origin on clonal propagation of *Aegle marmelos* (L.) a multipurpose tree species. Indian J Biotechnol 2015; 14: 566-73.

[55] Gupta A, Thomas T, Khan S. Phytopharmacological potentials and micropropagation of *Aegle marmelos*–A review. Pharmaceut Biosci J 2018; 24: 52-6.

[56] Ajithkumar D, Seeni S. Rapid clonal multiplication through *in vitro* axillary shoot proliferation of *Aegle marmelos* (L.) Corr., a medicinal tree. Plant Cell Rep 1998; 17(5): 422-6. [http://dx.doi.org/10.1007/s002990050418] [PMID: 30736583]

[57] Murashige T. Plant propagation through tissue cultures. Annu Rev Plant Physiol 1974; 25(1): 135-66. [http://dx.doi.org/10.1146/annurev.pp.25.060174.001031]

[58] Islam R, Hossain M, Joarder OI, Karim MR. Adventitious shoot formation on excised leaf explants of *in-vitro* grown seedlings of *Aegle marmelos* Corr. J Hortic Sci 1993; 68(4): 495-8.

[http://dx.doi.org/10.1080/00221589.1993.11516377]

[59] Islam R, Hossain M, Hoque A, Joarder OI. Somatic embryogenesis and plant regeneration in *Aegle marmelos*: A multipurpose social tree. Curr Sci 1996; 71(4): 259-60.

[60] Arumugam S. Regeneration of plantlets from immature leaflets derived callus cultures of *Aegle marmelos* (L.) Corr. Plant Tissue Cult 1998; 8: 77-82.

[61] Hazeena MS, Sulekha GR. Callus induction and plantlet regeneration in *Aegle marmelos* (L.) Corr. Using cotyledon explants. J Trop Agric 2008; 46(1-2): 79-84.

[62] Thangavel K, Sasikala M, Maridass M, Nadu T. *In vitro* antibacterial potential of *Aegle marmelos* (L.) callus extract. Pharmacol Online 2008; 3: 190-6.

[63] Ramanathan T, Satyavani K, Gurudeeban S. *In vitro* plant regeneration from leaf primordia of gum-bearing tree *Aegle marmelos*. E-Int Sci Res J 2011; 3(1): 47-50.

[64] Pati R, Chandra R, Chauhan UK, Mishra M, Srivastava N. *In vitro* clonal propagation of bael (*Aegle marmelos* Corr.) CV. CISH-B1 through enhanced axillary branching. Physiol Mol Biol Plants 2008; 14(4): 337-46.
[http://dx.doi.org/10.1007/s12298-008-0032-0] [PMID: 23572900]

[65] Pathirana C, Attanayake U, Dissanayake U, *et al.* Establishment of a micropropagation protocol for elite accessions of Bael (*Aegle marmelos* L. Corr.), a tropical hardwood species. Adv Agricul 2020; 10.

CHAPTER 7

Biotechnological Aspects for Micropropagation of *Artemisia absinthium* L.

Varsha S. Dhoran[1], **Vishal P. Deshmukh**[2] and **Varsha N. Nathar**[1,*]

[1] *Department of Botany, Sant Gadge Baba Amravati University, Amravati-444602, Dist-Amravati, Maharashtra, India*

[2] *Department of Botany, Jagadamba Mahavidyalaya, Achalpur city-444806, Dist-Amravati, Maharashtra, India*

Abstract: *Artemisia absinthium* L. or 'wormwood', commonly known as 'Dawna', is a small perennial herb with a dark fragrance due to glandular trichomes present all over the plant. Medicinal properties of *A. absinthium* are known in most of Asia, South America, and Europe. Essential oil, along with other phytoconstituents, like flavonoids, phenolic acids, tannins, and lignans, imparts medicinal potential to this species. It revealed antibacterial, antitumor, antimalarial, antioxidant, anthelmintic, antipyretic, antidepressant, antiulcer, antiprotozoal, hepatoprotective, neurotoxic and neuroprotective action. Due to its wide range of disease curing potential, *A. absinthium* germplasm is always under the pressure of overexploitation and loss of habitat. To cope with the higher industrial demand of this plant, the use of biotechnological techniques related to micropropagation can provide the best alternative. *In vitro* propagation using any explants has been extensively studied for the conservation of its plant genetic resources. Other micropropagation methods, such as callus culture, cell suspension, and organogenesis, have been adapted with the aim of secondary metabolite extraction and artemisinin enhancement. Modern biotechnological tools such as *Agrobacterium*-mediated gene transformation are mainly applied to hairy root and shoot cultures to optimize the biosynthesis of artemisinin. The present review throws light on various biotechnological studies carried out on *A. absinthium,* presenting the respective outcomes.

Keywords: Artemisinin, Biotechnology, *In vitro* propagation, Medicinal and Aromatic Plants, Micropropagation.

INTRODUCTION

Artemisia is an important genus belonging to the family Asteraceae (Compositae), having reputation for complex taxonomic problems. The generic name of this

* **Corresponding author Varsha N. Nathar:** Department of Botany, Sant Gadge Baba Amravati University, Amravati-444602, Dist-Amravati, Maharashtra, India; Tel: 91-0721-2662358; Ext. 287;
E-mail: varshanathar@sgbau.ac.in

T. Pullaiah (Ed.)

genus evolved from 'Artemis', a Greek Goddess, Diana [1]. Worldwide, the genus *Artemisia* is represented by more than 500 species, of which a maximum of 200 species were reported from China and 45 species documented in India [2]. In India, this genus was introduced around 800 A.D. by Arabian traders, amongst them *Artemisia absinthium* L. is used extensively in Materia medica [3]. *A. absinthium* is a perennial shrub measuring up to 80 cm in height, however at some locations, 1.5 m tall plants were reported [4]. The plant body is densely covered with hairs/glandular trichomes rich in essential oil, imparting a dense smell to the whole plant [5]. There are reports of 0.25-1.32% essential oils from stem and dry leaves of *A. absinthium* along with artemisinin, anabsin, artabsin, absinthin, matricin and anabsinthin. Compounds like flavonoids, phenolics, terpenes, and many other biologically active ingredients were profoundly found in *A. absinthium* [6, 7]. *A. absinthium* (Wormwood) is globally distributed in almost all continents. It was reported dominantly from Europe to North Asia, the Middle East, North and South America, and rare reports from Africa and Australia. In India, it is mostly confined to Kashmir Valley, Himachal Pradesh, and some hilly parts of South India above 1500 m [8]. In India, *A. absinthium* is known by many vernacular names like 'Afsanteen' in Urdu, 'Daman vishesh' 'Pranthaparna', 'Suparna' in Sanskrit, 'Saparna' 'Supreema', in Marathi, Majri, Karmala, Majtari, Mastiyarah in Hindi and as a 'Tethwen' in Kashmiri [1, 8]. As per medicinal importance, *A. absinthium* have antiulcer, neuroprotective, neurotoxic, anthelmintic, antiprotozoal, anti-inflammatory and anti-feedant properties [9 - 15]. Multiplication of *A. absinthium* is achieved by seed or by cutting and division, however, heterogeneous plantlets were seen in the seed germinated populations, and vegetative propagation is a quite slow process of multiplication [16]. Roots of *A. absinthium* are used for vegetative propagation. However, excessive irrigation damages the roots and leads to rotting [5]. In *A. absinthium*, true plant production is very critical as in field grown plants, biotic and abiotic stresses make plants more vulnerable [17]. Furthermore, ever-increasing demand for *A. absinthium* leads to overexploitation of its genetic resources leading to pressure on plant propagation [18]. To overcome this situation, *in vitro* propagation techniques provide an important alternative way. Micropropagation helps to produce infection-free genetically identical clones in very large numbers [19]. In recent years, *in vitro* propagation methodology has been extensively used for mass production and conservation of rare and endangered plant genetic resources [20]. The present review provides in-depth information on various biotechnological aspects successively used in micropropagation of *A. absinthium*.

BIOTECHNOLOGICAL ASPECTS OF *A. ABSINTHIUM*

Explants Sources

Leaves dissected from *A. absinthium* aseptic plantlets were successfully used for callus induction [21 - 25]. Foliage, stem, and nodal segment explants of *A. absinthium* were found effective in callogenesis and organogenesis [26 - 28]. Shoot induction through the callus of leaves was also reported by Koul and Lone [29] and Nin *et al.* [30]. Genetic transformation was quite successful in the callus cells grown from *A. absinthium* leaves and flowering tips [31], however, hairy root formation was achieved through aseptic shoots [32]. Soft tissues like shoot tips can be proliferated further in shoots and roots [33].

Culture Conditions

Light is an important physical factor that has the potential to modify plant architectural development [24]. The light intensity of 50 µmol m^{-2} s^{-1} photon flux density (PFD) for 16 h was applied by Koul and Lone [29]. In another study, Mannan *et al.* [34] used cool white fluorescent light of intensity 2000 lux for 16 h. While performing *Agrobacterium* mediated gene transfer, with cool white fluorescent light, 35 µmol m^{-2} s^{-2} was used [32]. A report of 40-50 µmol m^{-2} s^{-1} PFD light intensity for 12 h by cool white fluorescent was provided by Shekhawat and Manokari [27]. Cell suspension cultures placed at 40 µmol m^{-2} sec^{-1} revealed fluctuation in biomass [35]. The application of various spectral lights revealed morphogenic and biochemical variations in *A. absinthium* callus [24]. The cultures of *A. absinthium* were kept at 25 ± 1^0C [21]. Cell suspension cultures were also kept at 25 ± 1^0C [23]. Relative humidity was maintained at 60% [29]. pH of the culture medium was adjusted to 5.8 before heating using HCl or NaOH solutions, and the culture medium was subjected to autoclave for 20 min at 121°C and 108 kPa [33].

Tissue Surface Sterilization

In surface sterilization, initially *A. absinthium* explants were washed under running tap water to remove any surface attached material [21]. Mild detergent [26, 36] and Tween-20 (5% for 2 min) [33] were used to remove surface impurities from explants. Some researchers opt for antifungal agents like Benlate© solution. Lê *et al.* [37] washed leaf shoots with 0.1% Benlate© solution for 15 min. followed by soaking in 0.8% sodium hypochlorite for 15 min. In another experiment, Shekhawat and Manokari [27] used 0.1% aqueous solution of Bavistin for 5 min as a systemic disinfectant, followed by 0.1% mercuric chloride treatment. Explants were exposed to 70% ethanol for 30 minutes, followed by 0.01% mercuric chloride solution, later rinsed with sterile distilled water [19, 29,

31, 34, 24]. Kour *et al.* [38], in their experiment, used mercuric chloride solution to treat explants that were again subjected to sodium hypochlorite (0.5%) solution for 3 min and lastly washed the explants multiple times with sterile distilled water.

Culture Medium and Plant Growth Regulators (Hormones)

Among the many available media, Murashige and Skoog (MS) medium and B5 were extensively used for Micropropagation [34]. In the case of *A. absinthium*, MS0 (Murashige and Skoog basal medium) is prominently used for seed germination [21, 22]. For callogenesis and organogenesis, the MS medium proved the most significant [31, 34]. In the case of cell suspension culture, liquid MS media proved the most significant, however, some researchers also used B5 media [23, 32]. Plant hormones (also called Plant Growth Regulators) are chemical entities that have the potential to enhance or hamper plant growth. Leaf explant derived from the aseptic plant of *A. absinthium* was used by Ali *et al.* [21], who designed an experiment in which MS medium was supplemented with Thidiazuron (TDZ) (0.5 -5.0 mg/L) alone or along with Naphthalene Acetic Acid (NAA) (1 mg/L) or Indole-3-Acetic-Acid (IAA) (1.0 mg/L). The above combination produced significant callus, which, on further subculture, enhanced biomass. A combination of 1.0 mg/L TDZ and 1.0 mg/L NAA assisted in the successful development of cell suspension cultures [22]. Murashige and Skoog 10 medium, along with BAP (0.3 - 1.0 mg/L) and IAA or IBA (0.25-1.0 mg/L), was used to induce massive sprouting in *A. absinthium* nodal segments [26]. BAP, NAA, IAA, 2,4-Dichlorophenoxyacetic acid (2,4-D), and Kinetin (Kn) either alone or in combination were used for induction of *A. absinthium* callus and significantly enhanced shoot and root number and length [38]. Fig. (**1**) presents the *in vitro* regeneration of *A. absinthium* [38].

Source of Carbon

Among the many forms of available carbohydrates, sucrose is the best form of carbon and energy source used in tissue culture media for cell proliferation and secondary metabolite synthesis [39]; however, many alternatives, like glucose, and maltose, were used as per the experiment designed [34, 40, 29]. In the case of *A. annua* shoot culture, an increase in sucrose concentration to 30 g/L leads to get the highest artemisinin [39], and its elevation up to 70 g/L revealed enhanced artemisinin in hairy root culture [41]. 30g/L glucose produced the maximum number of roots and the longest induced roots [40]. According to Grech-Baran and Pietrosiuk [42], sucrose as a carbon source solely depends on the kind of culture; biomass growth in sucrose culture was comparable to fructose but was notably superior to glucose. So far, records using other carbon sources like galactose, lactose, mannose, melibiose, and cellobiose were lacking. Further research towards other alternative carbon sources needs to be done in the near future.

Fig. (1). (**A**) Leaf explants on MS medium without plant growth regulator (PGR); (**B**) GR); callus induction from leaf explants; (**C**) *in vitro* shoot regeneration from callus on MS medium; (**D**) shoot elongation; (**E**) *in vitro* shoot induction from nodal explant; (**F**) multiple shoot regeneration; (**g**) *in vitro* rooting on MS medium supplemented with IBA. (Image retrieved from B. Kour, G. Kour, S. Kaul, M. K. Dhar, "*In vitro* mass multiplication and assessment of genetic stability of *in vitro* raised Artemisia absinthium L. plants using ISSR and SSAP molecular markers", Advances in Botany, vol. 2014, Article ID 727020, 7 pages, 2014. https://doi.org/10.1155/2014/727020 and reproduced with the permission of Editor).

Source Nitrogen

In the case of the genus *Artemisia*, the ratio nitrate to ammonia [NO_3-/NH_4+] manipulated the artemisinin production and a 1:3 ratio of ammonia to nitrate was optimum for artemisinin production [42]. In some species, low total nitrogen concentration along with a high ratio of nitrate to ammonia in the culture medium helps in enhanced artemisinin concentration in hairy roots [39].

Additives

Agar is the most commonly used additive for solidification of culture medium for proper shooting rooting regeneration [29]; besides this, the use of Difco-Bacto agar is also reported in the hairy root culture of *A. absinthium* [32]. 50 mg/L ascorbic acid and 25 mg/L each of adenine sulphate, L-arginine, and citric acid were also used as an additive to fortify the medium for micropropagation of *A. absinthium* [27]. Elicitors like methyl jasmonate and jasmonic acids were used along with gibberellic acid to secondary metabolite accumulation and manipulate growth kinetics [23]. Basic MS medium was enriched with thiamine, pyridoxine, nicotinic acid, and myinositol for the multiplication of *A. absinthium* [37]. Although the addition of streptomycin sulphate antibiotics reduces bacterial infection, its addition during subculturing of *A. absinthium* revealed declined shoot formation [33].

In vitro Propagation/Micropropagation

Rapid and true-to-type propagation of *A. absinthium* can assist in designing and implementing conservation strategies, and also its bulk production significantly helped out in coping with market demand.

Callus Culture

Callogenic response from all explants initiated at the margins or from injury [28]. Green, brownish-yellow, compact, and friable callus was induced and proliferated from leaf explants of *A. absinthium* using TDZ alone and in combination with IAA and NAA on MS medium [21]. Tariq *et al.* [24] also reported an optimum response in callogenesis with the combination of TDZ (2.0 mg/L) + NAA (1.0 mg/L). The addition of methyl jasmonate and jasmonic acid to MS media enhances the biomass of calli [23]. Best callogenesis was observed with the use of leaf explants and 0.1 NAA + 2 mg/L BAP + sucrose (30g/L) [40]. The use of the BAP (1.0 mg/L) + 2, 4-D (0.5 mg/L) combination showed the highest 90% of responding frequency compared to the BAP and IAA combination [29]. In the case of *Artemisia* spp. GA_3 was of prime importance, having a role in the artemisinin synthesis pathway, which correlated with flowering [43]. The best results of GA_3 and BAP combination are due to the role of GA_3 in the breakdown of stored starch and proteins into sugars and amino acids useful for seedling growth [44, 45] and BAP stimulates lateral bud growth by boosting cell division in shoot meristems, controlling the development of vascular tissues and promotes the development of shoots from undifferentiated tissues of cultured tissues [44]. The 2,4-D + NAA combination revealed the finest callogenic response compared to other combinations [25].

Regeneration of Shoots

Cytokinins (BAP and Kn) have proved a record for cell elongation and division, which stimulate shoots from the explants [27]. BAP and Kn were the most successful cytokinins applied for *A. absinthium* shoot regeneration. Most excellent shoot induction was found on BAP (0.5 mg/L) + NAA (1.0 mg/L) [28]. Ghassemi and Nayeri [40] showed 0.5 mg/L BAP + 0.1 mg/L NAA + 30 g/L sucrose combination produces the maximum number and longest shoot in leaf explants. Using shoot tips, explants reported multiple shoots production with NAA in combination with BA or Kn, as presented by Aslam *et al.* [46]. A higher concentration, 1.0 mg/L of BAP, reported a decrease in the number of shoots, reflecting it was quite excessive for *A. absinthium* [26]. Callus grown on BAP (3.5 mg/L) and 2, 4-D (0.5 mg/L) along with 0.8% agar produced maximum percentages of response for shoot induction and shoots length was elongated by transferring shoots to MS salts plus GA_3 (1.0 mg/L) [29]. MS medium supplemented with 4.5 mg/L BAP and 0.5 mg/L NAA was found to be ideal for shoot induction from the callus; in addition to this, MS supplemented with only BAP at lower concentrations led to shoot formation; the number of shoots was quite less [38]. As per Shekhawat and Manokari [27], in the culture medium, a higher concentration of BAP and Kn than NAA could induce shoot organogenesis. It was quite evident that BAP has been proven as a beneficial hormone for the production of shoots in other members of *Artemisia* [47]. B5 medium with either 0.1 mg/L BAP or 0.1 mg/L Kn revealed the best direct shoot regeneration response using leaf explants [34]. According to Nin *et al.* [33], a low BA level and higher agar concentration depreciate the proliferation rate while the shoot vigour increases. 0.5 mg/L BAP and 0.25 mg/L Kn were found to be optimum for *A. absinthium* shoot induction [27].

Organogenesis of Root

2 mg/l of NAA and IAA was most significant for root induction in *Artemisia* [46]. Treatment of *A. absinthium* leaf cultures with 2 mg/L NAA + 30 g/L glucose reported the maximum number of root induction, and the longest induced roots were obtained in leaf explants culture with 0.1 mg/L NAA + 30 g/L glucose

treatment [40]. 1.0 mg/L and 0.50 mg/L IBA treatment showed maximum values of the number of roots and average main root length [26]. The application of silver nanoparticles significantly increases the root length up to 28 days of treatment compared to copper and gold nanoparticles [19]. MS medium with 0.5 mg/L IBA successfully produced roots from *in vitro* developed shoots [29]. Similarly, Kour *et al.* [38] reported the use of half-strength MS medium + 0.5 mg/L IBA as the most promising rooting combination, but a further increase in IBA leads to declined root numbers. Low 2, 4-D concentration triggered adventitious roots from 86% of all explants of *A. absinthium* [31]. Lê *et al.* [37] demonstrated the growth of rooting of newly developed shoots on hormone free medium. 0.025 mg/L NAA+ 0.1mg/L BAP in Gamborg's B5 medium, along with vitamins, was found to be the best combination for developing roots from regenerated shoots [34]. MS medium without IBA and with a minute amount of streptomycin sulfate and 1% sucrose helped to produce the highest (40%) rooting, as observed by Nin *et al.* [33]. Leaf explants maintained on MS medium with 1.81 μM 2, 4-D helped to direct the promotion of root primordial [30]. Shekhawat and Manokari [27] reported 100% *in vitro* rooting of regenerated shoots on one-fourth strength MS medium + 2.0 mg/L IBA along with this *ex vitro* rooting of *in vitro* regenerated shoots observed with treatment of 200 mg/L of IBA for 5 min. Although 1 or 2 roots of *A. absinthium* were reported on full MS and ½ MS with 0.25 mg/L IAA, still root induction is found problematic due to the development of callus at the base of shoots by inhibiting rooting.

Cell Suspension Culture

A. absinthium suspension culture reported rapid growth denoted by 6, 21, and 42 days of lag, log, and stationary phases with milky white, green, and brownish colours, respectively. While accumulating biomass in cell culture, gallic acid, caffeic acid, and catechin were also elicited [21]. In cell suspension cultures of *A. absinthium,* biomass accumulation and secondary metabolites production was altered using differential effects of light. Light grown cultures depicted around 4 times enhancement in biomass production and enhancement in total phenolic and total secondary metabolite contents compared to dark grown cultures [35]. In another study, Ali *et al.* [23] reported the inhibitory effect of elicitors *viz.*, methyl jasmonate, jasmonic acid, and GA on *A. absinthium* suspension cultures. The adverse results were mostly confined to the decline accumulation of dry biomass and undersized log phase of culture; however, treatment of suspension cultures with 1.0 mg/L of methyl jasmonate, jasmonic acid and GA showed an increase in total phenolic, total flavonoids contents and the highest radical scavenging activity in cultures.

Hairy Root and Shoot Induction

Artemisia species has a range of bioreactor configurations that have been reported to develop hairy root and shoot cultures; however, most bioreactors were widely used for the extraction of artemisinin [42]. In *A. absinthium*, *Agrobacterium tumefaciens* (MTCC532) mediated tumour induction reported 0.087 gm weight of dry extract and 700 cells/ml, additionally, the presence of mannopine confirmed by SDS PAGE [31]. Exposure of *A. absinthium* shoots to *Agrobacterium rhizogenes* strains 1855 and LBA 9402 resulted in hairy roots confirmed with southern blotting and PCR. Hormone free B5 agar-solidified medium with kanamycin or rifampicin suited best for the growth of subcultured transformed root lines. Additionally, 40 g/L sucrose was found best for enhancement in biomass production [32].

Hardening and Acclimatization

According to Mannan *et al.* [34], rooted plants were transferred in pots and kept in a growth chamber for 2 to 3 weeks before being exposed to the outer environment. Lê *et al.* [37] placed micropropagated plants on a horticulture substrate with compost, peat, sand, and perlite® (1:1:1:1 ratio) and later exposed to moisture saturated conditions; it ultimately reported a 95% survival rate. Nin *et al.* [30] transferred rooted shoots to the greenhouse in sterile peat pots and finally transplanted to the field. The gradual transformation of *in vitro* regenerated plants to perforated nursery polybags (Fig. **2**) having organic manure, red soil, and garden soil (1:1:1) at low humidity and high temperature for 40 days revealed successful acclimatization in the greenhouse showing 87% and 95% survival rates for *in vitro* and *ex vitro* rooted plantlets [27]. In another approach, Zia *et al.* [28] adopted a 3:1 ratio of soil and peat moss under high humid conditions until the leaves matured and later transferred to greenhouse for hardening.

CONCLUSION

A. absinthium is a well-known medicinal and aromatic plant having a wide range of phytoconstituents like polyphenols, flavonoids, *etc.* 'Artemisinin' is derived from *A. absinthium,* pharmacologically known worldwide as a drug of choice to cure malaria and other diseases. This plant has been under the radar of researchers to explore various aspects. Extensive work has been done on various biotechnological aspects related to micropropagation in this plant. This chapter provides a current and in-depth account of micropropagation studies. Additionally, *Agrobacterium* mediated gene transfer techniques for the production of hairy root culture to extract secondary metabolites and for the enhancement of biomass were also considered as an approach to conserve natural wild germplasm.

Fig. (2). (**a**) *Ex vitro* rooting to the *in vitro* regenerated shoots in Soilrite. (**b**), (**c**), and (**d**) Different stages in the hardening of the plantlets in the greenhouse. (**e**) Field transferred plant of A. absinthium. (Image retrieved from Mahipal S. Shekhawat, M. Manokari, "Efficient *in vitro* propagation by *ex vitro* rooting methods of Artemisia absinthium L., an ethnobotanically important plant", Chinese Journal of Biology, vol. 2015, Article ID 273405, 8 pages, 2015. https://doi.org/10.1155/2015/273405 and reproduced with permission of Editor).

REFERENCES

[1] Shah NC. *Artemisia* spp. ethnobotany, economics, taxonomy, and in Ayurveda, in India. J Nat Ayurvedic Med 2022; 6: 1-10.

[2] Huang YP, Ling YR. Economic Compositae in China. In: Caligari PDS, Hind DJN, Eds. Proceedings of the International Compositae Conference. 415-22.

[3] Wahid AH. A survey of drugs. Delhi: Hamdard Research Institute 1957; pp. 1-28.

[4] Ahamad J, Mir SR, Amin S. Pharmacognostic review on *Artemisia absinthium.* Int Res J Pharm 2019; 10(1): 25-31.
[http://dx.doi.org/10.7897/2230-8407.10015]

[5] Amidon C, Barnett R, Cathers J, Chambers B, Hamilton L, *et al.* Artemisia—An essential guide from the herb society of America. Kirtland, OH, USA: The Herb Society of America 2014.

[6] Kordali S, Cakir A, Mavi A, Kilic H, Yildirim A. Screening of chemical composition and antifungal and antioxidant activities of the essential oils from three Turkish *Artemisia* species. J Agric Food Chem 2005; 53(5): 1408-16.
[http://dx.doi.org/10.1021/jf048429n] [PMID: 15740015]

[7] Singh R, Verma P, Singh G. Total phenolic, flavonoids and tannin contents in different extracts of *Artemisia absinthium.* J Intercult Ethnopharmacol 2012; 1(2): 101-4.
[http://dx.doi.org/10.5455/jice.20120525014326]

[8] Beigh YA, Ganai AM. Potential of Wormwood (*Artemisia absinthium* Linn.) herb for use as additive in livestock feeding: A review. Pharma Innov 2017; 6: 176-87.

[9] Shafi N, Khan GA, Ghauri EG. Antiulcer effect of *Artemisia absinthium* L. in rats. Pak J Sci Ind Res 2004; 47: 130-4.

[10] Li Y, Ohizumi Y. Search for constituents with neurotrophic factor-potentiating activity from the medicinal plants of paraguay and Thailand. Yakugaku Zasshi 2004; 124(7): 417-24.
[http://dx.doi.org/10.1248/yakushi.124.417] [PMID: 15235225]

[11] Vogt DD. Absinthium: A nineteenth-century drug of abuse. J Ethnopharmacol 1981; 4(3): 337-42.
[http://dx.doi.org/10.1016/0378-8741(81)90002-7] [PMID: 7029147]

[12] Tariq KA, Chishti MZ, Ahmad F, Shawl AS. Anthelmintic activity of extracts of *Artemisia absinthium* against ovine nematodes. Vet Parasitol 2009; 160(1-2): 83-8.
[http://dx.doi.org/10.1016/j.vetpar.2008.10.084] [PMID: 19070963]

[13] Fernández-Calienes Valdés A, Mendiola Martínez J, Scull Lizama R, Vermeersch M, Cos P, Maes L. *In vitro* anti-microbial activity of the Cuban medicinal plants *Simarouba glauca* DC, *Melaleuca leucadendron* L and *Artemisia absinthium* L. Mem Inst Oswaldo Cruz 2008; 103(6): 615-8.
[http://dx.doi.org/10.1590/S0074-02762008000600019] [PMID: 18949336]

[14] Lee HG, Kim H, Oh WK. Tetramethoxyhydroxyflavone p-7F downregulates inflammatory mediators *via* the inhibition of nuclear factor kappa B. Ann N Y AcadSc 2004; 1030: 555-68.

[15] Jarić S, Popović Z, Mačukanović-Jocić M, *et al.* An ethnobotanical study on the usage of wild medicinal herbs from Kopaonik Mountain (Central Serbia). J Ethnopharmacol 2007; 111(1): 160-75.
[http://dx.doi.org/10.1016/j.jep.2006.11.007] [PMID: 17145148]

[16] Nin S, Bennici A. Biotechnology in Agriculture and Forestry.Transgenic Crops III. Berlin Heidelberg: Springer-Verlag 2001; 48.

[17] Geng S, Ye HC, Li GF. Flowering of *Artemisia annua* L. test tube plantlets and Artemisinin production with shoot cluster induced from flower organs explants. Chin J Appl Environ Biol 2001; 7: 201-6.

[18] Tiwari V, Tiwari KN, Singh BD. Comparative studies of cytokinins on *in vitro* propagation of *Bacopa monniera.* Plant Cell Tissue Organ Cult 2001; 66(1): 9-16.
[http://dx.doi.org/10.1023/A:1010652006417]

[19] Hussain M, Raja NI, Mashwani ZU, Iqbal M, Sabir S, *et al. In vitro* seed germination and biochemical profiling of *Artemisia absinthium* exposed to various metallic nanoparticles. 3 Biotech 2017; 7: 101.

[20] Sanyal R, Nandi S, Pandey S, *et al.* Biotechnology for propagation and secondary metabolite production in *Bacopa monnieri.* Appl Microbiol Biotechnol 2022; 106(5-6): 1837-54.
[http://dx.doi.org/10.1007/s00253-022-11820-6] [PMID: 35218388]

[21] Ali M, Abbasi BH, Ihsan-ul-haq . Production of commercially important secondary metabolites and antioxidant activity in cell suspension cultures of *Artemisia absinthium* L. Ind Crops Prod 2013; 49: 400-6.
[http://dx.doi.org/10.1016/j.indcrop.2013.05.033]

[22] Ali M, Abbasi BH. Thidiazuron-induced changes in biomass parameters, total phenolic content, and antioxidant activity in callus cultures of *Artemisia absinthium* L. Appl Biochem Biotechnol 2014;

172(5): 2363-76.
[http://dx.doi.org/10.1007/s12010-013-0663-7] [PMID: 24371002]

[23] Ali M, Abbasi BH, Ali GS. Elicitation of antioxidant secondary metabolites with jasmonates and gibberellic acid in cell suspension cultures of *Artemisia absinthium* L. Plant Cell Tissue Organ Cult 2015; 120(3): 1099-106.
[http://dx.doi.org/10.1007/s11240-014-0666-2]

[24] Tariq U, Ali M, Abbasi BH. Morphogenic and biochemical variations under different spectral lights in callus cultures of *Artemisia absinthium* L. J Photochem Photobiol B 2014; 130: 264-71.
[http://dx.doi.org/10.1016/j.jphotobiol.2013.11.026] [PMID: 24362323]

[25] Yatto GM, Nathar VN, Aziz WN, Gaikwad NB. Induction of callus and preliminary phytochemical profiling from callus of *Artemisia absinthium* L. and *Artemisia pallens* Wall. Int J Curr Trends Sci Technol 2018; 8: 20236-41.

[26] González HR, Sosa IH, Rodríguez Ferradá CA, Amitas MMR. *In vitro* propagation of *Artemisia absinthium*. L. in Cuba. Rev Cubana Plant Med 2003; p. 3.

[27] Shekhawat MS, Manokari M. Efficient *In vitro* propagation by *ex vitro* rooting methods of *Artemisia absinthium* L., an ethnobotanically important plant. Chin J Biol 2015; 8.

[28] Muhammad Z, Muhammad FC, Chaudhary MF. Hormonal regulation for callogenesis and organgenesis of *Artemisia absinthium* L. Afr J Biotechnol 2007; 6(16): 1874-8.
[http://dx.doi.org/10.5897/AJB2007.000-2281]

[29] Koul B, Lone OA. Optimization of *in vitro* regeneration of *Artemisia absinthium* L. (Worm wood) using leaf explants. J Chem Pharm Res 2016; 8: 303-10.

[30] Nin S, Morosi E, Schiff S, Bennici A. Callus cultures of *Artemisia absinthium* L.: Initiation, growth optimization and organogenesis. Plant Cell Tissue Organ Cult 1996; 45(1): 67-72.
[http://dx.doi.org/10.1007/BF00043430]

[31] Lata AJM, Dalei J, Sahoo D. Effect of different growth regulators on *in vitro* callus culture of *Artemisia absinthium* and genetic transformation by *Agrobacterium* mediated gene transfer. Int J Multidiscip Res Dev 2014; 1: 25-8.

[32] Nin S, Bennici A, Roselli G, Mariotti D, Schiff S, Magherini R. *Agrobacterium*-mediated transformation of *Artemisia absinthium* L. (wormwood) and production of secondary metabolites. Plant Cell Rep 1997; 16(10): 725-30.
[http://dx.doi.org/10.1007/s002990050310] [PMID: 30727627]

[33] Nin S, Schiff S, Bennici A, Magherini R. *In vitro* propagation of *Artemisia absinthium* L. Adv Hortic Sci 1994; 8: 145-7.

[34] Mannan A, Syed TN, Yameen MA, Ullah N, Ismail T, *et al.* Effect of growth regulators on *in vitro* germination of *Artemisia absinthium*. Sci Res Essays 2012; 7: 1501-7.
[http://dx.doi.org/10.5897/SRE11.1894]

[35] Ali M, Abbasi BH. Light-induced fluctuations in biomass accumulation, secondary metabolites production and antioxidant activity in cell suspension cultures of *Artemisia absinthium* L. J Photochem Photobiol B 2014; 140: 223-7.
[http://dx.doi.org/10.1016/j.jphotobiol.2014.08.008] [PMID: 25169773]

[36] Simon MJ, Damasceno S, Nogueira AM, Guizardi PS, Silva NCB. *In vitro* germination of seeds of losna (*Artemisia absinthium* L.) in different culture media. Annals... XIV Latin American Meeting Scientific Initiation and X Latin American Postgraduate Meeting.

[37] Lê CL, Julmi C, Tschuy F. Multiplication *in vitro* de l'absinthe (*Artemisia absinthium* L.). Rev Suisse Vitic Arboric Hortic 2007; 39: 263-7.

[38] Kour B, Kour G, Kaul S, Dhar MK. *In vitro* mass multiplication and assessment of genetic stability of *In vitro* raised *Artemisia absinthium* L. plants using ISSR and SSAP molecular markers. Adv Bot

2014; 7.

[39] Liu CZ, Guo C, Wang Y, Ouyang F. Factors influencing artemisinin production from shoot cultures of *Artemisia annua* L. World J Microbiol Biotechnol 2003; 19(5): 535-8. [http://dx.doi.org/10.1023/A:1025158416832]

[40] Ghassemi B, Nayeri FD. Interaction of four important factors affecting callogenesis and organogenesis of *Artemisia absinthium*. J Plant Mol Biols 2021. [http://dx.doi.org/10.22058/JPMB.2021.66838.1137]

[41] Liu CZ, Wang YC, Ouyang F, Ye HC, Li GF. Production of artemisinin by hairy root cultures of *Artemisia annua* L. Biotechnol Lett 1997; 19(9): 927-9. [http://dx.doi.org/10.1023/A:1018362309677]

[42] Grech-Baran M, Pietrosiuk A. *Artemisia* species *in vitro* cultures for production of biologically active secondary metabolites. BioTechnologia 2012; 4: 371-80. [http://dx.doi.org/10.5114/bta.2012.46591]

[43] Gulati A, Bharel S, Jain SK, Abdin MZ, Srivastava PS. *In vitro* micropropagation and flowering in *Artemisia annua.* J Plant Biochem Biotechnol 1996; 5(1): 31-5. [http://dx.doi.org/10.1007/BF03262976]

[44] Graham LE, Graham JM, Wilcox LW. Plant Biology. 2nd. United State of America: Sheri L. Snavely. Pearson Education, inc 2006; p. 670.

[45] Taylor OJ, Green NPO, Stout GW. Biological Science. United Kingdom: Cambridge University Press 1997; p. 984.

[46] Aslam N, Muhammad Zia , Chaudhary MF. Callogenesis and direct organogenesis of *Artemisia scoparia.* Pak J Biol Sci 2006; 9(9): 1783-6. [http://dx.doi.org/10.3923/pjbs.2006.1783.1786]

[47] Liu CZ, Murch SJ, EL-Demerdash M, Saxena PK. Regeneration of the Egyptian medicinal plant *Artemisia judaica* L. Plant Cell Rep 2003; 21(6): 525-30. [http://dx.doi.org/10.1007/s00299-002-0561-x] [PMID: 12789426]

CHAPTER 8

Conservation of Medicinal Plant Bramhi- *Bacopa monnieri* (L.) Wettstein Through *in vitro* Cultures

Kiranmai Chadipiralla[1,*], **Boddupalli Krishna Jaswanth**[1] and **Pichili Vijaya Bhaskar Reddy**[2]

[1] *Department of Biotechnology, Vikrama Simhapuri University, Nellore, Andhra Pradesh-524324, India*

[2] *Department of Life Science & Bioinformatics, Assam University Diphu Campus, Diphu-782462, Assam, India*

Abstract: *Bacopa monnieri* (L.) Wettstein is a medicinal herb from the family Plantaginaceae widely known as 'water hyssop' or 'brahmi'. The therapeutic potential of plants is due to the presence of many bioactive secondary metabolites, majorly brahmine, herpestine, alkaloids, and saponins (bacosides), which are responsible for pharmacological effects including neuroprotective, hepatoprotective, gastroprotective, antioxidant, anti-inflammatory, and antimicrobial properties. Vegetative cultivation of *Bacopa* on a large scale has its limitations due to the lack of viability of seeds during propagation and the unpredictable nature of the production of phytochemicals for commercial purposes, which can be overcome by tissue culture mechanism. Over the past few decades, many studies on the tissue culture of *Bacopa* in establishing a standardized protocol were reported. This chapter deals with *de novo* organogenesis of the root and shoot along with the callus induction and somatic embryogenesis from different explants of *B. monnieri* on MS basal nutrient medium supplemented with Plant Growth Regulators.

Keywords: *Bacopa monnieri*, Micropropagation, Plant growth regulator.

INTRODUCTION

Bacopa monnieri, a renowned medicinal herb, belongs to the family Plantaginaceae. The plant is a short-lived annual herb that is common in wet habitats and along waterways, and originates from tropical and subtropical Asia. It thrives on plains and slopes near flowing water and wetlands and is especially prolific during monsoon [1]. As an aquatic herb, it has been utilised both for therapeutic use and as a decorative plant in aquariums and ponds. It thrives in ma-

* **Corresponding author Kiranmai Chadipiralla:** Department of Biotechnology, Vikrama Simhapuri University, Nellore, Andhra Pradesh-524324, India; E-mails: cdpkiranmai@gmail.com and cdpkiranmai@vsu.ac.in

T. Pullaiah (Ed.)

rshy environments, lakeshores, shorelines, and seaside locations along canals and channels, where the formation of dense mats is frequently seen [2].

It can be found all over the world, including in Asia, Africa, the Americas, Australia, the Eastern Mediterranean, the Arabian Peninsula, and the Caribbean [2 - 4]. It is reported as having been introduced into Japan, Singapore, Spain, Portugal, Andaman Islands, California, Marquesas, and Cayman Islands, among other countries [2, 5 - 9]. *Bacopa monnieri* is most popularly known as Brahmi, water hyssop but has many international and local common names because it is widely distributed all over the world, such as coastal water-hyssop and herb-of-grace [2]. According to the Indian Medicinal Plants Database, there are 183 vernacular names for *B. monnieri* in 9 different languages in India.

Many Ayurvedic remedies, including Brahmi rasayana, Brahmivati, Brahmighrit, and Sarasvatarisht, use *B. monnieri* as a useful ingredient [10]. Ayurveda considers *Bacopa* as a booster for the nervous system, which aids in enhancing memory, focusing and treating mental disorders [11, 12]. An important traditional Medhya Rasayana medicine in Ayurveda is made from *B. monnieri* [13]. Both children's and teenager's cognitive and behavioral traits were enhanced by the herb [14]. Besides, plant exhibits anti-cancer, analgesic, antioxidant, antipyretic, pro-cognitive, neuropsychiatric, neuroprotective, anti-neuroinflammatory, anti-inflammatory, anti-bacterial, anti-fungal and anticonvulsant properties [13, 15 - 27].

The plant is used in India and Pakistan for gastrointestinal stimulation, a heart restorative, and to assist respiratory function during bronchoconstriction [28]. In the Indian Materia Medica and Traditional Chinese Medicine, *B. monnieri* has been recommended as a treatment for several mental health issues, including insomnia, psychosis, anxiety, impaired cognition, and depression [29 - 31]. Additionally, the plant extract provided defense against opioid and tacrolimus-mediated renal damage [32, 33]. Furthermore, the plant has been shown to have anti-anhedonian, vasodilator, hippocampus-strengthening, anti-cytotoxic, anti-genotoxic and hormetic properties [21, 34 - 36].

In *B. monnieri,* important bioactive compounds categorized under tannin, phlobetannin, saponin (bacosides A and B), steroid, flavonoid, cardiac glycoside, phenol, and alkaloid (nicotine and herpestine) were reported under phytochemical screening investigations [27]. The presence of numerous triterpenoid saponins, including bacosides A, B, C, and D, often known as "memory chemicals," has been linked with the pharmacological properties of *Bacopa* in improving comprehension and remembrance [37 - 39].

Due to the high therapeutic potential and wide usage of *B. monnieri,* it should be cultivated separately as the plant propagates vegetatively and reproduces asexually, which makes it difficult to recognise or detect in the field. Despite its rapid growth, invasiveness into the native environment, adaptability, and toleration under abiotic stress, it negatively impacts aquatic habitats by causing damage to the flora and fauna along with the water. In terrestrial environments, it destroys stream banks, wetlands and lakeside edges, and shorelines [2].

It can grow as a weed in rice fields and beneath date palms for small-scale irrigated fields [3, 8, 40], which does not compensate for structured mass cultivation. Apart from these, biotic or natural enemies for plant include *Anartia jatrophae*, the white peacock butterfly, that feeds its caterpillars on *B. monnieri*. Damage from the tobacco cutworm *Spodoptera litura* has been noted in greenhouse setups [41]. Several nematodes belonging to the genus *Meloidogyne* also live on *B. monnieri* as a host [40], which are major threats for the plant under its natural habitat. Because of their limited vitality (two months), *B. monnieri* seeds are regarded as poor propagules, and the seedlings frequently wither during the development of the secondary node which makes it challenging to grow from seeds. The whole plant is utilised for a variety of medical applications, creating a commercial demand for it that leads to the exploitation of plant species by various enterprises, including the adulteration of plant products [42].

Plant tissue culture offers an alternative approach for cultivation and conservation through *in vitro* propagation, which is much appropriate and sustainable compared to other methods. Additionally, it provides adequate samples for further research and development to explore a variety of additional advantages of *B. monnieri*, as well as contamination-free, carefully chosen plants of true type with quantification of desired phytochemicals that can meet industrial demand in the preparation of drugs.

MICROPROPAGATION OF *BACOPA MONNIERI*

Micropropagation of *B. monnieri* requires a stepwise protocol which includes,

Source and Selection of Explants

For direct organogenesis of shoots, the use of microshoots, leaves, or internodal explants removed from the base of the plant *B. monnieri* has been proven to be particularly efficient [43]. Additionally, direct somatic embryos were discovered utilizing leaf explants made from microshoots [44]. For the establishment of callus cultures, axillary buds, younger nodes, shoot tips, and young leaves cut from the young shoots have been employed [45, 46]. It was also reported that for the generation of bioactive compounds like bacosides, nodal parts and leaves were used as explants [47]. Begum and Mathur [48] utilized cotyledonary and epicotyledonary nodes of *Bacopa* axenic culture for shoot induction. Apical portions of healthy twigs with 4-5 nodal lengths of *Bacopa* were used for shooting and rooting by Joshi *et al.* [49] and callus induction by Karatas *et al.* [50]. Tips of the shoot were employed as explants, grown in the MS medium, and provided with $CaCl_2$ and sodium alginate for successful synthetic seed generation in *B. monnieri* [51].

Surface Sterilization

The collected *B. monnieri* explant was thoroughly washed under running water for 15-30 minutes in order to eliminate the surface-adhering muck and dust in the beginning [10, 52 - 54]. For the initial removal of microorganisms from the surface of the explants, moderate liquid detergents such as 1-2% Cetavlon (10 min) and 5% v/v Labolene (15 min) [52, 55] can be used. 70% (w/v) ethanol (30 sec to 5 mins) is used as disinfectant [56, 57]. Mostly, 3% (w/v) sodium hypochlorite (15 min) [54, 57] or 2% (w/v) Tween 20 (15-20 mins) [58, 59] were used as surfactants directly after the initial water wash or will be used after the treatment with the disinfectants and detergents mentioned above. It is followed by rinsing the explants 3- 4 times with sterilized distilled water.

For a few times the explants were also treated with the fungicide 0.1% (w/v) Bavistin® [60] and 0.2% (w/v) Teepol [61, 62] to avoid the fungal residues, which leads to the more common way of fungal contamination which was again followed by a thorough rinse with sterile distilled water for 3-4 times. Finally, the explants were treated with an aqueous solution containing 0.1% (w/v) mercuric chloride ($HgCl_2$) for 3-10 mins which was carried out in the laminar airflow chamber, followed by 3-5 washes with sterile (autoclaved) distilled water to remove the remnant sterilant traces [56, 58, 62].

Culture Media

In *B. monnieri,* full strength Murashige and Skoog (MS) medium was used for shoot [63] and root induction, and somatic embryogenesis [64, 65]; for the biosynthesis of bacosides [47, 66] and their enhancement [67], liquid MS medium was used; and for root induction, half strength MS medium was used [49, 60]. It was observed that the multiplication rate of explants decreases with a decrease in the strength of MS media [48] and was used for hardening of plantlets [68]. Gamborg's (B_5) medium supplemented with different elicitors was used for callus induction and bacoside production [69]. Apart from these, 80% response of shoot explants with a low multiplication rate was observed on a cyanobacterial medium supplemented with Kinetin (Kn) (2 mg/L) [70].

Plant Growth Regulators (PGRs)

In *B. monnieri,* terminal shoots with 4-5 nodal length were regenerated into shoots on MS media supplemented with Benzyl amino purine (BA) (2 mg/L) which were extended to 100% rooting with Naphthalene acetic acid (NAA) (1 mg/L) and somatic embryoids were also developed after four weeks of culture with lower concentrations of BA (0.1 mg/L) and Indole acetic acid (IAA) (0.2 mg/L) was reported [58]. Later, Ahuja *et al.* [71], Mohapatra and Rath [61] and many other researchers reinforced that the combinations of BA (1-2 mg/L) and IAA (0.5-2 mg/L) added to the MS media for successful shoot generation from nodal explants.

Developement of an embryogenic callus with clear globular, torpedo structures was identified on the MS media supplemented 2,4-D (0.5 mg/L) from leaves of *Bacopa* from two months of culture [64]. For root induction in *B. monnieri,* the best combination for MS media is with NAA (0.5-1.5 mg/L) was supported by Banerjee and Shrivastava [72]; Pandiyan and Selvaraj [73] and with Indole butyric acid (IBA) (0.25-2 mg/L) was reported [48, 50, 68, 74 - 77].

Some researchers have reported multiple shoot induction from MS media supplemented with algal extracts such as *Aulosira fertilissima* extract to produce 56 shoots/explant [52] and *Sargassum wightii* liquid extract to produce 25 shoots/explant [78] as a substitute for PGRs from the nodal explants of *Bacopa.* Sarkar and Jha (2017) experimented with PGR free basal medium, resulted in varying frequencies in direct shoot regeneration, which opened for a new discussion in strengthening of nutrient media with PGRs for *in vitro* propagation of *B. monnieri* [43].

Callus Induction

Callus is a group of unorganized cell mass developed in response to the cut or injury of explants. It is friable (greenish yellow or creamy white) or hard and compact (whitish green or greenish pink) in nature. As *B. monnieri* is concerned the whole plant is used for various medicinal purposes. The production of callus aids in generating a completely new plant *via* somatic embryo or direct organ development, which is much more precise for the development of desired organs when compared to usual agricultural practices. All the callus cells do not necessarily express totipotency for organ initiation but can be supported by the supplementation of appropriate PGRs for their differentiation. Callus extracts can be directly utilised for the extraction of bioactive compounds. A standard protocol for the induction of callus would eliminate the related problems in the callus production from different explants of *B. monnieri.*

Callus initiation is the prerequisite and foremost step in the micropropagation of medicinal plants. According to recent publications, a combination of PGRs includes predominantly one or two auxins and cytokinins, along with the growth nutrient media, promotes callus induction. Ali *et al.* [54] mentioned the highest calli growth from the MS semi-solid media supplemented with growth regulatory hormones 2,4-Dichlorophenoxyacetic acid (2,4-D) (2.5 mg/L), and NAA (2.5 mg/L) was recorded at 94.22% and 82.43% from leaf explants, 71.14% and 62.15% from nodal explants and, 65.21% and 52.14% with internodal explants of *B. monnieri* (Fig. **1A**).

Intercalary meristems present near the nodal and internodal regions proliferate into the callus in response to the hormonal concentration provided. Abiotic stress (salinity) can also affect callus production apart from the hormonal supplementation reported by Dogan [79], as callus formation was decreased with an increase in the salt concentration where dark green callus (100%) was obtained from MS medium supplemented with NaCl (0-60 mM), light green, brownish callus (55.55%) was obtained with NaCl (80 mM) and brownish, black, yellowish callus (33.33%) was obtained from NaCl (100mM) [79] shown in the (Figs. **1B, C**). Development of organogenic callus from the leaves, internodes, and nodes of *B. monnieri* on the MS medium supplemented with 2,4-D (1 mg/L) after 17 days and with BA (5 mg/L) after 50 days of culture was reported by Samanta *et al.* [59] Fig. (**1D**), which shows a prolonged period of exposure and high concentration of cytokinins when compared to low concentration of auxin with a short period of time will ensure in calli regeneration.

Callus induction (75%) from the leaf explants on the MS media containing BA (0.5 mg/L) along with NAA (0.5 mg/L) and from nodal explants on the MS medium with an increased concentration of BA (4.0 mg/L) and NAA (0.5 mg/L) in *B. monnieri* was observed [16]. Compact callus formation from the internodal segments of *B. monnieri* on the MS media supplemented with 2,4-D (1.5 mg/L) and friable callus was observed from the leaf margin in the MS medium with NAA (1.5 mg/L) along with plant regeneration from nodal and internodal segments was reported [80].

Callus was observed with 50-70% of explants responding to the media (MS) with different combinations of auxin (IAA, NAA, IBA, 2,4-D) and cytokinin (BA, Kn) within 20 days of culturing period [56]. The initiation of callus within 15 days of inoculation of leaf explant of *B. monnieri* was when MS medium supplemented with 6 BA (2 ppm) and IAA (1 ppm), IAA (1.5 ppm) and 6 BA (1.5 ppm), Kn (1.5 ppm) and IAA (1.5 ppm) but the development of yellowish green callus was observed in next 45 days after sub-culturing onto the MS media provided with the combination of 6 BA (2 ppm), IAA (1 ppm) and $ZnSO_4$ (600 μM) was reported [81].

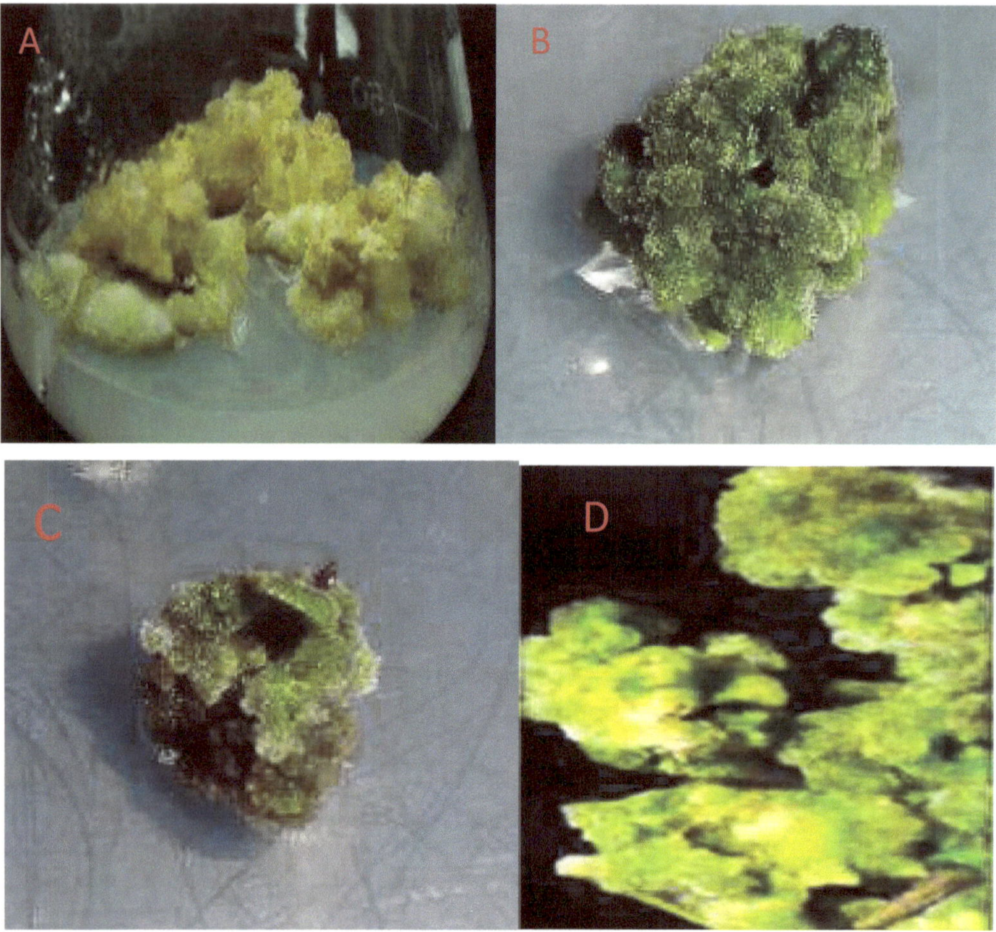

Fig. (1). A. Callus from MS (semi-solid) with 2,4-D (2.5 mg/L) + NAA (2.5 mg/L) (Source [54]:). **B**. Dark green callus, **C**. Light green, brownish callus, obtained from the MS media supplemented with NaCl concentrations (Source [79]:). **D**. Organogenic callus from 2,4-D (1 mg/L) and BA (5 mg/L) (Source: Samanta *et al.* [59]).

Talukdar [58] confirmed the growth of callus from nodal explants of *B. monnieri* in the MS medium supplemented with higher concentrations of 2,4-D (2 mg/L) in contradiction with Tiwari *et al.* [65] and Showkat *et al.* [46] than in lower concentrations of 2,4-D (0.5 mg/L). The callus turned friable by continuous sub-culturing on the same medium, which was initially semi-hard and weighed 35.60 g after 2 months of culture, which was a little more than the initial measurement, which was 15.20 g after 1 month of culture [46, 58, 65].

Somatic Embryogenesis

Using a single somatic cell or a cluster of somatic cells, a plant or embryo is created artificially through a process known as somatic embryogenesis [82]. It contrasts with totipotency, in which cells are flexible in differentiating into any type of cell or tissue, contributing to the development into a completely new plant through the process known as direct organogenesis. Whereas the embryos, which are produced from somatic cells through indirect organogenesis, are bipolar and can grow into tissues or organs (shoots and roots) [82].

Somatic embryogenesis is the development of embryos from vegetative callus from the explants of *B. monnieri*, which was previously reported by Tiwari *et al.* [65]. The benefits of micropropagation include encouraging both direct and indirect organ formation from explants. Both direct and indirect embryogenesis from somatic cells is possible, in addition to the direct development of organs from embryos. Direct embryogenesis occurs when the embryos develop from the explant cells directly, while indirect embryogenesis occurs when the embryos develop from the callus. In addition to contributing to biotechnological development, somatic embryogenesis has been used as a model to comprehend the physiological and biochemical events that take place throughout plant developmental processes [83].

According to the published articles, somatic embryogenesis in *B. monnieri* was observed in the MS media supplemented with different concentrations of 2,4-D alone and with the combination of kinetin and BA. Somatic embryogenic calli with 84% of callus developed into heart-shaped somatic embryos were observed from the MS solid medium supplemented with 2,4-D (2.0 mg/L) along with Kn (1.5 mg/L) by Ali *et al.* [54], shown in Fig. (**2**).

Fig. (2). Heart-shaped somatic embryoid development (Ali *et al.* [54]).

Parale *et al.* [84] also reported the development of the globular, heart, and torpedo stage somatic embryos from the nodular callus after 21 days of culture in the MS media containing an equal ratio of 2,4-D (20 µM) and Kn (20 µM) with 12.0 ± 0.3 embryos (per 100g of callus) which showed an advancement from the earlier experiments with time duration of developing embryoids after 60 days from culture by using single hormone 2,4-D (0.5 mg/L) by Sharath *et al.* [64]. The embryo induction and maturation on subculturing were reported in the MS media supplemented with 2,4-D (1mg/L) by Samanta *et al.* [59]. Khilwani *et al.* [44] reported somatic embryogenesis was influenced by sucrose and PGR concentrations where leaf explants were differentiated into somatic embryos on the MS medium supplemented with BA (12.5 µM) and 2,4-D (1.0 µM) with 47.1% (maximum frequency), and with sucrose (250 mM), which showed 77% of frequency in development.

Shoot Organogenesis

One of the critical benefits of micropropagation, the preservation of genetic diversity, and biomass output is the ability to multiply shoots from a single explant. Numerous scholars advocated nodal explant's effectiveness in generating numerous shoots [85]. On an isolated explant, such as a leaf or stem, shoot development can occur directly through direct organogenesis or indirectly through callus production. The cytokinin signalling system is a possible target for regulating *in vitro* plant regeneration and *de novo* shoot organogenesis. The cytokinins to auxins ratio in the culture media significantly affects how much the cells proliferate.

The majority of the explants exhibited shoot regeneration from the MS medium supplemented with BA alone Fig. (**3**) [49] and in combination with NAA at low concentrations [50, 86, 88] however, multiple shoot production from a single explant was less successful. Tiwari *et al.* [88] demonstrated a considerable increase in axillary shoot induction (50 axillary shoots/explants) from nodal explants due to the combined effects of an antibiotic TMP (Trimethoprim) and a fungicide BVN (Bavistin Methyl benzimidozole carbamate) in MS media. In addition, adventitious shoot buds (30) were generated from these nodal explants, but the histological proof was not presented. Therefore, it is unknown whether these adventitious shoot buds were produced *de novo* or resulted from the proliferation of existing meristems.

Fig. (3). **A**. Formation of shoot buds of entire wounded leaf all over from upper surface. **B**. Development of microshoots was more number from 8 weeks of culture (Source: Joshi *et al.* [49]). **C**. Regeneration and elongation of shoots from nodal explants 5% sucrose (C) and at **D**. 3% sucrose (Source: Srivastava *et al.* [95]).

Fortification of Kn (2 mg/L) alone to the medium resulted in the production of shoots, 72 shoots/explant [89], and even more (126 shoots/explant) compared to the combinations of Kn/BA (98 shoots/explant) and Kn/IAA (84 shoots/explant) [90]. According to Chauhan and Shirkot [62], the medium supplemented with equal proportions of BA (0.5 mg/L) and Kn (0.5 mg/L) yields the greatest sprouting (87.5%) from axillary shoot buds of *Bacopa*. In contrast, a decrease in the rate of sprouting was observed when the concentration of BA was decreased in conjunction with the addition of other plant growth regulators (NAA), and the length of multiple microshoots (7) was observed in media supplemented with a constant concentration of Kn and a slight increase in the concentration of BA (1.0 mg/L).

Behera *et al.* [75] reported that gibberellic acid (GA_3) supplemented at a concentration of 1.0 mg/L to MS media resulted in the most significant number of shoots (114.2 shoots/nodes); after they were transferred to media supplemented with BA (3.0 mg/L), which resulted in a lower number of shoot regeneration (6.5 shoots/nodes). TDZ (N-phenyl-N'- 1,2,3-thiadiazol-5-ylurea), which acts like cytokinin [91], has been found in vast numbers of plant regeneration *in vitro* [92]. However, its method of action and function during the inductive phase of *in vitro* morphogenesis are unknown [93]. In *Bacopa*, after 4-8 weeks of culture, TDZ was found to be much less appropriate for multiple shoot organogenesis [42, 94] when compared to the same media supplemented with NAA [60].

Numerous studies have proposed that 3% sucrose as a carbon source [49, 74, 87, 95 - 97]and 5% sucrose [95] is optimal for inducing numerous shoots.

Root Organogenesis

De novo root organogenesis is the regeneration of adventitious roots from detached or damaged plant organs or tissues. In normal circumstances, organ regeneration is believed to depend on endogenous hormones. The optimum sorts and quantities of plant hormones in the media are essential to induce adventitious roots in tissue culture [98]. However, when moved to *ex vitro* circumstances, the benefit of plant tissue culture may be realized by root organogenesis hardening and adaptation of the *in vitro* developed plant.

In *Bacopa* initially, root induction was reported by Tiwari *et al.* [65], where 100% of shoots regenerated roots and were rooted in root-inducing media (RIM) containing NAA (1 mg/L), and they expanded their rooting studies. Within two weeks, Tiwari *et al.* [42] found that *in vitro*-derived shoots on growth regulator free MS media demonstrated a superior rooting response on a medium (half and full-strength MS) containing IBA Fig. (**4**), which was later supported by Karatas *et al.* [50], Sharma *et al.* [55], Mohapatra and Rath [61], Jain *et al.* [68], Behera [75], Karatas *et al.* [76], Croom *et al.* [77], Jain *et al.* [86], Tiwari *et al.* [88], Srivastava *et al.* [95] and Rao *et al.* [99].

Binita *et al.* [100] experimented on liquid MS media supplemented with both auxin (1.1 μM BA) and cytokinin (0.2 μM IAA), where rooting was reported to be successful. Later, Banerjee and Shrivastava [72], in establishing an improved protocol, also suggested NAA (1.5 mg/L) for root induction, where 12 ± 1.73 roots/shoot was observed. Rooting was increased to 15 ± 2.20 roots/shoot with a 100% survival rate achieved by Banerjee and Modi [52], on 40 ml of MS + 60 ml of cyanobacterial medium (*Aulosira fertilissima* extract) provided along with calcium chloride (0.44 gm/L) and Kn (1.0 mg/L).

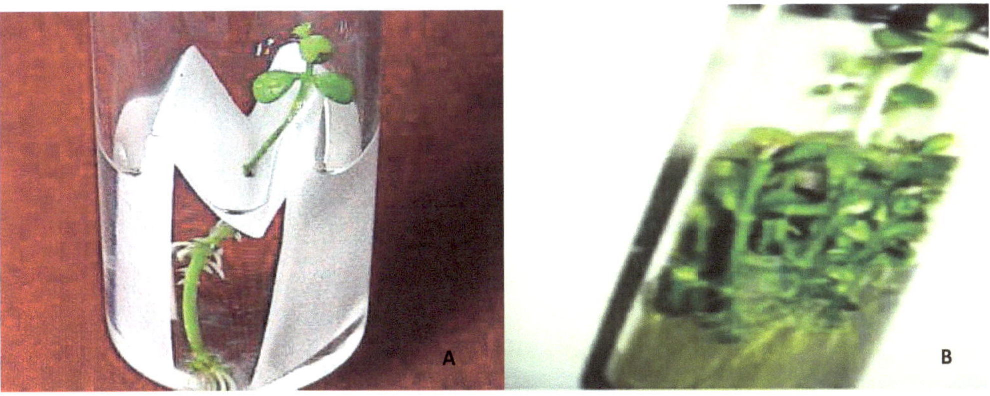

Fig. (4). A. Induction of roots in half-strength MS liquid medium + Sucrose (1%) + IBA (2 µM) with a filter paper bridge culture (Source: Joshi *et al.* [49]). **B**. Root induction of regenerated shoots from nodal explants (Source: Srivastava *et al.* [95]).

Ceasar *et al.* [60] reported a more significant number of roots (16.5) with root length (8.7 cm) in half strength MS basal medium fortified with IBA (1.0 mg/L) and phenolic compound, phloroglucinol (0.5 mg/L), since phloroglucinol exhibits both cytokinin-like and auxin-like action. When given to a rooting medium with auxin, phloroglucinol induces rooting further [101].

In comparison to other carbohydrates, sucrose has been proposed as a carbon source for roots by a few studies. Begum *et al.* [48] and Joshi *et al.* [49] suggested 1% sucrose, whereas Kaur *et al.* [74], Rout *et al.* [87] and Mishra *et al.* [102] suggested 2% sucrose, along with the combination of auxins, IAA/NAA and IBA to the MS media (RIM) for optimal root induction.

Acclimatization

Numerous studies reported that *Bacopa monnieri* plantlets produced *in vitro* were effectively transferred to the field and acclimatized after rooting. According to the researchers, the success rates in acclimatization of plantlets onto sterilized soilrite [88] were 85% [103], 95% [65, 89] and 100% [42], whereas the survival rate was 90% when the plants were transferred to the sand along with garden soil [68, 71] and 98% in sand, soil and farmyard manure [16, 55, 100] 100% in garden soil, vermiculite and sand [52, 72], where all the components were reported to be mixed in equal proportions. Surprisingly, 100% of the plants transferred were acclimatized in MS basal salts containing vermicompost [60]. Later, most researchers preferred transferring the plants to shade net [86], mist [48], greenhouse [49, 99] and outside the net-house [16] for acclimatization before transferring to the field Fig. (**5**).

Fig. (5). A. *Bacopa monnieri* plantlets which are *in vitro* cultivated were transferred onto an equal mixed ratio of soil, sand, and farmyard manure (FYM) and maintained in a shade net-house. **B**. *Bacopa* plants which are acclimatized were growing in the pot, outside the net-house (Source: Ranjan *et al.* [16]).

CONCLUSION

Bacopa monnieri is a commonly found ingredient in various Ayurvedic and herbal formulations that are increasingly being used for improving neurocognitive functions and memory enhancement in recent years. Owing to its high therapeutic potential benefits, *B. monnieri* is a highly sought after plant by pharmaceutical companies for research as well as the development of novel drug molecules. Due to the growing demand in the market, *Bacopa* has gained importance for natural health products, especially those that help with brain health and memory. At present, the commercialization of *B. monnieri* is facing several challenges, like the lack of standardization of the plant materials due to the presence of multiple varieties and the variability of the active compounds. These challenges can be solved using biotechnology and plant tissue culture techniques involving micropropagation of *B. monnieri* which can be achieved directly through the shoot and root organogenesis and indirectly through callus regeneration from nodal explants in a nutrient-rich growth medium supplemented with PGRs (auxins and cytokinins). *In vitro* propagation of *B. monnieri* can be a useful method for producing large numbers of genetically identical plants for commercial or research purposes.

REFERENCES

[1] Vikaspedia.in. vikaspedia Domains. 2017. Available from: https://vikaspedia.in/agriculture/crop-production/package-of-practices/medicinal-and-aromatic-plants/bacopa-monnieri

[2] Rojas-Sandoval J. *Bacopa monnieri* (water hyssop). Invasive Species Compendium 2018; 112638.

[3] Lansdown RV, Knees SG, Patzelt A. *Bacopa monnieri*. The IUCN Red list of Threatened Species 2013; T164168A17722668.

[4] USDA, Agricultural Research Service, National Plant Germplasm System. Germplasm Resources

Information Network (GRIN Taxonomy). Beltsville, Maryland: National Germplasm Resources Laboratory 2022.

[5] Jaime FagúndezDíaz. Manuel BarradaBeiras. Invasive plants of Galicia: biology, distribution and control methods [Internet]. Xunta de Galicia, [Santiago de Compostela]. 2007. Available from: https://www.worldcat.org/title/plantas-invasoras-de-galicia-bioloxia-distribucion-e-me-odos-de-control/oclc/434478933

[6] The Ecological Society of Japan. Provisional list of alien species naturalized in Japan [Internet]. APASD. National Institute for Agro-Environmental Sciences. 2007. Available from: http://www.naro.affrc.go.jp/archive/niaes/techdoc/apasd/Bacopa%20monnieri%20-B.html

[7] Roy D, Alderman D, Anastasiu P, *et al.* DAISIE : Inventory of alien invasive species in Europe. Version 17 Research Institute for Nature and Forest (INBO) 2020 Checklist dataset 2020. [http://dx.doi.org/10.15468/ybwd3x]

[8] Global Invasive Species Database. Species profile: *Bacopa monnieri*. 2022. Available from: http://www.iucngisd.org/gisd/speciesname/Bacopa+monnieri

[9] Pagad S, Genovesi P, Carnevali L, Schigel D, McGeoch MA. Introducing the global register of introduced and invasive species. Sci Data 2018; 5(1): 170202. [http://dx.doi.org/10.1038/sdata.2017.202] [PMID: 29360103]

[10] Sharma N, Singh R, Pandey R. *In vitro* propagation and conservation of *Bacopa monnieri* L. Protocols for *in vitro* cultures and secondary metabolite analysis of aromatic and medicinal plants. 2nd. New York, NY: Humana Press 2016; pp. 153-71.

[11] Brimson JM, Brimson S, Prasanth MI, Thitilertdecha P, Malar DS, Tencomnao T. The effectiveness of *Bacopa monnieri* (Linn.) Wettst. as a nootropic, neuroprotective, or antidepressant supplement: Analysis of the available clinical data. Sci Rep 2021; 11(1): 596. [http://dx.doi.org/10.1038/s41598-020-80045-2] [PMID: 33436817]

[12] Lopresti AL, Smith SJ, Ali S, Metse AP, Kalns J, Drummond PD. Effects of a *Bacopa monnieri* extract (Bacognize®) on stress, fatigue, quality of life and sleep in adults with self-reported poor sleep: A randomised, double-blind, placebo-controlled study. J Funct Foods 2021; 85: 104671. [http://dx.doi.org/10.1016/j.jff.2021.104671]

[13] Abdul Manap AS, Vijayabalan S, Madhavan P, *et al. Bacopa monnieri*, a neuroprotective lead in Alzheimer disease: A review on its properties, mechanisms of action, and preclinical and clinical studies. Drug Target Insights 2019; 13. [http://dx.doi.org/10.1177/1177392819866412] [PMID: 31391778]

[14] Kean J, Downey L, Stough C. Systematic overview of *Bacopa monnieri* (L.) Wettst. dominant poly-herbal formulas in children and adolescents. Medicines 2017; 4(4): 86. [http://dx.doi.org/10.3390/medicines4040086] [PMID: 29165401]

[15] Nemetchek MD, Stierle AA, Stierle DB, Lurie DI. The Ayurvedic plant *Bacopa monnieri* inhibits inflammatory pathways in the brain. J Ethnopharmacol 2017; 197: 92-100. [http://dx.doi.org/10.1016/j.jep.2016.07.073] [PMID: 27473605]

[16] Ranjan R, Kumar S, Singh AK. An efficient *in vitro* propagation protocol of local germplasm of *Bacopa monnieri* (L.) found in Bihar: A plant with wide variety of medicinal properties. J Pharmacogn Phytochem 2018; 7(1): 1803-7.

[17] Brimson JM, Prasanth MI, Plaingam W, Tencomnao T. *Bacopa monnieri* (L.) Wettst. Extract protects against glutamate toxicity and increases the longevity of *Caenorhabditis elegans*. J Tradit Complement Med 2020; 10(5): 460-70. [http://dx.doi.org/10.1016/j.jtcme.2019.10.001] [PMID: 32953562]

[18] Castelli V, Melani F, Ferri C, *et al.* Neuroprotective activities of bacopa, lycopene, astaxanthin, and vitamin B12 combination on oxidative stress-dependent neuronal death. J Cell Biochem 2020; 121(12): 4862-9.

[http://dx.doi.org/10.1002/jcb.29722] [PMID: 32449987]

[19] Jeyasri R, Muthuramalingam P, Suba V, Ramesh M, Chen JT. *Bacopa monnieri* and their bioactive compounds inferred multi-target treatment strategy for neurological diseases: A cheminformatics and system pharmacology approach. Biomolecules 2020; 10(4): 536.
[http://dx.doi.org/10.3390/biom10040536] [PMID: 32252235]

[20] Kiani AK, Miggiano GAD, Aquilanti B, *et al.* Food supplements based on palmitoylethanolamide plus hydroxytyrosol from olive tree or *Bacopa monnieri* extracts for neurological diseases. Acta Biomed 2020; 91(13-S) (13): e2020007.
[PMID: 33170159]

[21] Micheli L, Spitoni S, Di Cesare Mannelli L, Bilia AR, Ghelardini C, Pallanti S. *Bacopa monnieri* as augmentation therapy in the treatment of anhedonia, preclinical and clinical evaluation. Phytother Res 2020; 34(9): 2331-40.
[http://dx.doi.org/10.1002/ptr.6684] [PMID: 32236999]

[22] Cheema AK, Wiener LE, McNeil RB, *et al.* A randomized phase II remote study to assess *Bacopa* for Gulf War Illness associated cognitive dysfunction: Design and methods of a national study. Life Sci 2021; 282: 119819.
[http://dx.doi.org/10.1016/j.lfs.2021.119819] [PMID: 34256038]

[23] Datta S, Ramamurthy PC, Anand U, *et al.* Wonder or evil?: Multifaceted health hazards and health benefits of *Cannabis sativa* and its phytochemicals. Saudi J Biol Sci 2021; 28(12): 7290-313.
[http://dx.doi.org/10.1016/j.sjbs.2021.08.036] [PMID: 34867033]

[24] Dutta T, Anand U, Saha SC, *et al.* Advancing urban ethnopharmacology: A modern concept of sustainability, conservation and cross-cultural adaptations of medicinal plant lore in the urban environment. Conserv Physiol 2021; 9(1): coab073.
[http://dx.doi.org/10.1093/conphys/coab073] [PMID: 34548925]

[25] Paul S, Chakraborty S, Anand U, *et al.* *Withania somnifera* (L.) Dunal (Ashwagandha): A comprehensive review on ethnopharmacology, pharmacotherapeutics, biomedicinal and toxicological aspects. Biomed Pharmacother 2021; 143: 112175.
[http://dx.doi.org/10.1016/j.biopha.2021.112175] [PMID: 34649336]

[26] Sharma M, Gupta PK, Gupta P, Garabadu D. Antinociceptive activity of standardized extract of *Bacopa monnieri* in different pain models of zebrafish. J Ethnopharmacol 2022; 282: 114546.
[http://dx.doi.org/10.1016/j.jep.2021.114546] [PMID: 34418512]

[27] Jain P, Sharma HP, Basri F, Priya K, Singh P. Phytochemical analysis of *Bacopa monnieri* (L.) Wettst. and their anti-fungal activities. Indian J Tradit Knowl 2017; 16(2): 310-8.

[28] Saha PS, Sarkar S, Jeyasri R, Muthuramalingam P, Ramesh M, Jha S. *In vitro* propagation, phytochemical and neuropharmacological profiles of *Bacopa monnieri* (L.) Wettst.: A review. Plants 2020; 9(4): 411.
[http://dx.doi.org/10.3390/plants9040411] [PMID: 32224997]

[29] Moskwa J, Naliwajko SK, Markiewicz-Żukowska R, *et al.* Chemical composition of Polish propolis and its antiproliferative effect in combination with *Bacopa monnieri* on glioblastoma cell lines. Sci Rep 2020; 10(1): 21127.
[http://dx.doi.org/10.1038/s41598-020-78014-w] [PMID: 31913322]

[30] Halder S, Anand U, Nandy S, *et al.* Herbal drugs and natural bioactive products as potential therapeutics: A review on pro-cognitives and brain boosters perspectives. Saudi Pharm J 2021; 29(8): 879-907.
[http://dx.doi.org/10.1016/j.jsps.2021.07.003] [PMID: 34408548]

[31] Anand U, Tudu CK, Nandy S, *et al.* Ethnodermatological use of medicinal plants in India: From ayurvedic formulations to clinical perspectives : A review. J Ethnopharmacol 2022; 284: 114744.
[http://dx.doi.org/10.1016/j.jep.2021.114744] [PMID: 34656666]

[32] Shahid M, Subhan F, Ullah I, Ali G, Alam J, Shah R. Beneficial effects of *Bacopa monnieri* extract on opioid induced toxicity. Heliyon 2016; 2(2): e00068. [http://dx.doi.org/10.1016/j.heliyon.2016.e00068] [PMID: 27441247]

[33] Oyouni AA, Saggu S, Tousson E, Mohan A, Farasani A. Mitochondrial nephrotoxicity induced by tacrolimus (FK-506) and modulatory effects of *Bacopa monnieri* (Farafakh) of Tabuk Region. Pharmacognosy Res 2019; 11(1): 20-4. [http://dx.doi.org/10.4103/pr.pr_100_18]

[34] Kamkaew N, Paracha TU, Ingkaninan K, Waranuch N, Chootip K. Vasodilatory effects and mechanisms of action of *Bacopa monnieri* active compounds on rat mesenteric arteries. Molecules 2019; 24(12): 2243. [http://dx.doi.org/10.3390/molecules24122243] [PMID: 31208086]

[35] Promsuban C, Limsuvan S, Akarasereenont P, Tilokskulchai K, Tapechum S, Pakaprot N. *Bacopa monnieri* extract enhances learning-dependent hippocampal long-term synaptic potentiation. Neuroreport 2017; 28(16): 1031-5. [http://dx.doi.org/10.1097/WNR.0000000000000862] [PMID: 28885486]

[36] Dogan M, Emsen B. Anti-cytotoxic-genotoxic influences of *in vitro* propagated *Bacopa monnieri* L. Pennell in cultured human lymphocytes. Eurasian J Biol Chem Sci 2018; 1(2): 48-53.

[37] Dey A, Hazra AK, Nongdam P, *et al.* Enhanced bacoside content in polyamine treated *in-vitro* raised *Bacopa monnieri* (L.) Wettst. S Afr J Bot 2019; 123: 259-69. [http://dx.doi.org/10.1016/j.sajb.2019.03.012]

[38] Banerjee S, Anand U, Ghosh S, *et al.* Bacosides from *Bacopa monnieri* extract: An overview of the effects on neurological disorders. Phytother Res 2021; 35(10): 5668-79. [http://dx.doi.org/10.1002/ptr.7203] [PMID: 34254371]

[39] Nandy S, Das T, Tudu CK, *et al.* Unravelling the multi-faceted regulatory role of polyamines in plant biotechnology, transgenics and secondary metabolomics. Appl Microbiol Biotechnol 2022; 106(3): 905-29. [http://dx.doi.org/10.1007/s00253-021-11748-3] [PMID: 35039927]

[40] Aguilar NO. *Bacopa monnieri* (L.) Pennell.In: van Valkenburg, J.L.C.H. and Bunyapraphatsara, N. (Editors) [Internet]. Prosea : plant resources of southeast asia. PROSEA Foundation, Bogor, Indonesia. 2001. Available from: https://www.prota4u.org/prosea/view.aspx?id=931

[41] Jyoti T, Dwijendra S, Shalini M. Outbreak of *Spodoptera litura* on brahmi-an important medicinal plant. Insect Environ 1997; 2(4): 134.

[42] Tiwari V, Tiwari KN, Singh BD. Comparative studies of cytokinins on *in vitro* propagation of *Bacopa monnieri*. Plant Cell Tissue Organ Cult 2001; 66(1): 9-16. [http://dx.doi.org/10.1023/A:1010652006417]

[43] Sarkar S, Jha S. Morpho-histological characterization and direct shoot organogenesis in two types of explants from *Bacopa monnieri* on unsupplemented basal medium. Plant Cell Tissue Organ Cult 2017; 130(2): 435-41. [http://dx.doi.org/10.1007/s11240-017-1231-6]

[44] Khilwani B, Kaur A, Ranjan R, Kumar A. Direct somatic embryogenesis and encapsulation of somatic embryos for *in vitro* conservation of *Bacopa monnieri* (L.) Wettst. Plant Cell Tissue Organ Cult 2016; 127(2): 433-42. [http://dx.doi.org/10.1007/s11240-016-1067-5]

[45] Bhusari S, Wanjari R, Khobragade P. Cost effective *in vitro* clonal propagation of *Bacopa monnieri* L. Penell. Intern J Indigenous Med Plants 2013; 46(2): 1239-44.

[46] Showkat P, Zaidi Y, Asghar S, Jamaluddin S. *In vitro* propagation and callus formation of *Bacopa monnieri* (L.) Penn. Plant Tissue Cult Biotechnol 1970; 20(2): 119-25. [http://dx.doi.org/10.3329/ptcb.v20i2.6890]

[47] Praveen N, Naik PM, Manohar SH, Nayeem A, Murthy HN. *In vitro* regeneration of brahmi shoots using semisolid and liquid cultures and quantitative analysis of bacoside A. Acta Physiol Plant 2009; 31(4): 723-8.
[http://dx.doi.org/10.1007/s11738-009-0284-5]

[48] Begum T, Mathur M. *In vitro* regeneration of *Catharanthus roseus* and *Bacopa monnieri* and their survey around Jaipur district. Int J Pure App Biosci 2014; 2(4): 210-21.

[49] Joshi AG, Pathak AR, Sharma AM, Singh S. High frequency of shoot regeneration on leaf explants of *Bacopa monnieri.* Environ Exp Biol 2010; 8: 81-4.

[50] Karatas M, Aasim M, Dogan M, Khawar KM. Adventitious shoot regeneration of the medicinal aquatic plant water hyssop (*Bacopa monnieri* L. Pennell) using different internodes. Arch Biol Sci 2013; 65(1): 297-303.
[http://dx.doi.org/10.2298/ABS1301297K]

[51] Pramanik B, Sarkar S, Bhattacharyya S, Gantait S. meta-Topolin-induced enhanced biomass production *via* direct and indirect regeneration, synthetic seed production, and genetic fidelity assessment of *Bacopa monnieri* (L.) Pennell, a memory-booster plant. Acta Physiol Plant 2021; 43(7): 107.
[http://dx.doi.org/10.1007/s11738-021-03279-1]

[52] Banerjee M, Modi P. Micropropagation of *Bacopa monnieri* using cyanobacterial liquid medium. Plant Tissue Cult Biotechnol 1970; 20(2): 225-31.
[http://dx.doi.org/10.3329/ptcb.v20i2.6917]

[53] Mahender A, Mallesham B, Srinivas K, *et al.* A rapid and efficient method for *in vitro* shoot organogenesis and production of transgenic *Bacopa monnieri* L. mediated by *Agrobacterium tumefaciens. In vitro.* Cell Dev Biol Plant 2012; 48(2): 153-9.
[http://dx.doi.org/10.1007/s11627-011-9421-0]

[54] Ali D, Alarifi S, Pandian A. Somatic embryogenesis and *in vitro* plant regeneration of *Bacopa monnieri* (Linn.) Wettst., a potential medicinal water hyssop plant. Saudi J Biol Sci 2021; 28(1): 353-9.
[http://dx.doi.org/10.1016/j.sjbs.2020.10.013] [PMID: 33424317]

[55] Sharma S, Kamal B, Rathi N, *et al. In vitro* rapid and mass multiplication of highly valuable medicinal plant *Bacopa monnieri* (L.) Wettst. Afr J Biotechnol 2010; 9(49): 8318-22.

[56] Jat BL, Rashmi P, Dilip G, Bhat TS, Rawat RS. *In vitro* production of bacosides in tissue cultures of *Bacopa monneri.* World J Pharm Res 2016; 5(7): 1087-107.

[57] Kalita BC, Doley P, Mitra PK, Tag H, Das AK. Optimization of MS media for micropropagation of an medicinal plant-*Bacopa monnieri* (L.). World J Pharm Pharm Sci 2018; 7(6): 1014-22.

[58] Talukdar A. Biosynthesis of total bacosides in the callus culture of *Bacopa monnieri.* L. Pennel from North-East India. Int J Curr Microbiol Appl Sci 2014; 3(3): 140-5.

[59] Samanta D, Mallick B, Roy D. *In vitro* clonal propagation, organogenesis and somatic embryogenesis in *Bacopa monnieri* (L.) Wettst. Plant Sci Today 2019; 6(4): 442-9.
[http://dx.doi.org/10.14719/pst.2019.6.4.600]

[60] Antony Ceasar S, Lenin Maxwell S, Bhargav Prasad K, Karthigan M, Ignacimuthu S. Highly efficient shoot regeneration of *Bacopa monnieri* (L.) using a two-stage culture procedure and assessment of genetic integrity of micropropagated plants by RAPD. Acta Physiol Plant 2010; 32(3): 443-52.
[http://dx.doi.org/10.1007/s11738-009-0419-8]

[61] Mohapatra HP, Rath SP. *In vitro* studies of *Bacopa monnieri* : An important medicinal plant with reference to its biochemical variations. Indian J Exp Biol 2005; 43(4): 373-6.
[PMID: 15875724]

[62] Chauhan R, Shirkot P. Micropropagation of endangered medicinal plant *Bacopa monnieri* (L.) Pennell.

J Pharmacogn Phytochem 2020; 9(2): 1614-20.

[63] Shrivastava N, Rajani M. Multiple shoot regeneration and tissue culture studies on *Bacopa monnieri* (L.) Pennell. Plant Cell Rep 1999; 18(11): 919-23. [http://dx.doi.org/10.1007/s002990050684]

[64] Sharath R, Krishna V, Sathyanarayana BN, Prasad BM, Harish BG. High frequency regeneration through somatic embryogenesis in *Bacopa monnieri* (L.) Wettest, an important medicinal plant. Med Aromat Plant Sci Biotechnol 2007; 1: 138-41.

[65] Tiwari V, Deo Singh B, Nath Tiwari K. Shoot regeneration and somatic embryogenesis from different explants of Brahmi (*Bacopa monniera* (L.) Wettst.). Plant Cell Rep 1998; 17(6-7): 538-43. [http://dx.doi.org/10.1007/s002990050438] [PMID: 30736632]

[66] Sharma M, Ahuja A, Gupta R, Mallubhotla S. Enhanced bacoside production in shoot cultures of *Bacopa monnieri* under the influence of abiotic elicitors. Nat Prod Res 2015; 29(8): 745-9. [http://dx.doi.org/10.1080/14786419.2014.986657] [PMID: 25485652]

[67] Parale A, Barmukh R, Nikam T. Influence of organic supplements on production of shoot and callus biomass and accumulation of bacoside in *Bacopa monniera* (L.) Pennell. Physiol Mol Biol Plants 2010; 16(2): 167-75. [http://dx.doi.org/10.1007/s12298-010-0018-6] [PMID: 23572966]

[68] Jain A, Pandey K, Benjamin D, Meena AK, Singh RK. *In vitro* approach of medicinal herb: *Bacopa monnieri*. Int J Innov Res Sci Eng Technol 2014; 3(5): 12088-93.

[69] Koul A, Mallubhotla S. Elicitation and enhancement of bacoside production using suspension cultures of *Bacopa monnieri* (L.) Wettst. 3 Biotech 2020; 10(6): 1-4.

[70] Ghasolia B, Shandilya D, Maheshwari R. Multiple shoot regeneration of *Bacopa monnieri* (L.) using cyanobacterial media—a novel approach and effect of phytoregulators on *in vitro* micropropagation. Int J Recent Biotechnol 2013; 1(2): 27-33.

[71] Ahuja A, Gupta KK, Khajuria RK, *et al.* Production of bacosides by multiple shoot cultures and *In vitro* regenerated plantlets of selected cultivar of *Bacopa monnieri* (L.) Wettst. Plant biotechnology and its application in plant tissue culture 2005; 155-62.

[72] Banerjee M, Shrivastava S. An improved protocol for *In vitro* multiplication of *Bacopa monnieri* (L.). World J Microbiol Biotechnol 2008; 24(8): 1355-9. [http://dx.doi.org/10.1007/s11274-007-9612-3]

[73] Pandiyan P, Selvaraj T. *In vitro* multiplication of *Bacopa monnieri* (L.) Pennell from shoot tip and nodal explants. Agric Technol Thail 2012; 8(3): 1099-108.

[74] Jaspreet K, Kalpana N, Manu P. *In vitro* propagation of *Bacopa monnieri* (L.) Wettst : A medicinally priced herb. Int J Curr Microbiol Appl Sci 2013; 2(8): 131-8.

[75] Behera S, Nayak N, Barik DP, Naik SK. An efficient micropropagation protocol of *Bacopa monnieri* (L.) Pennell through two-stage culture of nodal segments and *ex vitro* acclimatization. J Appl Biol Biotechnol 2015; 3(3): 1-2.

[76] Karataş M, Aasim M. Efficient *in vitro* regeneration of medicinal aquatic plant water hyssop (*Bacopa monnieri* L. Pennell). Pak J Agric Sci 2014; 51(3): 667-72.

[77] Croom LA, Jackson CL, Vaidya BN, Parajuli P, Joshee N. Thin Cell Layer (TCL) culture system for herbal biomass production and genetic transformation of *Bacopa monnieri* L. Wettst. Am J Plant Sci 2016; 7(8): 1232-45. [http://dx.doi.org/10.4236/ajps.2016.78119]

[78] Govindan P, Ebenezer RS, Christdas EJ, Harshavardhan S. Comparative analysis on the effect of seaweed liquid extracts and commercial plant growth regulators on *in vitro* propagation of *Bacopa monnieri*. Int J Res Biosci 2016; 5(2): 1-9.

[79] Dogan M. Effect of salt stress on *in vitro* organogenesis from nodal explant of *Limnophila aromatica*

(Lamk.) Merr. and *Bacopa monnieri* (L.) Wettst. and their physio-morphological and biochemical responses. Physiol Mol Biol Plants 2020; 26(4): 803-16. [http://dx.doi.org/10.1007/s12298-020-00798-y] [PMID: 32255941]

[80] Mehta A. Effect of plant growth regulators on callus multiplication and *in vitro* plant regeneration in *Bacopa monnieri* L. Int J Med Plants Res 2017; 6(5): 337-45.

[81] Ahmed A, Rahman ME, Tajuddin T, *et al.* Effect of nutrient medium, phytohormones and elicitation treatment on *in-vitro* callus culture of *Bacopa monnieri* and expression of secondary metabolites. Nat Prod J 2014; 4(1): 13-7. [http://dx.doi.org/10.2174/221031550401140715142624]

[82] Sahoo JP. Organogenesis and somatic embryogenesis - *In vitro* mutant selection for biotic and abiotic stresses. Conference: Class PresentationAt: Department of Agril Biotech, CA, OUAT, BBSRAffiliation: Orissa University of Agriculture & Technology 2018. [http://dx.doi.org/10.13140/RG.2.2.26278.57928/1]

[83] Quiroz-Figueroa FR, Rojas-Herrera R, Galaz-Avalos RM, Loyola-Vargas VM. Embryo production through somatic embryogenesis can be used to study cell differentiation in plants. Plant Cell Tissue Organ Cult 2006; 86(3): 285-301. [http://dx.doi.org/10.1007/s11240-006-9139-6]

[84] Parale A, Sangle S. Induction of somatic embryogenesis in *in vitro* cultures of *Bacopa monniera* (L.) Pennell. Int J Botany Stud 2020; 5(3): 627-9.

[85] Dey A, Hazra AK, Nandy S, Kaur P, Pandey DK. Selection of elite germplasms for industrially viable medicinal crop *Bacopa monnieri* for bacoside A production: An HPTLC-coupled chemotaxonomic study. Ind Crops Prod 2020; 158: 112975. [http://dx.doi.org/10.1016/j.indcrop.2020.112975]

[86] Richa J, Bheem P, Manju J. *In-vitro* regeneration of *Bacopa monnieri* (L.): A highly valuable medicinal plant. Int J Curr Microbiol Appl Sci 2013; 2(12): 198-205.

[87] Rout JR, Sahoo SL, Ray SS, Sethi BK, Das R. Standardization of an efficient protocol for *in vitro* clonal propagation of *Bacopa monnieri* L. : An important medicinal plant. Agric Technol Thail 2011; 7: 289-99.

[88] Tiwari KN, Tiwari V, Singh J, Singh BD, Ahuja P. Synergistic effect of trimethoprim and bavistin for micropropagation of *Bacopa monniera.* Biol Plant 2012; 56(1): 177-80. [http://dx.doi.org/10.1007/s10535-012-0038-x]

[89] Naik PM, Patil BR, Kotagi KS, Kazi AM, Lokesh H, Kamplikoppa SG. Rapid one step protocol for *in vitro* regeneration of *Bacopa monnieri* (L.). J. Cell Tissue Res 2014; 14(2): 4293.

[90] Umesh TG, Sharma A, Rao NN. Regeneration potential and major metabolite analysis in nootropic plant-*Bacopa monnieri* (L.) Pennell. Asian J Pharm Clin Res 2014; 7(1): 134-6.

[91] Huetteman CA, Preece JE. Thidiazuron: A potent cytokinin for woody plant tissue culture. Plant Cell Tissue Organ Cult 1993; 33(2): 105-19. [http://dx.doi.org/10.1007/BF01983223]

[92] Deepa AV, Anju M, Dennis Thomas T. The applications of TDZ in medicinal plant tissue culture. Thidiazuron: From urea derivative to plant growth regulator. In: Ahmed N, Faisal M, Eds. Thidiazuron: From Urea Derivative to Plant Growth Regulator. Springer 2018; pp. 297-316. [http://dx.doi.org/10.1007/978-981-10-8004-3_15]

[93] Dinani ET, Shukla MR, Turi CE, Sullivan JA, Saxena PK. Thidiazuron: Modulator of morphogenesis *in vitro*. Thidiazuron: From urea derivative to plant growth regulator. In: Ahmed N, Faisal M, Eds. Thidiazuron: From Urea Derivative to Plant Growth Regulator. Springer 2018; pp. 1-36. [http://dx.doi.org/10.1007/978-981-10-8004-3_1]

[94] Faisal M, Alatar AA, El-Sheikh MA, Abdel-Salam EM, Qahtan AA. Thidiazuron induced *in vitro* morphogenesis for sustainable supply of genetically true quality plantlets of Brahmi. Ind Crops Prod

2018; 118: 173-9.
[http://dx.doi.org/10.1016/j.indcrop.2018.03.054]

[95] Srivastava P, Tiwari KN, Srivastava G. Effect of different carbon sources on *in vitro* regeneration of Brahmi *Bacopa monnieri* (L.) An important memory vitalizer. J Med Plants Stud 2017; 5(3): 202-8.

[96] Menezes RS, Soares AT, Marques Júnior JG, *et al.* Culture medium influence on growth, fatty acid, and pigment composition of *Choricystis minor* var. *minor*: A suitable microalga for biodiesel production. J Appl Phycol 2016; 28(5): 2679-86.
[http://dx.doi.org/10.1007/s10811-016-0828-1]

[97] Hegazi GA. *In vitro* preservation of *Bacopa monnieri* (L.) Pennell as a rare medicinal plant in Egypt. J Basic Appl Sci Res 2016; 6(12): 35-43.

[98] Chen X, Qu Y, Sheng L, Liu J, Huang H, Xu L. A simple method suitable to study de novo root organogenesis. Front Plant Sci 2014; 5: 208.
[http://dx.doi.org/10.3389/fpls.2014.00208] [PMID: 24860589]

[99] Rao S, Rajkumar P, Kaviraj C, Parveen PA. Efficient plant regeneration from leaf explants of *Bacopa monniera* (L.) Wettst.: A threatened medicinal herb. Ann Phytomed 2012; 1: 110-7.

[100] Binita BC, Ashok MD, Yogesh TJ. *Bacopa monnieri* (L.) Pennell: A rapid, efficient and cost effective micropropagation. Plant Tissue Cult Biotechnol 2005; 15(2): 167-75.

[101] Teixeira da Silva JA, Dobránszki J, Ross S. Phloroglucinol in plant tissue culture. *In Vitro* Cell Dev Biol Plant 2013; 49(1): 1-16.
[http://dx.doi.org/10.1007/s11627-013-9491-2]

[102] Mishra SK, Tiwari KN, Shivna PL, Mishra AK. Micropropagation and comparative phytochemical, antioxidant study of *Bacopa monnieri* (L.) Pennell. Res J Pharm Biol Chem Sci 2015; 6(6): 902-12.

[103] Tiwari V, Tiwari KN, Deo Singh B. Shoot bud regeneration from different explants of *Bacopa monniera* (L.) Wettst. by trimethoprim and bavistin. Plant Cell Rep 2006; 25(7): 629-35.
[http://dx.doi.org/10.1007/s00299-006-0126-5] [PMID: 16482428]

CHAPTER 9

A Review of *Momordica charantia* L.: Regeneration *via* Organogenesis *versus* Embryogenesis

Mala Agarwal[1,*]

[1] *Department of Botany, B.B.D. Government College, Chimanpura (Shahpura), Jaipur, Rajasthan, India*

Abstract: *Momordica charantia* L., commonly known as bitter melon/gourd, is a slender tendril-climbing annual vine of the family Cucurbitaceae. Bitter melon grows in tropical areas, including parts of the Amazon, Asia, and the Caribbean, and is cultivated throughout South America. It is a common food of the tropics used in the treatment of many diseases and is also known for its potent hypoglycemic actions. A steroidal sapogenins known as charantin, insulin-like peptides, and alkaloids have been reported to have hypoglycemic or other actions of potential benefit in diabetes mellitus. The present chapter gives a comprehensive review of the tissue culture of *Momordica charantia*. There are two ways of regeneration, direct organogenesis and indirect organogenesis; both take place through the production of adventitious buds and somatic embryogenesis. The present review gives a complete *in vitro* regeneration protocol of *M. charantia*.

Keywords: Clonal propagation, Differentiation, Explants, *In vitro* regeneration, Organogenesis, Regeneration, Somatic embryogenesis.

INTRODUCTION

Momordica charantia L., commonly known as bitter melon/bitter gourd, is a tendril climbing plant belonging to the family Cucurbitaceae. It is distributed throughout the globe with predominance in tropical and subtropical areas [1, 2]. It is widely grown in India and other parts of the Indian subcontinent, Southeast Asia, China, Africa, the Caribbean, and South America for food and medicine [3]. Fruits of *M. charantia* are used as daily food, whereas the fruits, leaves, roots, and seeds of bitter melon are used as traditional medicine in Southeast Asia and Indo-China [4]. *M. charantia* has culinary use and is also a vegetable [5, 6]. *M. charantia* possesses many biological activities. Roots are acrid, astringent, and bitter, and leaves are antipyretic, emetic as well as purgative and bitter. Fruits are

* **Corresponding author Mala Agarwal:** Department of Botany, B.B.D. Government College, Chimanpura (Shahpura), Jaipur, Rajasthan, India; E-mail: agarwal.mala@yahoo.co.in

T. Pullaiah (Ed.)

acrid, anthelmintic, anti-diabetic, anti-inflammatory, appetizer, bitter, depurative, digestive, purgative, stimulant, stomachic, and thermogenic [7]. Momordicin, a compound of bitter melon, gives the plant a characteristic of bitter taste and also has stomachic properties [8].

Momordica charantia has a very important place in Ayurveda and is recommended for treating many diseases like anemia, bronchitis, blood diseases, cholera, diarrhea, dysentery, fever, hepatitis, itch, ulcer, measles, and sexual tonic and as a cure for gonorrhea [9]. Fruits are used as a traditional medication to cure various diseases like rheumatism, gout, worms, colic, diseases of the liver, and spleen [10]. It is also found useful in the treatment of cancer and diabetes mellitus [11].

It contains an array of biologically active chemicals, including triterpenes, proteins, steroids, alkaloids, saponins, flavonoids, and acids due to which plant possesses anti-fungal, anti-bacterial, anti-parasitic, anti-viral, anti-fertility, anti-tumorous, hypoglycemic, and anti-carcinogenic properties [12 - 16]. This plant is blessed with therapeutic potential [2, 17 - 19], such as antioxidant [20 - 23], antimicrobial [2, 17, 21, 24 - 26], anthelmintic [27, 28], anti-diabetic [2, 20, 29, 30], anti-inflammatory [30, 31], antihyperglycemic [2, 30, 32], anticancer [4, 28, 30, 33], antimicrobial [43] and nutritional as antilipolytic [20, 34] due to the presence of bioactive compounds. It is a potent hypoglycemic agent due to alkaloids and insulin-like peptides and a mixture of steroidal sapogenins known as charantin [32].

Chemical Composition

Chemically, the plant is enriched with secondary metabolites like triterpenoids [20, 35-37], saponins [38], polypeptides [39], flavonoids [23, 40, 41], alkaloids [42, 44] and sterols [22, 45 - 47] which are distributed throughout the entire plant. The seed is not edible; it contains extractable oils, mostly a conjugated triene *cis*-9, trans-, trans-ctt [9, 11, 13], and conjugated omeroflinolenic acid, known as α-elestearic acid (α-ESA). The anti-cancer and anti-obesity properties are due to the presence of ESA [48], while phenolic compounds phenylpropanoids, flavonoids, triterpenes, and carotenoids [40, 49] are responsible for pharmaceutical properties. The main bioactive compounds of the fruit of *M. charantia* are carbohydrates, proteins, lipids, and more [35, 47, 50].

Antimicrobial properties of the plant are due to the presence of charantin which is a Cucurbitane-typetriterpenoid [51]. Charcnte is a mixture of two steroidal saponins, β-sitosterol glycoside and stigma sterol glycoside [52]. Charantin is present mainly in fruits, leaves, and roots [27, 35, 36, 44]. However, cucurbitane-type triterpenoids are found in the entire plant. Antimicrobial activity is also linked to [53, 54] α-momorcharin (leaf and seed) and MAP30 (fruit and seed). MAP30 and α-momorcharin show antimicrobial properties as they are ribosome-inactivating proteins [35, 54].

PLANT TISSUE CULTURE OF *MOMORDICA CHARANTIA* L.

Initiation of adventitious buds and somatic embryo formation are two ways of regeneration [55]. Callus formation is an important stage in the plant regeneration process as the callus can give rise to shoots, roots, and both roots as well as multiple shoots. The process of regeneration mediated by callus is also known as indirect regeneration.

Callus Formation in *Momordica charantia* L.

Thiruvengadam *et al.* [55] reported about the callus formation in *M. charantia.* They stated that in Murashige and Skoog (MS) medium supplemented with 1.0 mg/l 2,4-Dichlorophenoxy acetic acid (2,4-D) well-organized friable calluses were formed by approximately 90% of leaf explants. Compact and hard green calluses were produced at various concentrations of hormones like 6-Benzy--aminopurine (BAP) and Kn, which turned embryogenic due to plant growth regulator stress. BAP is useful for the development of good texture callus [56]. In *Momordica dioica,* 1.0 mg/l BAP + 0.1 mg/naphthalene acetic acid (NAA) combination produced soft, green, light, and friable calli [57]. The greenish compact callus was produced on the combination of 7.7 μM NAA + 2.2 μM Thiadiazuron (TDZ) from leaf explants of *M. charantia.*

Indirect Organogenesis

Indirect Organogenesis through Somatic Embryos

Thiruvengadam *et al.* [58] developed a system for the somatic embryogenic suspension culture of *M. charantia.* They reported the formation of friable calluses in 30-day-old leaves on semi-solid MS medium [59] supplemented with 1 mg/l 2,4-D. On sub-culturing the callus in liquid medium with 1.5 mg/l 2,4-D, a large number of globular embryos (about 24.6%) was noticed and by removing 2,4-D during later stages, they turned to heart/torpedo stages. The germination of these embryos was on basal MS media. They studied the effect of media, carbohydrates, and amino acids on somatic embryogenesis *via* the formation of cell clusters, which then enlarged to pro-embryos, and gave rise to mature embryos within a period of 2 weeks. A high-frequency induction, maturation, and development of somatic embryos were on MS medium with 50 mg/l polyvinyl pyrrolidone (PVP) and 40 mg/l glutamine. About 6.2% phenotypically normal young plantlets were developed from friable callus.

Indirect Organogenesis Through Adventitious Buds

Thiruvenagadam *et al.* [60] described *in vitro* organogenesis from leaf explants derived callus of *M. charantia*. The immature leaf explants of 15-day-old *in vitro* seedlings and mature leaf explants of 45-day-old *vivo* plants were used for callus generation. The explants were grown on Murashige and Skoog (MS) medium [59] with Gamborg (B5) [61] vitamins containing 30 g/l sucrose, 2.2 g/l Gelrite, and 7.7 μM NAA with 2.2μM TDZ. Regeneration of adventitious shoots from callus (30–40 shoots per explant) was achieved on MS medium containing 5.5 μM TDZ, 2.2 μM NAA, and 3.3 μM silver nitrate ($AgNO_3$). 1.0 cm long shoots were excised from the callus and elongated in MS medium with 3.5 μM GA_3 and were rooted in MS medium supplemented with 4.0 μM IBA. Plantlets with roots were hardened in the greenhouse, and transferred to soil and a 90% survival rate was achieved. These protocols gave the production of forty plantlets in ninety-eight days culture period from leaf explants.

Triruvenagadam *et al.* [62] reported *in vitro* regeneration from internodal explants of *M. charantia via* indirect organogenesis. They reported high-frequency shoot regeneration from organogenic callus. They used internodal explants from 30 days old *in vivo* grown plants and found that around 97.5% of the explants produced green, compact nodular organogenic callus in Murashige and Skoog (MS) [59] plus Gamborg *et al.* [61] (B_5) medium containing 5.0 μM 2,4-D and 2.0 μM TDZ by transferring two times in eleven-day interval [62]. When callus was transferred to MS medium supplemented with .0 μM TDZ, 1.5 μM 2, 4-D, and 0.07 mM L-glutamine, it produced adventitious shoots having 96.5% shoot induction frequency and 48% adventitious shoots per explants from callus after eighty days of culture. They reported that shoot production was better on L-glutamine from L-asparagine. Callus produced more shoots on high cytokinin: low auxin ratio and more callus on low cytokinin: high auxin ratio. Elongation of the regenerated shoots was carried out on the same medium and rooting was done on MS medium with 3.0 μM IBA. Rooted plants were acclimatized and subsequently established in soil with a survival rate of ninety-five percent.

Effect of Callus Type and Seedling Age on Regeneration

Manye *et al.* [63], who studied *in vitro* regeneration of *M.charantia*, reported the production of yellow-green, fragile yellow, and green calluses from *M. charantia* seedlings. They reported that in the differentiation of adventitious bud from callus, phytohormone as well as the type of callus is significant. Yellow-green callus produced differentiation of adventitious buds on MS medium + BA 4.0 mg/l + Kinetin (Kn) 2.0 mg/l. The frequency was about 66.7%. No adventitious bud was formed on the yellow callus. On the green callus, the frequency of differentiation was low (< 15.0%) even in optimum conditions. The MS medium with zeatin 5.0 mg/l and Kn 0.5 mg/l was favorable for bud proliferation, after three weeks proliferation coefficient was five to six-fold. The *in vitro* rooting of the shoot was best on ½MS+ zeatin 0.02 mg/l or ½MS media with six to seven new roots production in 3 weeks. A seventy percent survival rate was achieved for plantlets. The *in vitro* morphogenesis of *M. charantia* is affected by the age of seedlings, hormone combinations and concentration ratio, and different genotypes [64]. Seedling age was significant for direct adventitious bud regeneration; ten-day-old seedlings on MS medium with BA 0.5 mg/l + NAA 0.2 mg/l and MS + zeatin 2.0 mg/l +NAA 0.1 mg/l were optimum for multiple shoots production. Different genotypes had little effect on clustered shoot production, however, the variety Lvubaoshi is the most effective. The selection of age of seedling explants with consideration of the colour of cotyledon is important for explant selection for induction of clustered shoots in *M. charantia.*

Effects of Explants and Hormones on Regeneration

In tissue culture medium cytokinin concentration is determinal for shoot induction response. Cytokinin activates totipotent cells of the callus for shoot production and functions as a signaling molecule, besides cytokinin activates preexisting machinery for somatic cells like leaf, stem, cotyledon, *etc.*, in the direct organogenesis, and stimulates the growth in meristematic cells of shoot apex. Many a time cytokinins stimulate the explants for multiple shoot production. Cytokinins like Kn and BAP stimulate the growth and development in cultures [65], they promote cell division more effectively with an auxin combination. They decrease apical dominance and retard ageing and thus induce adventitious shoot formation at higher concentrations. The morphogenetic response was low from cotyledonary node explant which is a regeneration frequency of 50.0% on the same hormonal combinations shoot tip explants and gave maximum hundred percent responses. BAP alone can induce shoot differentiation in *M. charantia,* NAA in combination with Indole-3-butyric acid (IBA) also gives good results [66]. The nodal segments give the best response for the production of multiple shoots in MS medium with BAP and NAA [67]. This combination (BAP+NAA) is the best combination for adventitious multiple shoot production in teasle gourd also [68]. Sultana and Bari showed that the full-strength MS medium with 0.1 mg/l NAA gave the best rooting with 1.38±1.06 average roots per flask and 62.5% acclimatization.

Differentiation and callogenesis are determined by the endogenous and exogenous levels of growth regulators however, the habituation property of the callus hinders the organogenic response in *M.charantia* [69].

The effect of various combinations and concentrations of growth hormones auxins and cytokinins in MS medium on different explants (leaf, stem, and cotyledonary of *M. charantia*, for callus formation and direct and indirect organogenesis were analysed. All three explants like leaf, stem, and cotyledon on MS medium with 1.0- 1.5 mg/l BAP + 1.5 mg/l NAA + 1.0 mg/l 2,4-D, gave the best result for callus formation. 2,4-D proved best for callus formation from all three explants but was not very responsive at low levels (0.1 and 1.0 mg/l). Maximum callus formation was from leaf explants, in comparison to stem and cotyledons, on MS medium with BAP, NAA, and Kn. Plant growth regulators are also important for various callus textures and morphology. The calluses were compact, hard brownish, yellowish green at all concentrations of 2, 4-D. The second subculture was unresponsive for shoot production and indirect organogenesis as the callus hardened followed by habituation; therefore, no organogenesis occurred at any concentration /combination tested except leaves with BAP and TDZ/ NAA.

Best shoot induction was from shoot tip explants with BAP and NAA (1.0+0.2 mg/l) in MS medium; the average number of shoots per flask was 2.75±0.71 with a maximum length of 1.74±0.69 cm. An average number of shoots was 1.50±1.69 with 0.800±0.899 cm length in BAP with TDZ. Shoot induction response was low when only BAP was used. Shoot tip and cotyledonary node explants gave the best response for shoot induction with 1.0 mg/l BAP + 0.1 TDZ and 1.5 mg/l BAP+0.2mg/l.

When regenerated shoots were transferred to MS medium both full and half strength with different auxin concentrations resulted in root formation. The best rooting response from regenerated shoots was observed at half-strength MS medium supplemented with 0.5 mg/l NAA with a 2.63±1.30 average number of roots giving 87.5% response and half-strength MS without any growth regulator that gave 100% rooting response.

Tang *et al.* [70] investigated the effects of TDZ, silver nitrate ($AgNO_3$) in adventitious buds induction from stems of *M. charantia* and found that TDZ was necessary for bud development and the higher concentration of it could induce adventitious buds efficiently, while 0.1 mg/l was the best concentration to induce adventitious buds. Bud formation was significantly affected by $AgNO_3$ and triacontanol. The best results were obtained at a concentration of 2.0 mg/l separately.

Effect of Seedling Age, Genotype, and Hormones on Direct Regeneration

In Gazipaşa and Silifke genotypes of *M. charatia*, a study by Saglam [71] on *in vitro* and *in vivo* germination with subsequent callus formation and regeneration, it was found that during *in vivo* conditions, germination was 20 and 50% in the pots, however during *in vitro* conditions, germination was higher in the Gazipaşa genotype compared to Silifke genotype. In callus culture studies, leaf and stem explants from 9-10-day old plantlets, cultured in nutrient media having various concentrations of 2,4-D (2, 4, 6, and 8 mg/l) were used. The *in vitro*-grown explants gave a hundred percent callus formation, this rate was found to be eighty percent in explants obtained *in vivo*. In both genotypes and leaf and stem explants taken *in vitro*, 6 mg/l of 2,4-D concentration was the best. Maximum callus formation in leaf explants was at 2 mg/l 2,4-D. In both the explants the most effective concentration of auxin was 2 mg/l under *in vivo* conditions. Maximum callus formation was at 4 and 8 mg/l 2,4-D in the Silifke genotype and the most effective concentration of auxin in the stem explants of the Gazipaşa genotype was 2 mg/l. The calluses of the Silifke genotype were less dispersed from those of the genotype Gazipaşa. Callus formation was found to be 73% and 67% from *in vitro* and *in vivo* of the Silifke genotype, whereas in the Gazipaşa genotype it was 20% from *in vivo* explants. However, in both culture conditions in the control group, callus formation was not detected from leaf explants. In both *in vivo* and *in vitro* conditions, callus formation was higher in the leaf explants from the stem explants. However, indirect somatic embryogenesis from callus derived from stem explants of the Silifke genotype produced some plantlets. Besides this, one plantlet was formed by the callus formed by the indirect somatic embryogenesis from stem explants of the Silifke genotype which was not from the leaf explants. This plantlet was transferred to the pot, subsequently to the field, and adaptation was maintained.

Direct Organogenesis

Huda *et al.* [72] studied *in vitro* plant production through apical meristem culture of *M. charantia* and observed growth of meristem on semisolid MS medium with 0.05 mg/l Kn + 0.1 mg/l GA_3. After three weeks these meristems were transferred to MS medium with BA, Kn, IBA, NAA, and Indole-3-acetic acid (IAA) singly/in combination for shoot elongation and root initiation. Shoot elongation was on MS medium with 1.0 mg/l BA + 0.1 mg/l IBA + 0.3 mg/l Gibberellic acid (GA_3). MS medium with 0.5 mg/l IBA and 0.1 mg/l NAA gave the best rooting. *In vitro* plantlets (10 weeks old) were transferred in soil and were acclimatized gradually.

Effect of Seedling Age, Genotype, and Hormones on Direct Regeneration

Chao *et al.* [73] studied the *in vitro* regeneration of Changbai, Dabai, and Youlv varieties (cotyledonary nodes) of Balsam pear (*M. charantia*). They investigated the effects of different genotypes of Balsam pear, hormone combinations, seedling age, and dark period and $AgNO_3$ concentrations. They found an eight-day-old seedling of *M. charantia* most suitable for the regeneration of multiple buds with the induction rate of shoot regeneration frequency of 78.63%, in comparison to six- and ten-day-old seedlings. Besides this rate of multiple buds in eight-day-old seedlings, the combination of BAP and IBA was the best for the multiple bud induction.

The most suitable medium for the multiple bud induction in Balsam pear was MS medium [59] with 2.5 mg/l of BA and 0.1 mg/l of IBA. Cotyledonary nodes of *M. charantia* cotyledonary nodes cultured on MS medium swelled and turned green after 2-3 days with the appearance of callus at the lower edges and after one-week formation of multiple buds started.

The average bud formation was highest (up to 4.42) on MS medium with 2.0 mg/l BA and 0.1 mg/l NAA, but were weak and failed to elongate. When explants were cultured on MS medium supplemented with 2.5 mg/l BA and 0.1 mg/l IBA, multiple buds were large in number and vigorous. In addition, the induction rate of these different hormone types and proportions of MS medium reached as high as 90.26%. So these different hormone types and proportions of MS medium were the optimum medium in the regeneration of Balsam pear. Induction frequency varied with the genotypes, the induction rate was found to be highest (up to 80.65%) in Youlv Balsam pear, followed by Changbai Balsam pear (78.79%) and lowest (7.74%) in Dabai Balsam pear. However, the average multiple bud production was higher in Changbai in comparison to Balsam pear cultivars. They found that buds of "Changbai" Balsam pear were found to be more vigorous and easy to grow among the three varieties of Balsam pear, therefore this genotype is optimum in this respect.

Two to three cm long regenerated buds excised and transferred to various media for rooting. The best root development was obtained in the rooting medium containing half MS medium supplemented with 0.4 mg/l IBA with eighty to ninety per cent rooting rate in about two weeks.

Seedling age is determinal for the production of adventitious shoot regeneration with high frequency [64, 67, 74, 75] because different physiological states of explants affect the regeneration capacity of shoots [76, 77] and is determinal for the success of *in vitro* regeneration in Balsam pear. Genotype is also important for organogenesis in Balsam pear tissue culture [64, 75, 78, 79]. Auxins and cytokinins have considerable effects on the redifferentiation and dedifferentiation of plant cells [64, 66, 67, 79, 80].

A plantlet obtained by tissue culture can be established in the field only by rooting of regenerated buds. In many plants like *M. charantia*, IBA is found to be the best rooting medium [66].

In Vitro Regeneration/Clonal Propagation of Momordica Charantia

Munsur *et al.* [81] observed that the combination of 2,4-D (1.0 mg/l) and BAP (1.0 mg/l) produced the highest callus frequency in nodal (93.75%), leaf (78.75%), and roots (75.00%) segments. In the case of root tips, 2,4-D (1.0 mg/l) + BAP (0.5 mg/l) produced the highest callus (72.5%). Nodal segments required a minimum number of 8.25 days while root tips required a maximum of 32 days for callus induction. A combination of BAP (2.5 mg/l + NAA (0.6 mg/l) produced the highest callus frequency (86.25%) in nodal segments followed by root segments (85.00%). Among the combinations 2,4-D (1.0 mg/l and BAP 1.0 mg/l were the most suitable for producing greenish friable callus and BAP 2.0 or 2.5 mg/l + NAA (0.30 or 0.60 mg/l) was suitable for callus induction. A combination of 2,4-D (1.0 mg/l) + BAP (1.0 mg/l) exhibited 75.00% and 56.0% shoot direct regeneration from nodal segments and leaf segments, respectively. Nodal segments had 70% shoot regeneration *via* callus on medium BAP (2.0 mg/l +IBA 0.5 mg/l + GA_3 (0.2 mg/l). Leaf segments and root tips had an equal percentage (65.0) of shoot regeneration upon culture on medium with BAP 2.0 mg/l and NAA (0.3 mg/l).

Daniel *et al.* [82] established a protocol for *in vitro* regeneration of *M. charantia.* They used MS medium with different concentrations of cytokinin (BAP, Kn) and auxin (IBA, IAA, NAA) for *in vitro* clonal propagation of *M. charantia* [59]. Maximum callus formation was from leaf explants rather than stem and cotyledons. A green, compact callus was formed at different concentrations of BAP and Kn. Nodal segments of *M. charantia* on MS medium with BAP and NAA were found to be most suitable for multiple shoot regeneration.

In one of the studies, Agarwal and Kamal [66] showed that the shoot tip, nodal, internodal, and leaf explants from *in vitro*-grown seedlings were cultured on modified MS (Fig. **1A**). Direct organogenesis was observed at a cytokinin concentration of 0.5 mg/l and 2 mg/l BAP, while on IBA/NAA, the root formation was observed (Fig. **1B**). Both roots and shoots were formed from all the explants (shoot tip, internodal, nodal, leaf) on MS media containing BAP + IBA/NAA with best response on 2 mg//l NAA + 0.5 mg BAP and 4 mg/l IBA + 2 mg/l BAP (Fig. **2**).

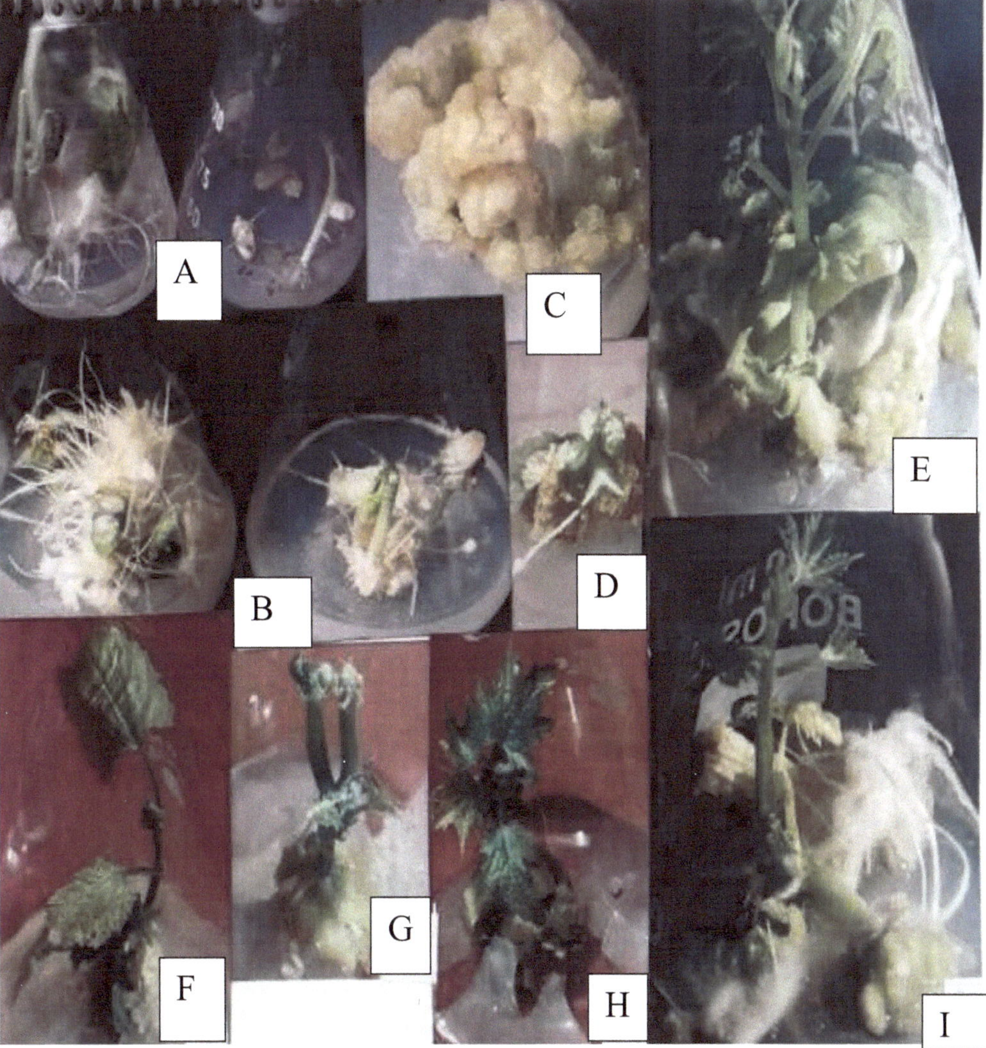

Fig. (1). Micropropagation of *Momordica charantia* L. Indirect Organogenesis.
A) *In vitro* regeneration of seedling. **B**) Profuse Rooting. **C**) Formation of callus on explants. **D**) Formation of callus and rooting on *in vitro* leaf explants. **E**) Regeneration of shoots from callus. **F**) Regeneration of leaves and buds from callus. **G**) Shoot formation on callus. **H**) Shoot and leaves regeneration from callus. **I**) Regeneration of roots and shoots from callus.

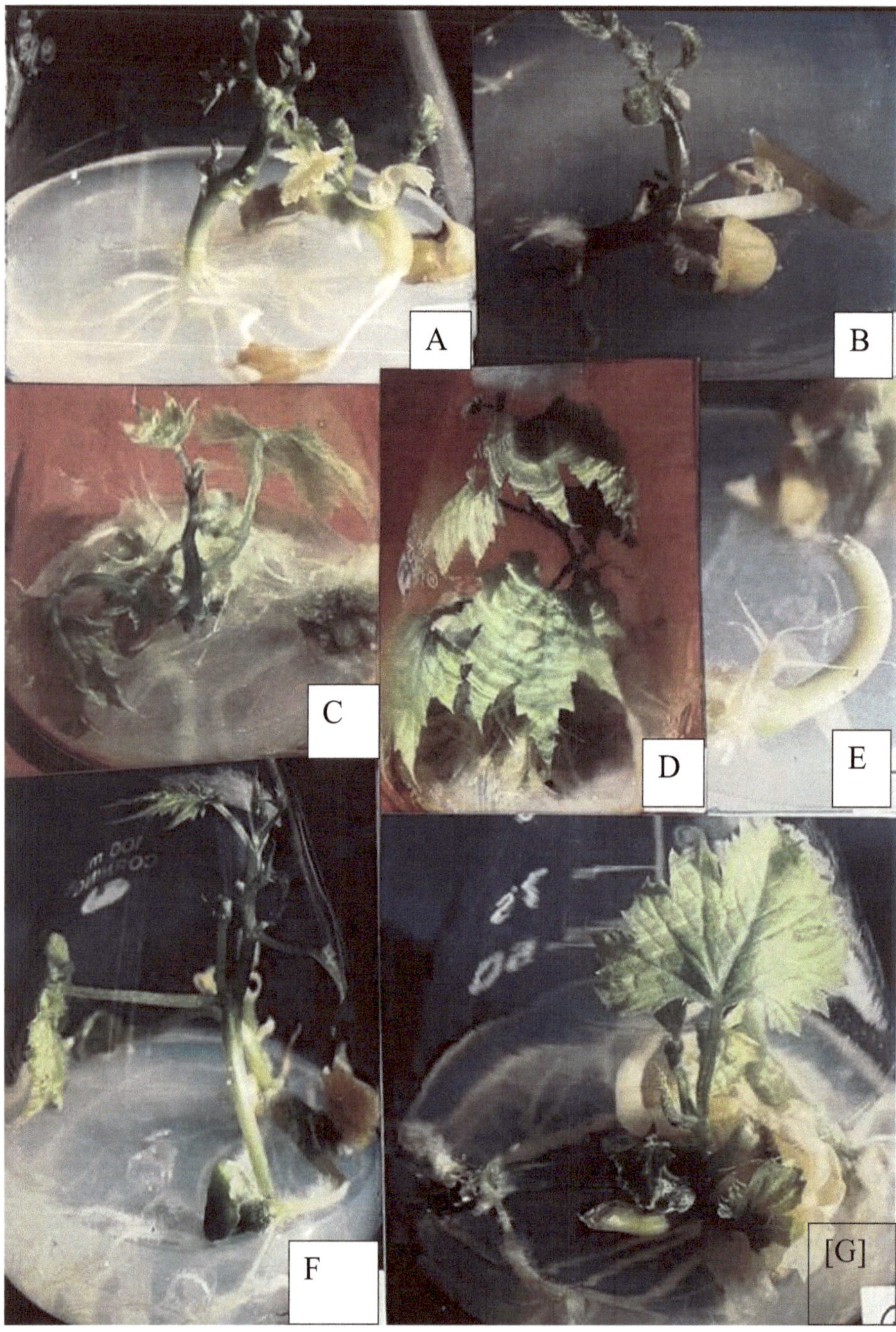

Fig. (2). Micropropagation of *Momordica charantia* L.Direct Organogenesis
A, B) formation of shoot buds and rooting on nodal explant. **C**) Formation of leaves. **D**) Fully expended leaves on *in vitro* explants. **E**) Rooting on nodal explant. **F, G**) *In vitro* plant regeneration.

Fig. (3). Micropropagation of *Momordica charantia* L. Multiple shoots regeneration.
A) Initiation of shoots from explant. **B**) Multiple shoots from callus. **C**) Direct multiple shoot formation. **D**) Initiation of multiple shoots from nodal explant. **E**) Multiple shoots with rooting.

Indirect organogenesis was shown when callus differentiation was obtained on an MS medium fortified with BAP, IBA, and 2,4-D respectively. Profuse callusing was at 2 mg/l of 2,4-D. A combination of NAA + BAP + 2,4-D was most effective for callus formation with the best response on 2 mg/l NAA + 0.5 mg/l BAP + 2 mg/l 2,4-D (Figs. **1C**, **D**). Regeneration of shoots from callus was on BAP (Figs. **1E-H**), regeneration of shoots as well as root was obtained on medium containing BAP + IBA/NAA with best response at 2 mg//l NAA + 0.5 mg BAP and 4 mg/l IBA + 2 mg/l BAP (Fig. **1I**). Shoots were 55 mm in length with 2-3 nodes and 1-2 fully expanded leaves. There was indirect regeneration from callus formed on different explants, as well as direct regeneration from explants taken from *in vitro* grown seedlings.

In this study, development of multiple shoots was observed on BAP; a low concentration of BAP (0.5mg/l) resulted in high shoot production with minimal callus (Figs. **1E**, **F**), on a much higher concentration of BAP shoot production was moderate, and was also associated with increased callus (Fig. **1G**). On finding that BAP induces more callus MS medium without hormones, it was used for further production of multiple shoots; on this medium multiple shoots as well as roots were formed (Fig. **1H**). The shoots attained a length of about 60 mm with 2-3 nodes and 1-2 fully expanded leaves. Therefore, multiple shoots with roots were obtained on MS medium without hormones (MSO) (Fig. **3**).

The roots were developed on the shoots regenerated from callus, seedling explants, and multiple shoots when they were transferred to MS medium with different concentrations of rooting hormones, 3 mg/L IBA was found to be most suitable, and the roots were developed in ten days (Fig. **1I**). Plantlets were hardened in greenhouse and transferred to pots with a 40% survival rate (Figs. **1J**, **3**). This confirms the previous studies [66] that multiple shoots can be formed on MS medium without hormones and endogenous and exogenous level of growth regulators is important for callus formation and for differentiation also (Fig. **3**). All the reagents and hormones used were of Central Drug House Private Limited, New Delhi, India.

Mishra *et al.* [83] studied *in vitro* propagation of *M. charantia* and found that using different concentrations of auxins and cytokinins the *in vitro* development of *M. charantia* on MS medium [59] gives hundred percent germination of the seeds having normal hypocotyls. Basal MS media was best for seed germination and MS media with BAP and IBA was suited for multiplication of shoot, callus formation, and rooting of shoot. About 1.5-2.0 cm long shoot apex was inoculated on MS medium with different concentrations of BAP (1 mg/l, 2 mg/l, 4 mg/l) and they produced a good number of shoots. *In vitro* regenerated shoots were inoculated on MS medium with different concentrations of IBA (0.1 and 1 mg/l) and in two weeks profuse rooting occurred with little callusing.

Hairy Root Cultures for the Production of Secondary Metabolites

Thiruvengadam *et al.* [84] studied the production of secondary metabolites from hairy root cultures of Indian *M. charantia* (IMC) and Korean *M. charantia* (KMC) and their antioxidant and antimicrobial activities. The *Agrobacterium rhizogenes* strains (KCTC 2703 and KCTC 2704) were used for the induction of hairy roots and explants from *in vitro* seedlings like cotyledons, hypocotyls, roots, leaves, and nodes were inoculated on it. Polymerase chain reaction (PCR), reverse transcription-PCR, and sequencing using rolC-specific primers were used to establish transgenic clones of hairy roots. Hairy roots cultured in MS liquid medium with 3% sucrose produced the highest accumulation of biomass [95.11 g/fresh mass (FM) and 10.61 g/l dry mass (DM) in IMC and 93.58 g/l FM and 10.12g/l DM in KMC]. The growth of hairy roots was best on MS basal medium with 3% sucrose in comparison with other culture media (B5, NN, and N6), for the production of biomass and phenolic compounds. Phenolic compounds as well as antioxidant and antimicrobial activities were more in hairy roots in comparison to untransformed control roots. The hairy roots of IMC were superior to KMC for biomass, phenolic compounds, and biological activities.

CONCLUSION AND FUTURE PROSPECTS

Medicinal plants and products obtained from them have been used since a long time to cure a number of diseases of human beings and are considered an important therapeutic aid for humankind. *M. charantia* is a potent medicinal plant extensively studied throughout the world for its medicinal properties to cure diseases. This may be due to the fact that the plant has more than 225 different medicinal constituents having medicinal effects either separately or together.

Modern bio techniques like plant cell culture culturing facilitate provides commercial processing of medicinally important plants and the bioactive compounds they provide. This can be a source of a continuous, reliable source of natural products as well as synthesis of bioactive secondary metabolites, running in controlled environmental conditions. Molecular biology with new techniques to produce transgenic cultures as well as to affect the expression and regulation of biosynthetic pathways can be a significant step for making cell cultures acceptable and applicable for the commercial production of secondary metabolites.

REFERENCES

[1] Jia S, Shen M, Zhang F, Xie J. Recent advances in *Momordica charantia*: Functional components and biological activities. Int J Mol Sci 2017; 18(12): 2555.

[http://dx.doi.org/10.3390/ijms18122555] [PMID: 29182587]

[2] Basch E, Gabardi S, Ulbricht C. Bitter melon (*Momordica charantia*): A review of efficacy and safety. Am J Health Syst Pharm 2003; 60(4): 356-9. [http://dx.doi.org/10.1093/ajhp/60.4.356] [PMID: 12625217]

[3] Jiraungkoorskul W, Poolperm S. An update review on the anthelmintic activity of bitter gourd, *Momordica charantia*. Pharmacogn Rev 2017; 11(21): 31-4. [http://dx.doi.org/10.4103/phrev.phrev_52_16] [PMID: 28503051]

[4] Kumar DS, Sharathnath KV, Yogeswaran P, *et al.* A medicinal potency of *Momordica charantia*. Int J Pharm Sci Rev Res 2010; 1(2): 95.

[5] Karaman K, Dalda-Şekerci A, Yetişir H, Gülşen O, Coşkun ÖF. Molecular, morphological and biochemical characterization of some Turkish bitter melon (*Momordica charantia* L.) genotypes. Ind Crops Prod 2018; 123: 93-9. [http://dx.doi.org/10.1016/j.indcrop.2018.06.036]

[6] Perez JL, Jayaprakasha GK, Patil BS. Metabolite profiling and *in vitro* biological activities of two commercial bitter melon (*Momordica charantia* Linn.) cultivars. Food Chem 2019; 288: 178-86. [http://dx.doi.org/10.1016/j.foodchem.2019.02.120] [PMID: 30902279]

[7] Jadhav D. Medicinal plants of Madhya Pradesh and Chhattisgarh. Daya Publishing House 2008; pp. 213-4.

[8] Taylor L. Technical Data Report for Bitter melon (Momordica charantia) Herbal Secrets of the Rainforest. 2nd edition., Austin: Sage Press 2002.

[9] Kirtikar KR, Basu BD. Indian Medicinal Plants. Compositae, Vol 2. Dehradun: International Book Distributors 1987; pp. 1420-3.

[10] Nazimuddin S, Naqvi SS. Flora of Pakistan No154. Cucurbitaceae Deptt Botany University Karachi 1984; p. 56.

[11] Duke JA. Handbook of medicinal herbs. Boca Raton FL: CRC Press 1985; pp. 315-6.

[12] Beloin N, Gbeassor M, Akpagana K, *et al.* Ethnomedicinal uses of *Momordica charantia* (Cucurbitaceae) in Togo and relation to its phytochemistry and biological activity. J Ethnopharmacol 2005; 96(1-2): 49-55. [http://dx.doi.org/10.1016/j.jep.2004.08.009] [PMID: 15588650]

[13] Grover JK, Yadav SP. Pharmacological actions and potential uses of *Momordica charantia*: A review. J Ethnopharmacol 2004; 93(1): 123-32. [http://dx.doi.org/10.1016/j.jep.2004.03.035] [PMID: 15182917]

[14] Ng TB, Chan WY, Yeung HW. Proteins with abortifacient, ribosome inactivating, immunomodulatory, antitumor and anti-AIDS activities from Cucurbitaceae plants. Gen Pharmacol 1992; 23(4): 575-90. [http://dx.doi.org/10.1016/0306-3623(92)90131-3] [PMID: 1397965]

[15] Scartezzini P, Speroni E. Review on some plants of Indian traditional medicine with antioxidant activity. J Ethnopharmacol 2000; 71(1-2): 23-43. [http://dx.doi.org/10.1016/S0378-8741(00)00213-0] [PMID: 10904144]

[16] Zafar R, Neerja . *Momordica charantia*--a review. Hamdard Med 1991; 34(3): 49-61. [PMID: 11613982]

[17] Bukhari SA, Farah N, Mustafa G, Mahmood S, Naqvi SAR. Magneto-primingim proved nutraceutical potential and antimicrobial activity of *Momordica charantia* L. without affecting nutritive value. Appl Biochem Biotechnol 2019; 188(3): 878-92. [http://dx.doi.org/10.1007/s12010-019-02955-w] [PMID: 30729394]

[18] Supe U, Daniel P. HPLC method for analysis of bioactive compound from *Momordica charantia.* Am J Agric Environ Sci 2015; 15(11): 2196-200.

[19] Gupta M, Sharma S, Gautam A, Bhaduria R. *Momordica charantia* Linn. (Karela): Nature's silent healer. Int J Pharm Sci Rev Res 2011; 11(1): 32-7.

[20] Wang S, Li Z, Yang G, Ho CT, Li S. *Momordica charantia*: A popular health-promoting vegetable with multifunctionality. Food Funct 2017; 8(5): 1749-62. [http://dx.doi.org/10.1039/C6FO01812B] [PMID: 28474032]

[21] Shafie MH, Samsudin D, Yusof R, Gan C-Y. Characterization of bio-based plastic made from a mixture of *Momordica charantia* bioactive polysaccharide and choline chloride/ glycerol based deep eutectic solvent. Int J Biol Macromol 2018; 118: 1183-92.

[22] Mala M, Hepsibah AH, Jothi GJ. Silver nano particles synthesis using *Coccinia grandis*(L.) Voigt and *Momordica charantia* L, its characterization and biological screening. J Bionanoscience 2017; 11(6): 504-13. [http://dx.doi.org/10.1166/jbns.2017.1480]

[23] Svobodova B, Barros L, Calhelha RC, *et al.* Bioactive properties and phenolic profile of *Momordica charantia* L. medicinal plant growing wild in Trinidad and Tobago. Ind Crops Prod 2017; 95: 365-73. [http://dx.doi.org/10.1016/j.indcrop.2016.10.046]

[24] Zubair MF, Atolani O, Ibrahim SO, *et al.* Chemical and biological evaluations of potent antiseptic cosmetic products obtained from *Momordica charantia* seed oil. Sustain Chem Pharm 2018; 9: 35-41. [http://dx.doi.org/10.1016/j.scp.2018.05.005]

[25] Abalaka ME, Onaolapo JA, Inabo HI, Olonitola OS. Antibacterial activity of chromatographically separated pure fractions of whole plant of *Momordica charantia* L (Cucurbitaceae). Adv Environ Biol 2010; 4(3): 509-14.

[26] Abid M, Chohan S, Mehmood MA, Naz S, Naqvi SAH. Antifungal potential of indigenous medicinal plants against *Myrothecium* leaf spot of bitter gourd (*Momordica charantia* L.). Braz Arch Biol Technol 2018; 60: e17160395.

[27] Swarna J, Ravindhran R. *Agrobacterium rhizogenes* - mediated hairy root induction of *Momordica charantia* Linn. and the detection of charantin, a potent hypoglycemic agent in hairy roots. Res J Biotechnol 2012; 7(4): 227-31.

[28] Wang S, Zheng Y, Xiang F, Li S, Yang G. Antifungal activity of *Momordica charantia* seed extracts toward the pathogenic fungus *Fusarium solani* L. Yao Wu Shi Pin Fen Xi 2016; 24(4): 881-7. [PMID: 28911628]

[29] Peter EL, Kasali FM, Deyno S, *et al. Momordica charantia* L. lowers elevated glycaemia in type 2 diabetes mellitus patients: Systematic review and meta-analysis. J Ethnopharmacol 2019; 231: 311-24. [http://dx.doi.org/10.1016/j.jep.2018.10.033] [PMID: 30385422]

[30] Lucas EA, Dumancas GG, Smith BJ, Clarke SL, Arjmandi BH. Health benefits of bitter melon (*Momordica charantia*). In: Watson RR, Preedy VR, Eds. Bioactive Foods in Promoting Health. San Diego: AcademicPress 2010; pp. 525-49. [http://dx.doi.org/10.1016/B978-0-12-374628-3.00035-9]

[31] Zeng Y, Guan M, Li C, *et al.* Bitter melon (*Momordica charantia*) attenuates atherosclerosis in apo-E knock-out mice possibly through reducing triglyceride and anti-inflammation. Lipids Health Dis 2018; 17(1): 251. [http://dx.doi.org/10.1186/s12944-018-0896-0] [PMID: 30400958]

[32] Palamthodi S, Lele SS. Nutraceutical applications of gourd family vegetables: *Benincasa hispida, Lagenaria siceraria* and *Momordica charantia.* Biomed Prev Nutr 2014; 4(1): 15-21. [http://dx.doi.org/10.1016/j.bionut.2013.03.004]

[33] Ji H, Zhang L, Li J, Yang M, Liu X. Optimization of ultra high pressure extraction of momordicosides from bitter melon. Int J Food Eng 2010; 6(6): 3. [http://dx.doi.org/10.2202/1556-3758.1962]

[34] Nerurkar PV, Lee YK, Nerurkar VR. *Momordica charantia* (bitter melon) inhibits primary human adipocyte differentiation by modulating adipogenic genes. BMC Complement Altern Med 2010; 10(1): 34.
[http://dx.doi.org/10.1186/1472-6882-10-34] [PMID: 20587058]

[35] Upadhyay A, Agrahari P, Singh DK. A review on salient pharmacological features of *Momordica charantia.* Int J Pharmacol 2015; 11(5): 405-13.
[http://dx.doi.org/10.3923/ijp.2015.405.413]

[36] Panlilio BG, Macabeo APG, Knorn M, *et al.* A lanostane aldehyde from *Momordica charantia.* Phytochem Lett 2012; 5(3): 682-4.
[http://dx.doi.org/10.1016/j.phytol.2012.07.006]

[37] Chen JC. Eight new cucurbitane glycosides, kuguaglycosides A-H,from the root of *Momordica charantia* L. Helv Chim Acta 2008; 91(5): 920-9.
[http://dx.doi.org/10.1002/hlca.200890097]

[38] Popovich DG, Li L, Zhang W. Bitter melon (*Momordica charantia*) triterpenoid extract reduces preadipocyte viability, lipid accumulation and adiponectin expression in 3T3-L1 cells. Food Chem Toxicol 2010; 48(6): 1619-26.
[http://dx.doi.org/10.1016/j.fct.2010.03.035] [PMID: 20347917]

[39] Wang S, Zhang Y, Liu H, *et al.* Molecular cloning and functional analysis of a recombinant ribosome-inactivating protein (alpha-momorcharin) from *Momordica charantia.* Appl Microbiol Biotechnol 2012; 96(4): 939-50.
[http://dx.doi.org/10.1007/s00253-012-3886-6] [PMID: 22262229]

[40] Cuong D, Kwon SJ, Jeon J, Park Y, Park J, Park S. Identification and characterization of phenylpropanoid biosynthetic genes and their accumulation in bitter melon (*Momordica charantia*). Molecules 2018; 23(2): 469.
[http://dx.doi.org/10.3390/molecules23020469] [PMID: 29466305]

[41] Thiruvengadam M, Praveen N, Maria John KM, Yang YS, Kim SH, Chung IM. Establishment of *Momordica charantia* hairy root cultures for the production of phenolic compounds and determination of their biological activities. Plant Cell Tissue Organ Cult 2014; 118(3): 545-57.
[http://dx.doi.org/10.1007/s11240-014-0506-4]

[42] Makhija M, Ahuja D, Nandy BC, Gautam S, Tiwari K, Awasthi A, *et al.* Evaluation and comparison of antibacterial activity of leaves, seeds and fruits extract of *Momordica charantia.* Res J Pharm Biol Chem Sci 2011; 2(2): 185-92.

[43] Ajitha B, Reddy YAK, Reddy PS. Biosynthesis of silver nanoparticles using *Momordica charantia* leaf broth: Evaluation of their innate antimicrobial and catalytic activities. J Photochem Photobiol B 2015; 146: 1-9.
[http://dx.doi.org/10.1016/j.jphotobiol.2015.02.017] [PMID: 25771428]

[44] Joseph B, Jini D. Antidiabetic effects of *Momordica charantia* (bitter melon) and its medicinal potency. Asian Pac J Trop Dis 2013; 3(2): 93-102.
[http://dx.doi.org/10.1016/S2222-1808(13)60052-3]

[45] Guevara AP, Lim-Sylianco CY, Dayrit FM, Finch P. Acylglucosyl sterols from *Momordica charantia.* Phytochemistry 1989; 28(6): 1721-4.
[http://dx.doi.org/10.1016/S0031-9422(00)97832-4]

[46] Park H, Moon B, Kim S. Antioxidant and α-glucosidase inhibitory activities of fresh bitter melon and change of charantin and lutein content upon brining and blanching treatments of pickling. Korean J Food Sci Technol 2018; 50(4): 430-6.

[47] Rohajatien U, Harijono H, Estiasih T, Sriwahyuni E. Bitter melon (*Momordica charantia* L) fruit decreased blood glucose level and improved lipid profile of streptozotocin induced hyperglycemia rats. Curr Res Nutr Food Sci 2018; 6(2): 359-70.

[http://dx.doi.org/10.12944/CRNFSJ.6.2.11]

[48] Chan FK, Hsu C, Li TC, Chen WH, Tseng KT, Chao PM. Bitter melon seed oil increases mitochondrial content in gastrocnemius muscle and improves running endurance in sedentary C57BL/6J mice. J Nutr Biochem 2018; 58: 150-7. [http://dx.doi.org/10.1016/j.jnutbio.2018.05.008] [PMID: 29957359]

[49] Saeed MK, Shahzadi I, Ahmad I, Ahma R, Shahzad K, Ashraf M, *et al.* Nutritional analysis and antioxidant activity of bitter gourd (*Momordica charantia*) from Pakistan. Pharmacologyonline 2010; 1: 252-60.

[50] Zhang F, Lin L, Xie J. A mini-review of chemical and biological properties of polysaccharides from *Momordica charantia.* Int J Biol Macromol 2016; 92: 246-53. [http://dx.doi.org/10.1016/j.ijbiomac.2016.06.101] [PMID: 27377459]

[51] Patel S, Patel T, Parmar K, Bhatt Y, Patel Y, Patel NM. Isolation, characterization and antimicrobial activity of charantin from *Momordica charantia* Linn. fruit. Int J Drug Dev Res 2010; 2(3): 629-34.

[52] Desai S, Tatke P. Charantin: An important lead compound from *Momordica charantia* for the treatment of diabetes. J Pharmacogn Phytochem 2015; 3(6): 163-6.

[53] Zhu F, Zhang P, Meng YF, *et al.* Alpha-momorcharin, a RIP produced by bitter melon, enhances defense response in tobacco plants against diverse plant viruses and shows antifungal activity *in vitro.* Planta 2013; 237(1): 77-88. [http://dx.doi.org/10.1007/s00425-012-1746-3] [PMID: 22983699]

[54] Ching-Dong C, Ping-Yuan L, Yo-Chia C, Han-Hsiang H, Wen-Ling S. Novel purification method and antibiotic activity of recombinant *Momordica charantia* MAP 30. 3 Biotech 2017; 7(1): 3.

[55] Thiruvengadam M, Rekha KT, Jayabalan N. An efficient *in vitro* propagation of *Momordica dioica* Roxb. ex Willd. Philipp Agric Sci 2006; 89: 165-71.

[56] Berg AJW, Zhang XP, Rhodes BB. Micropropagation of *Citrullus lanatus* (Thumb.) Matsum and Nakai (WaterMelon). In: Bajaj YPS, Ed. Biotechnology in Agriculture and Forestry High-Tech and Micro Propagation. Berlin, Heidelberg: Springer-Verlag 1997; Vol. 39.

[57] Nabi A, Rashid MM, Al-Amin M, Rasul MG. Organogenesis in teasle gourd (*Momordica dioica* Roxb.). Plant Tissue Cult 2002; 112: 173-80.

[58] Thiruvengadam M, Varisai Mohamed S, Yang CH, Jayabalan N, Chang M. Development of an embryogenic suspension culture of bitter melon (*Momordica charantia* L.). Scientia Horticulturae 2006; 109(2): 123-9.

[59] Murashige T, Skoog F. A revised medium for rapid growth and bioassays with tobacco tissue cultures. Physiol Plant 1962; 15(3): 473-97. [http://dx.doi.org/10.1111/j.1399-3054.1962.tb08052.x]

[60] Thiruvengadam M, Rekha KT, Yang CH, Jayabalan N, Chung I-M. High-frequency shoot regeneration from leaf explants through organogenesis in bitter melon (*Momordica charantia* L.). Plant Biotechnol Rep 2010; 4(4): 321-8. [http://dx.doi.org/10.1007/s11816-010-0151-2]

[61] Gamborg OL, Miller RA, Ojima K. Nutrient requirements of suspension cultures of soybean root cells. Exp Cell Res 1968; 50(1): 151-8. [http://dx.doi.org/10.1016/0014-4827(68)90403-5] [PMID: 5650857]

[62] Thiruvengadam M, Praveen N, Maria KM. Ill–Min Chug. *In vitro* regeneration from internodal explants of bitter melon (*Momordica charantia* L.) *via* indirect organogenesis. Afr J Biotechnol 2012; 11(32): 8218-24.

[63] Manye Y, Zhao M, Lan L, Chen F. Establishment of *in vitro* regeneration system of bitter melon (*Momordica charantia* L.). High Technol Lett 2004; 10(1): 44-8.

[64] Wang GL, Fan HY, Lin RY, Li XT. Establishment of *in vitro* regeneration system of *Momordica*

charantia L. with different genotype. Anhui Nongye Kexue 2008; 36(28): 12125-7.

[65] Pierik RLM. *In vitro* propagation of higher plants. Boston: Martinus Nizhoof Publisher 1987. [http://dx.doi.org/10.1007/978-94-009-3621-8]

[66] Agarwal M, Kamal R. *In vitro* clonal propagation of *Momordica charantia* L. Indian J Biotechnol 2004; 3: 426-30.

[67] Sultanas B, Miah MA. *In vitro* propagation of karalla (*Momordica charantia* L). J Biol Sci 2003; 1134-9.

[68] Hoque A, Islam R, Joarder OI. *In vitro* plantlets differentiation in kakrol (*Momordica dioica* Roxb.). Plant Tissue Cult 1995; 5: 119-24.

[69] Malik S, Zia M, Riaz-ur-Rehman , Chaudhary MF. *In vitro* plant regeneration from direct and indirect organogenesis of *Memordica charantia*. Pak J Biol Sci 2007; 10(22): 4118-22. [http://dx.doi.org/10.3923/pjbs.2007.4118.4122] [PMID: 19090290]

[70] Tang Y, Liu J, Li X, Li J, Li H. Promote adventitious buds induction from stem segments of Bitter melon (*Momordica charantia* L.). J Agric Sci 2011; 3(2).

[71] Saglam S. *In vitro* propagation of Bitter Gourd (*Momorica charantia* L.). Sci Bull Ser F Biotechnol 2017; XXI.

[72] Huda A, Sikdar B. *In vitro* plant production through apical meristem culture of Bitter gourd (*Momorica charantia* L.). Plant Tissue Cult Biotechnol 2006; 16((1): 31-6.

[73] Ma C, Tang Y, Li X, Wang L, Li H. *In vitro* induction of multiple buds from cotyledonary nodes of balsam pear (*Momordica charantia* L.). Afr J Biotechnol 2012; 11(3): 3106-15.

[74] Zhang Y, Zhou J, Wu T, Cao J. Shoot regeneration and the relationship between organogenic capacity and endogenous hormonal contents in pumpkin. Plant Cell Tissue Organ Cult 2008; 93(3): 323-31. [http://dx.doi.org/10.1007/s11240-008-9380-2]

[75] Ntui SVO, Thirukkumaran G. Efficient plant regeneration *via* organogenesis in "Egos" melon (*Colocynthis citrullus* L.). Sci Hortic 2009; 1119: 397-402.

[76] Lin YZ, Lou HY, Zhang ZZ. Induction of multiple buds from cotyledonary nodes of Balsam pear (*Momordica charantia* L.). Redai Zuowu Xuebao 2006; 27(2): 60-3.

[77] Li J, Li M. The System of *in vitro* culture and shoot regeneration from cotyledon of balsam pear of Cuifei. Northern Hort 2007; 110: 181-3.

[78] Wehner TC, Locy RD. *In vitro* adventitious shoot and root formation of cultivars and lines of *Cucumis sativus* L. Hort Sci 1981; 116(6): 759-60.

[79] Munsur MAZA, Haque MS, Nasiruddin KM, Hossain MS. *In vitro* propagation of bitter gourd (*Momordica charantia* L.) from nodal and root segment. Plant Tissue Cult Biotechnol 1970; 19(1): 45-52. [http://dx.doi.org/10.3329/ptcb.v19i1.4916]

[80] Song RLY, Gao F. Changes of endogenous hormones in *Momordica charantia* L. during *in vitro* Culture. Zhiwuxue Tongbao 2006; 23(2): 192-6.

[81] Munsur MAZA, Haque MS, Nasiruddin KM, Hossain MS. *In vitro* propagation of Bitter gourd (*Momordica charantia* L.) from nodal and root segments. Plant Tissue Cult Biotechnol 1970; 19(1): 45-52. [http://dx.doi.org/10.3329/ptcb.v19i1.4916]

[82] Daniel P, Supe U, Roymon MG. A review on phytochemical analysis of *Momordica charantia.* Intern J Pharm Biol Chem 2014; 3(1): 214-20.

[83] Mishra J, Tiwari KL, Jadhav SK. Micro propagation of *Momordica charantia*L. Intern J Biol Health Sci 2012; 1(1): 25-31.

[84] Thiruvengadam M, Praveen N, Maria John KM, Yang YS, Kim SH, Chung I-M. Establishment of *Momordica charantia* hairy root cultures for the production of phenolic compounds and determination of their biological activities. Plant Cell Tissue Organ Cult 2014; 118(3): 545-57. [http://dx.doi.org/10.1007/s11240-014-0506-4]

CHAPTER 10

In Vitro Protocols for Micropropagation of *Catharanthus roseus* (L.) G. Don

Govindugari Vijaya Laxmi[1,*] and **K. Dharmalingam**[1]

[1] *Department of Biotechnology, Chaitanya Bharathi Institute of Technology, Hyderabad, Telangana-500075, India*

Abstract: *Catharanthus roseus* (*C. roseus*) is an important alkaloid-yielding medicinal and ornamental plant belonging to the family Apocynaceae. The genus *Catharanthus* is well studied and reported to contain biologically active terpenoid indole alkaloids (TIAs) with over 130 compounds isolated and identified. It has great medicinal importance in treating various ailments to treat diseases as diabetes, malaria, menorrhagia, Hodgkin's disease, *etc.* In view of the immense importance in the pharmaceutical industry, micropropagation of *C. roseus* has been the best alternative for continuous source of plants and also for *in vitro* production of secondary metabolites. Various explants have been studied for micropropagation; however, nodal explants were the most suitable. For surface sterilization, 0.1% $HgCl_2$ or 70% ethanol, followed by sodium hypochlorite and Bavistin (carbendazim), was optimum to control the microbial contamination. Murashige and Skoog (MS) medium was the most widely used for its success rate. 2,4-D for callus initiation and BAP, along with zeatin and activated charcoal, were reported to be promising for regeneration of plantlets. The 100% acclimatization of plantlets on transfer to field depends on the soil mixture and environmental conditions and humidity in the initial stages of transfer from *in vitro* cultures.

Keywords: Acclimatization, *Catharanthus roseus*, *In vitro* Studies, Micropropagation, Organogenesis, Plant Regeneration, Somatic Embryogenesis.

INTRODUCTION

Catharanthus roseus (L.) G. Don, commonly known as Madagascar periwinkle, is an important medicinal plant. It belongs to the family Apocynaceae. It contains several commercially valuable secondary metabolites, making it the most demanding medicinal plant. The secondary metabolites are used in the treatment of various ailments and disorders like Hodkin's disease, lymphoblastic leukaemia, breast and skin cancer and cancerous tumours [1]. The periwinkle has been repor-

* **Corresponding author Govindugari Vijaya Laxmi:** Department of Biotechnology, Chaitanya Bharathi Institute of Technology, Hyderabad, Telangana-500075, India; Tel: +91-040-24193276; E-mail: drgvlaxmi_biotech@cbit.ac.in

T. Pullaiah (Ed.)

ted to contain a good source of commercial bioactive alkaloids, including vinblastine and vincristine, which have anti-cancer activities. *C. roseus* plant has been reported to contain several other important bioactive compounds, such as anthocyanins, flavanol glycosides, phenolic acids, saponins, steroids, and terpenoids, that exhibit antidiarrheal, antidiabetic, anti-hypoglycaemic, antimicrobial, wound healing, and antioxidant activities, respectively [1, 2].

The leaves of *C. roseus* contain secondary metabolites (vindoline, vinblastine, catharanthine, vincristine, *etc.*), whereas in stem and roots, ajmalicine, reserpine, serpentine, horhammericine, taborsonine, *etc.*, are present. Two alkaloids vincristine and vinblastine possess anticancer properties. Hence, it is used immensely in the pharmaceutical industry. The alkaloids of *C. roseus* comprise a group of about 130 terpenoid indole alkaloids (TIA). Vinblastine has now marketed for more than 40 years as an anticancer drug and has become a true lead compound for drug development [2]. Due to its low volume with high value in pharmaceutically important drugs, *in vitro* studies are the best alternatives for the production of these alkaloids. The conventional methods of propagation through seed germination were not encouraged due to its low germination rate (30%) and low vigour [3]. Plant tissue culture technology is a promising method to produce true-to-type plants without destroying the plant. Many researchers tried and tested various plant tissue culture techniques for clonal propagation on a large scale and also to improve the alkaloid content of *C. roseus*.

IN VITRO STUDIES IN *CATHARANTHUS ROSEUS*

The Explants

A large number of plants can be developed in a short period in a small space under controlled and aseptic conditions *via* micropropagation. The selection of explant plays an important role in plant tissue culture. Various explants of *C. roseus* have been used, such as nodal segment, axillary bud, shoot tip or apical bud, leaf, stem, anther, petiole, root, *etc*. The choice of explant is based on various criteria such as availability of material, response and objective [4]. With respect to optimum production of shoots and roots *in vitro*, the nodal segment (node) was most responsive [5, 6]. Apical meristem or shoot tip consisting of apical or axillary buds also proved to be a quick responsive explant for direct organogenesis [7]. Verma and Mathur [8] produced adventitious shoot buds and roots using *in vitro* grown leaf explants. Studies reported the use of hypocotyl, anthers and zygotic embryos as explants for callus induction and regeneration produced embryogenic callus from anthers [9, 10], however, hypocotyl explant was reported to be promising for somatic embryogenesis [11]. Leaf petiole when used as an explant produced callus and roots [12]. Epicotyl with increased vincristine with shooty teratoma was reported by Begam *et al.* [13]. Due to availability of lateral meristem, the nodal segments may be the best explants for mass multiplication through direct organogenesis [13, 14].

Surface Sterilization

The selection of surface sterilants is another crucial step for the establishment of aseptic cultures *in vitro*. It is a well-known fact that the concentration and the duration of treatment with sterilant vary for different species and explant to explant besides the load of contamination in the explant. The use of 70% (v/v) ethanol wash in *C. roseus* for 30 seconds to 1 minute, along with other sterilants, was found to be very common for surface sterilization of explants. However, some researchers reported to use higher concentration of ethanol (75-90%) and up to 2 minutes with 70% ethanol [15]. Ethanol wash is followed by treatment with commercial bleach or sodium hypochlorite (0.1–25%), and the duration of treatment was 5–45 min and sometimes along with a few drops of liquid detergent Tween-20 (polysorbate 20) or Tween-80 (polysorbate 80) or Triton-X was also reported.

The explants of *C. roseus* were also treated with mercuric chloride. Even though mercuric chloride is toxic to plants, it has been used by researchers in the range of 0.04–0.5% (w/v) for 2–5 min, maximum up to 15 min [16]. Besides, the explants were also treated with hydrogen peroxide, Labolene (Qualigen- Fisher, LR grade), Teepol (Reckitt Benckiser Pvt. Ltd., India), Dettol, *etc* [6]. In order to control contamination further with fungal species, researchers also treated the explants with fungicide Bavistin (Crystal crop protection Ltd., India) and/or antibiotic solution (Cefotaxime or Streptomycin). To summarize, the explants treated with multiple surface sterilants starting from 70% ethanol treatment followed by NaOCl along with detergent, Bavistin, mercuric chloride and sometimes 5% H_2O_2 were found to be more effective in preventing and controlling microbial contamination in *C. roseus* cultures [17].

Culture Medium

The selection of a suitable basal medium depends upon the objective and the type of plant species used in the experiment [18, 19]. Murashige and Skoog (MS) medium [20] is the most commonly used basal medium to this day and is reported to be also responsive to culturing of *C. roseus*. For organogenesis, a full-strength MS medium was utilized, reported to provide the required nutrients for shoot bud induction and shoot proliferation (Table **1**). Half strength MS medium was reported to be used for root induction in *C. roseus* [13, 14]. Other media such as woody plant medium (WPM) [19], Gamborg's B5 medium [21], and liquid Linsmaier and Skoog [22] medium were also tested.

Table 1. Factors influencing multiple shoot induction through direct organogenesis in *Catharanthus roseus*.

Explant	Sterilants Used	Media	Hormones (mg/l)	Additives (mg/l)	Solidifying Agent	Culture Conditions/Temp/Light/Photo Period	Response	References
IV-S	-	MS	0.05 NAA + 2.5 Kn	-	1% agar	25 °C, 1500, 12 h	Multiple shoots	[33]
Node	1% Teepol; 70% EtOH; 0.12% $HgCl_2$	MA	2 Kn	-	-	25 ± 2 °C, 16 h	Multiple shoots	[34]
Shoot segment	79% EtOH; 1% NaOCl (v/v) 20 min	MS	7 BA + 1 NAA	-	0.6% agar	26 ± 0.5 °C	Multiple shoots	[36]
Node	70% EtOH 30s; 7% (w/v) NaOCl 30 min	½ MS	1 BA 0.5 BA + 0.125 NAA	-	0.7% agar	25 ± 2 °C; 1000; 16 h	Multiple shoots Rooting	[37]
IV-S	70% EtOH; 5% NaOCl	MS	1BA	-	-	25 ± 2 °C, 16 h	Multiple shoots	[38]
Internode	70% EtOH-5 min; 20% NaOCl-15 min	WPM	5 µM BA +5 µM NAA	-	-	21 °C; 16 h	Multiple shoots	[39]
Nodal segment	0.5% cetrimide: 10 min; 1% Bavistin + Streptomycin sulfate: 30 min; 0.1% $HgCl_2$: 5 min; 70% EtOH	MS	0.1 NAA + 1 BA	500mg/l CH	0.62% agar	25 ± 2 °C; 16 h	Multiple shoots	[17]
Axillary bud	Teepol: 2 min; 70% EtOH: 1 min; 0.1% $HgCl_2$: 5 min	MS	4 NAA + 4 BA	-	0.8% agar	25±2 °C; 16 h	Shoot induction	[3]
Nodal segment	Tween80 + 0.4% NaOCl 25 min; 0.04% $HgCl_2$ 5 min+ Tween 20, 8 min	MS	5 µM BA +5 µM NAA	-	-	25±2 °C; 14 h	Multiple shoots	[14]
In vitro leaf	-	½ MS	7 BA + 3 NAA	-	0.4% phytagel	24 ± 2 °C; 3000; 16 h; NM	Adventitious shoots	[8]
Apical bud	Drops of Tween 20+1000ppm cefotaxim,10 min; 1000 ppm bavistin 10 min; 0.1% $HgCl_2$ 15 min	MS	1 BA + 0.2 NAA	0.25 g/l charcoal	-	25 ± 2 °C; 16 h	Multiple shoot	[30]
Nodal segment	Tween 20; 0.1% $HgCl_2$ 5 min	MS	0.5 BA + 1 NAA 5I BA	-	0.8% agar	25 ± 2 °C; 2000–2500; 16 h	Multiple shoot Roots	[31]
Nodal segment	0.1% $HgCl_2$	MS	6 BA+ 6 Kn 10 IBA	-	0.8–1.0% agar	25–28 °C; 16h	Multiple shoot Roots	[16]

(Table 1) cont.....

Explant	Sterilants Used	Media	Hormones (mg/l)	Additives (mg/l)	Solidifying Agent	Culture Conditions/Temp/Light/Photo Period	Response	References
Shoot tip	-	MS	2 BA + 0.2 NAA	10 thiamine HCl, 1 glycine	0.8% agar	25 ± 1 °C; 1000; 16 h;	Multiple shoot	[15]
In vitro nodal segment	70% EtOH 30s; 1% NaOCl 15 min	MS	0.5 BA+1NAA	250 tryptophan	0.8%	-	Multiple shoots	[5]
Nodal segment	Labolin; 0.1% $HgCl_2$ 15 min	MS	0.5 BA +1 NAA	-	0.8–1.0% agar	25 ± 2 °C; 2000–2500;16h	Multiple shoot	[13]
Nodal segment	H_2O_2 2 min; 2% Bavistin+3 drops Tween20-20 min; 0.1% $HgCl_2$ 2 min	MS	3 µM BA + 3 µM Kn	0.1 µm $AgNO_3$	7% agar	25 ± 1 °C; 12 h light/dark	Multiple shoot	[6]
In vitro axillary shoot	70% EtOH 30s; 0.1% $HgCl_2$ 2 min	MS	1 BA + 0.1 NAA	0.38 thymine HCl	0.8%	-	-	[7]
Nodal explants	0.1% Tween 20, (v/v) 12 min; EtOH 70% (v/v) 30s; 5% (v/v) + 6 drops of Tween 20-15 min; 70% EtOH 60 s	MS	2 µM TDZ+ 1 µM NAA	-	0.8% (w/v) plant agar	23 ± 1 °C 16/8 light/dark	Multiple shoots	[40]
Apical node	Labklin 5 drops in 100 ml 15 min; 1%Bavistin+1% Blitox -10 min; 0.1% $HgCl_2$ 3 min	MS	3 BAP+ 1 NAA 0.5 IBA	-	0.8% agar	25 ± 2 °C; 16/8 light/dark; 2000-3000 lux	Multiple shoots	[41]

EtOH: Ethanol; BA or BAP: 6-Benzylaminopurine; Kn: Kinetin; CH: Casein hydrolysate; NAA: α-naphthalene acetic acid; IBA: Indole-3-butyric acid; TDZ: Thidiazuron; WPM: Woody Plant Medium; -: not mentioned.

Carbon Source

The most commonly used sugar in the culture medium is sucrose. Most of the studies reported to use 3% sucrose as a carbohydrate source and an increase in callus biomass with 6% sucrose [23]. Other sugars like glucose and maltose were also used as carbon sources [24].

Physical Factors

The growth response to the cultured explant also depends on physical factors like temperature, relative humidity, photoperiod and light.

Temperature

The majority of the studies on *C. roseus* reported a set of a temperatures ranging from 23 to 28 °C optimum for uniform growth. However, a range of 20 to 22 °C and 27°C for micropropagation and 35°C for callus induction were also reported [25].

Relative Humidity

Variations in relative humidity (RH) in the culture room were reported to be detrimental to plantlets. An RH of 50-60% and a few studies with 80% have been reported [26].

Photoperiod

The duration of light to which a plant is exposed in 24 h has been studied by various labs. The majority of the studies on photoperiod reported a 16 h light and 8 h dark regime to be optimum to maintain the physiological response of *C. roseus in vitro.* A 12 h light and 12 h dark condition has also been reported [25].

Light Intensity

The quantity of light intensity focussed on cultures is an important parameter in *in vitro* studies. It is measured in lux or photosynthetic photon flux density (PPFD). A good light intensity was reported to maintain proper growth and development. Hence, a light intensity of 2000-3000 lux was reported by majority of studies. Some studies reported explants were initially cultured under dark conditions [27].

Plantlet Regeneration *in vitro*

Plant regeneration in *C. roseus* was reported *via* organogenesis and somatic embryogenesis. In organogenic pathway, the shoots or roots were observed either direct or indirect *via* the caulogenesis (*via* callus) phase.

Direct Organogenesis

In direct organogenesis, the shoots or roots are produced from the explant without the callus phase or caulogenesis. Direct or indirect regeneration of the explant in the medium is decided by the threshold concentrations of auxins and cytokinins within the explant and those added in the media. The available literature on *C. roseus* studies indicates a combination of cytokinin along with auxin known to effectively promote shoot induction and its proliferation. The nodal segments, shoot tips, and axillary buds from *C. roseus* seedlings and mature plants produced direct multiple shoot regeneration *in vitro* [6 - 8, 28 - 32]. Studies in *C. roseus* reported the use of a high proportion of the cytokinin to auxin ratio for the optimum number of shoots per explant [7, 8, 17, 30 - 34]. The overall studies on direct organogenesis report that Benzyl amino purine (BA) with Naphthalene acetic acid (NAA) was a promising combination. Mohammed *et al.* [35] reported BAP (0.5 mg/l) + NAA (1.0 mg/l) + 3% activated charcoal was very optimum and produced 35.10 ± 0.74 shoots per explant. The shoots were rooted in half or full-strength media, supplemented with IBA. Kumar *et al.* [30] reported rooting on MS media without hormones, *i.e.*, Basal media supplemented with 0.25 g/l activated charcoal (Table **1**).

Indirect Organogenesis

The plant regeneration occurs *via* the callus phase. The concentration of cytokinin to auxin leads to caulogenesis followed by morphogenesis of the callus. Previous studies in *C. roseus* indicate that auxin, in combination with cytokinin, in lower doses, induces callus from the explant. NAA or 2,4-D was commonly used for callus induction. NAA, along with kinetin (Kn), proved to be best for callus induction, as reported by Kim *et al.* [10] and several other researchers [24, 42, 43]. Some of the studies reported to use IAA in combination with NAA and Kn for caulogenesis [44 - 47]. 2,4-D (2,4-Dichlorophenoxyacetic acid) along with NAA + Kn, reported to produce 95% callogenesis. The callus induction from *in vitro* grown seedling- hypocotyl callused on NAA and BA supplemented media, and the callus regenerated into shoots with Zeatin, NAA and BAP in the regeneration media [48]. The use of auxins like dicamba or picloram was not reported in *C. roseus* for indirect regeneration studies. Therefore, callus induction was reported to be optimum in *C. roseus* when 2,4-D was used along with NAA and Kn. BAP promoted shoot induction and the IBA or NAA for rooting of the shoots (Table **2**).

Table 2. Factors influencing callus induction and plant regeneration in *C. roseus*.

Explant	Sterilants used	Media /carbohydrates	Hormones (mg/l)	Additives (mg/l)	Solidifying agent	Culture conditions/Temp/light /photoperiod	Response	References
Anther	-	MS	1 NAA + 0.1 Kn	0.4 mg/l Thiamine	0.4% gelrite	25 °C; 16 h;	Somatic embryos	[10]
In vitro-hypocotyl	1% cetrimide 5 min; 10% NaOCl: 20 min; 0.1% $HgCl_2$: 5 min; 70% EtOH:3 min	MS / 3% sucrose	2 NAA + 5BA; 0.1 NAA+5 BA + 1 Zeatin	1000 CH + 100 asparagine 100 asparagine + 100 glutamine	0.6% agar	25 ± 2°C; 1045; 10 h; 55%	Callus and Shoot regeneration	[48]

(Table 2) cont.....

Explant	Sterilants used	Media /carbohydrates	Hormones (mg/l)	Additives (mg/l)	Solidifying agent	Culture conditions/Temp/light /photoperiod	Response	References
Leaf	25% NaOCl: 5 min	MS/ 3% sucrose	1 NAA+ 0.1 Kn	-	0.8% agar	25 °C; 12 hr	Callus initiation and proliferation	[43]
Seedling segment	-	B5/ 3% sucrose	5.88 µM 2,4-D + 1.16 µM Kn + 1.34 µM NAA	-	0.8% agar	-	95% callusing	[49]
Leaf	-	MS/ 3% sucrose	1 2,4-D+ 1 IAA+ 0.5 Kn	1100 KNO_3 + 2 thiamine HCL + 0.1 riboflavin + 0.1 biotin + 0.1 folic acid	-	23±2 °C; dark ;	Callus induction	[50]
Stem and leaf	-	MS/ 4% sucrose	1 NAA + 1 IAA+ 0.5 Kn	-	-	23±2 °C; dark	Callus induction	[51]
Stem and leaf	-	MS/ 3% sucrose	5.37 µM NAA + 4.65 µM Kn	1100 KNO_3 + 2 thiamine HCL + 0.1 riboflavin + 0.1 biotin + 0.1 folic acid	-	23±2 °C; dark /45min/day fluorescent light	Compact callus induction	[52]
stem	-	MS/ 3% sucrose	5.37 µM NAA+4.65 µM Kn	1100 KNO_3 + 2 thiamine HCL+ 0.1 riboflavin + 0.1 biotin + 0.1 folic acid	-	23±2 °C; dark	Compact callus induction	[53]
Young stem	-	MS/3% sucrose	2 NAA + 2 IAA + 0.1 Kn	-	-	25 °C dark	Callus induction	[46]
Young stem	-	MS/3% sucrose	2 NAA + 2 IAA+ 0.1 Kn	-	-	25 °C dark	Callus induction	[47]
In vitro hypocotyl	70% EtOH 0.5% $HgCl_2$: 2 min; 5% H_2O_2	MS	1.5BA +1NAA	-	-	25±2 °C; 16h	Embryogenic callus induction	[54]

(Table 2) cont.....

Explant	Sterilants used	Media /carbohydrates	Hormones (mg/l)	Additives (mg/l)	Solidifying agent	Culture conditions/Temp/light /photoperiod	Response	References
Leaf petiole	5% NaOCl: 45 min	MS	0.1 NAA+0.1 Kn	-	-	35 °C; dark	Callus and root	[55]
Leaf	0.5% Bavistin: 20 min; 0.01% Tween 20: 20 min; 0.1% $HgCl_2$: 5 min	MS /3% sucrose	2 NAA + 0.2 Kn	-	-	25±2 °C; 16 h light	Callus induction	[42]
In vitro leaf	Labolene; savlon; 0.1% $HgCl_2$: 1 min	MS	1 2,4-D + 0.5 BA	-	0.8% agar	25°C; 1200; 12 h	Callus	[56]
Hypocotyl	Tap water + few drops of Teepol; 0.4% Bavistin 30 min	MS	1 BA + 1 NAA 1.5 BA+ 1 NAA+ 1.5 BA + 1 NAA	-	0.8% agar	26°C; dark; 16 hr	Callus Multiple shoots Rooting	[57]
Leaf	0.1% HgCl2:	MS	4 2,4-D 6 BA 10 IBA	-	0.8-1% agar	25-28°C /100/16h	Callus Buds Multiple shoots Rooting	[16]
Nodal segment	Liq. Detergent 10-20 min; 1% Bavistin 15-20min; 0.1% NaOCl 5-7 min	MS / 3% sucrose	2NAA+ 1 Kn	10 Thiamine	0.7% agar	27±2°C; 16h	callus	[24]

EtOH: Ethanol; BA or BAP: 6-Benzylaminopurine; Kn: Kinetin; 2,4-D: 2,4-Dichlorophenoxyacetic acid; CH: casein hydrolysate; NAA: α-naphthalene acetic acid; IBA: Indole-3-butyric acid; TDZ: Thidiazuron; GA3: Gibberellic acid; -: not mentioned.

Somatic Embryogenesis

The embryos are obtained *in vitro* from somatic tissues in a unique process in somatic embryogenesis. Plant regeneration *via* somatic embryogenesis is a desirable pathway. Studies from various laboratories have been conducted to develop somatic embryogenesis in *C. roseus*. Various factors that influence the production of somatic embryos from vegetative cells include genotype, explant types, the carbohydrates used, the growth conditions, the type of plant growth regulators *etc*. The use of lower concentrations of 2,4-D alone promoted embryogenesis [58, 59]. Some have been reported to produce somatic embryogenesis on media supplemented with BAP, NAA, GA_3 [60] and TDZ [61]. Most of the researchers reported to initiate and proliferate somatic embryos using BAP and NAA [62 - 66]. The use of triacontanol (TRIA) along with 2,4-D for callus induction and abscisic acid (ABA) for embryo maturation has been reported [27]. However, successful studies on direct embryogenesis have to be carried out (Table **3**).

Table 3. Factors influencing Somatic embryogenesis in *C roseus*.

Explant	Sterilants Used	Media	Hormones (mg/l)	Additives	Solidifying Agent	Culture Conditions/Temp./Light/Photo Period	Response	References
Anther	-	MS	1 NAA + 0.1 Kn	0.4 mg/l Thiamine	0.4% gelrite	25 °C; 16 h;	Embryogenic callus	[10]
Immature Zygotic embryos	0.4% NaOCl; 10 min	MS/ 3% sucrose	-	0.4 mg/l Thiamine	0.4% gelrite	25 °C; 3 W/m2; 16 h;	20% Somatic Embryos (SE)	[58]
In vitro hypocotyl	-	MS	1 2,4-D 1 NAA 1 GA_3 0.5 BA	-	-	20 °C dark 25°C ;16 h.	Embryogenic callus; proliferation, maturation, germination	[59]
In vitro hypocotyl	70% EtOH; 0.5% $HgCl_2$: 2 min; 5% H_2O_2	MS	1-2 2,4-D, NAA or CPA	-	-	25 ± 2 °C; 16 h	Somatic embryos	[62]
In vitro hypocotyl	70% EtOH 2 min; 0.5% $HgCl_2$: 2 min; 5% H_2O_2	MS/3% sucrose	6.78 μM 2,4-D+ 2.24 μM BA + 1.53 μM NAA 2.6 μM GA_3 2.25 μM BA	-	0.8% agar	27 ± 2 °C; 16h	Embryogenic callus, proliferation, maturation, germination	[63]
Mature zygotic embryos	70% EtOH 2 min; 1.5% sodium hypochlorite:10 min	MS+ B5 vitamins	7.5 μM of thidiazuron (TDZ)	-	4 g/l gelrite	25 ± 1°C	Somatic embryos	[61]
Hypocotyl	0.1% $HgCl_2$. 5 min; tween 20; 5 min	MS /3% sucrose	0.45 2,4-D+6.62 BA+1.44 GA3 5.37 NAA+ 5.71 IAA	-	0.8% agar	25 ± 2°C; 16h; 50-60%	Somatic embryos Root induction	[60]
In vitro hypocotyl	70% EtOH 2 min; 0.5% HgCl2: 2 min; 5% H_2O_2	MS/3% sucrose	1 2,4-D + 1 NAA + 0.5 BA	-	-	25 ± 2 °C; 16 h	SE Initiation, proliferation, maturation	[66]
In vitro hypocotyl	75% EtOH; 2 min; 5.25% NaOCl:10 min	MS/ 3% sucrose	1 2,4-D + 1 NAA + 0.1 Zeatin 5 BA+ 0.5 NAA	-	0.3% gelrite	25 ± 2 °C; 16h	Callus Embryogenic callus	[67]

(Table 3) cont.....

Explant	Sterilants Used	Media	Hormones (mg/l)	Additives	Solidifying Agent	Culture Conditions/Temp./Light/Photo Period	Response	References
Shoot tip	0/1% streptomycin: 20s; 70% EtOH 50s; 0.1% $HgCl_2$ 2min	MS /3% ½ MS	2 2,4-D+ TRIA 5 ABA	1 g/l CH+ 0.5 g/l glutamine+250 mg/l peptone + 0.2 g/l PABA+ 0.1 g/l Biotin.	0.7% agar; 0.8% agar; 0.7% agar	25±3°C; dark; 55-60%	Callus SE maturation, SE germination	[27]
IV-hypocotyl	-	MS	2 mg/l 2,4-D; sub cultured: 2.0 mg/l BAP+ 1.5 mg/l NAA; finally, to 2.0 mg/l GA_3	-	-	-	Green mature embryos	[68]

CPA: Chlorophenoxyacetic acid ; EtOH: Ethanol; BA or BAP: 6-Benzylaminopurine;KN: Kinetin; CH: casein hydrolysate; NAA: α-naphthalene acetic acid; IBA: Indole-3-butyric acid; TDZ: Thidiazuron; GA3: Gibberellic acid; ABA: Abscisic acid; -: not mentioned.

Media Additives

The addition of certain media additives in the form of vitamins, amino acids, proteins, natural juices, inorganic salts, *etc*., is reported to enhance the response *in vitro*. Casein hydrolysate was added to *in vitro* cultures in various concentrations ranging from 1 mg/l to 500 mg/l [36, 17]. An improved callus induction was reported with the addition of vitamins like biotin, folic acid, riboflavin, thiamine HCl, and inorganic salt KNO_3 to the MS medium supplemented with NAA and Kinetin [45, 69]. The addition of an anti-ethylene compound, $AgNO_3$, was reported to induce precocious flowering with multiple shoots by Panigrahi *et al.* [6]. Studies on *C. roseus in vitro* reported to add amino acids such as tryptophan, glycine, asparagine and glutamine in media. The addition of tryptophan in a concentration of 250 mg/l for shoot induction and 350 mg/l for root induction was reported [5]. For efficient callus induction, the addition of 1000 mg/l CH, 100 mg/l asparagine and 100 mg/l asparagine with 100 mg/l glutamine was used for shoot regeneration from callus [48].

Acclimatization

After successful plant regeneration *in vitro,* the plantlets with established root systems are transferred to a growth chamber for acclimatization to withstand the external environment. The success rate of acclimatization depends on factors like the type of soil mixture, temperature and humidity. A standardized acclimatization protocol has not been defined for *C. roseus,* but a variety of compositions, including vermiculite and peatmoss potting mixture along with farm soil or sand, has been reported. The acclimatization studies were reported with a survival rate of 60-100% using peat soil with a humidity of 98% [43, 54, 55]. Some researchers reported the acclimatization of plantlets in a mixture of sterile garden soil, sand and vermiculite in equal proportions [3]. Few studies reported to acclimatize the plantlets using only soil [16] or a combination of sterile soil rite initially and followed by transfer to soil rite: sand (1:1) and subsequently to normal soil resulted in 100% acclimatization [14, 30, 62, 63].

Secondary Metabolites

C. roseus is reported to contain a large number of terpenoid indole alkaloids. The secondary metabolites such as ajmalicine, catharanthine, horhammericine, leurosine, lochnerine, reserpine, serpentine, vindoline, vinblastine, vincristine and tabersonine were some of the medicinally important alkaloids. Various studies have been carried out in enhancing the secondary metabolites at elevated quantities *in vitro*. The cultures were subjected to various types of biotic and abiotic elicitors for enhanced production ajmalicine and catharanthine. Biotic elements like cerium and neodymium and abiotic elements: mannitol, KCl, sodium alginate, methyl jasmonate, copper, and zinc were studied. The studies were reported to demonstrate an increase in 3-10-fold production of the alkaloids *in vitro* using abiotic elicitors [38, 69, 71]. Another best alternative for enhanced production of important alkaloids was reported to be the induction of hairy roots. The explants such as leaf, shoot or hypocotyl of *in vitro* grown plantlets were infected with soil borne gram negative bacteria *Agrobacterium rhizogenes.* The t-DNA of its plasmid encodes genes for auxin and cytokinin, and on integration in the host plant, it induces hairy roots at the site of infection [70, 72, 73]. Goklany *et al.* [74] reported hairy root cultures elicited with a range of Methyl Jasmonate (MJ) dosages (0-1,000 μM). The terpenoid indole alkaloids production of secologanin, strictosidine, and tabersonine was investigated in *C. roseus* [74]. An increase of 150-370% of TIA was observed with the addition of 250 μM of MJ [64]. Some researchers reported a combination of malate and alginate treatment of *C. roseus* cultures for indole alkaloid production *via* the jasmonate pathway [28]. Production of catharanthine and ajmalicine using bioreactors was successful.

About 17 monomeric indole alkaloids were isolated, including vindoline, ajmalicine, lochnericine and tabersonine [75, 76].

CONCLUSION

C. roseus is one of the important medicinal plants producing secondary metabolites like ajmalicine, vincristine, and vinblastine. These alkaloids are most effectively used in treating cancer, hypertension and many more ailments. The production of these TIAs in wild-type plants and through chemical synthesis is not viable due to their limitations with respect to low content of alkaloids and chemical synthesis on a large scale, respectively [77]. Hence, the use of biotechnological investigations is reported to be the best approach in the production of TIAs *in vitro*. Protocols for somatic embryogenesis and direct and indirect regeneration of microclones have been optimised in various laboratories. In view of the huge requirement for safe plant-based pharmaceuticals, the development of efficient plant micropropagation techniques and methods for large-scale *in vitro* root cultivation is of great importance and desirable.

REFERENCES

[1] Greenwell M, Rahman PK. Medicinal plants: their use in anticancer treatment. Int J Pharmaceut Sci Res. 2015; 6(10):4103.

[2] Heijden R, Jacobs D, Snoeijer W, Hallard D, Verpoorte R. The Catharanthus alkaloids: Pharmacognosy and biotechnology. Curr Med Chem 2004; 11(5): 607-28. [http://dx.doi.org/10.2174/0929867043455846] [PMID: 15032608]

[3] Bakrudeen A, Subha Shanthi G, Gouthaman T, Kavitha M, Rao M. *In vitro* micropropagation of *Catharanthus roseus* : An anticancer medicinal plant. Acta Bot Hung 2011; 53(1-2): 197-209. [http://dx.doi.org/10.1556/ABot.53.2011.1-2.20]

[4] Salma U, Kundu S, Gantait S. Conserving biodiversity of a potent anticancer plant, *Catharanthus roseus* through *in vitro* biotechnological intercessions: Substantial progress and imminent prospects. Anticancer plants: natural products and biotechnological implements. Singapore: Springer 2018; pp. 83-107. [http://dx.doi.org/10.1007/978-981-10-8064-7_5]

[5] Rahmatzadeh S, Khara J, Kazemitabar SK. The study of *in vitro* regeneration and growth parameters in *Catharanthus roseus* L. under application of tryptophan. J Sci Kharazmi Univ 2014; 14(3): 249-60.

[6] Panigrahi J, Dholu P, Shah TJ, Gantait S. Silver nitrate-induced *in vitro* shoot multiplication and precocious flowering in *Catharanthus roseus* (L.) G. Don, a rich source of terpenoid indole alkaloids. Plant Cell Tissue Organ Cult 2018; 132(3): 579-84. [http://dx.doi.org/10.1007/s11240-017-1351-z]

[7] Sharma A, Mathur AK, Ganpathy J, Joshi B, Patel P. Effect of abiotic elicitation and pathway precursors feeding over terpenoid indole alkaloids production in multiple shoot and callus cultures of *Catharanthus roseus.* Biologia 2019; 74(5): 543-53. [http://dx.doi.org/10.2478/s11756-019-00202-5]

[8] Verma P, Mathur AK. Direct shoot bud organogenesis and plant regeneration from pre-plasmolysed leaf explants in *Catharanthus roseus.* Plant Cell Tissue Organ Cult 2011; 106(3): 401-8. [http://dx.doi.org/10.1007/s11240-011-9936-4]

[9] Tonk D, Mujib A, Maqsood M, Ali M, Zafar N. *Aspergillus flavus* fungus elicitation improves vincristine and vinblastine yield by augmenting callus biomass growth in *Catharanthus roseus.* Plant Cell Tissue Organ Cult 2016; 126(2): 291-303. [http://dx.doi.org/10.1007/s11240-016-0998-1]

[10] Kim S, Song N, Jung K, Kwak S, Liu J. High frequency plant regeneration from anther-derived cell suspension cultures *via* somatic embryogenesis in *Catharanthus roseus.* Plant Cell Rep 1994; 13(6): 319-22. [http://dx.doi.org/10.1007/BF00232629] [PMID: 24193829]

[11] Maqsood M, Mujib A, Siddiqui ZH. Synthetic seed development and conversion to plantlet in *Catharanthus roseus* (L.) G. Don. Biotechnology 2011; 11(1): 37-43. [http://dx.doi.org/10.3923/biotech.2012.37.43]

[12] Rahmatzadeh S, Khara J, Kazemitabar SK. The study of *in vitro* regeneration and growth parameters in *Catharanthus roseus* L. under application of tryptophan. J Sci Kharazmi Univ. 2014; 14(3):249-60.

[13] Begum T, Mathur M. *In vitro* regeneration of *Catharanthus roseus* and *Bacopa monnieri* and their survey around Jaipur district. Int J Pure App Biosci 2014; 2(4): 210-21.

[14] Pati PK, Kaur J, Singh P. A liquid culture system for shoot proliferation and analysis of pharmaceutically active constituents of *Catharanthus roseus* (L.) G. Don. Plant Cell Tissue Organ Cult 2011; 105(3): 299-307. [http://dx.doi.org/10.1007/s11240-010-9868-4]

[15] Al-Oubaidi HK, Mohammed-Ameen AS. Effect of benzyl adenine on multiplication of *Catharanthus roseus* L. *in vitro*. World J Pharm Pharm Sci 2014; 3(6): 2101-7.

[16] Rajora RK, Sharma NK, Sharma V. Effect of plant growth regulators on micropropagation of *Catharanthus roseus*. Int J Adv Biotechnol Res 2013; 4(1): 123-30.

[17] Srivastava T, Das S, Sopory SK, Srivastava PS. A reliable protocol for transformation of *Catharanthus roseus* through *Agrobacterium tumefaciens*. Physiol Mol Biol Plants 2009; 15(1): 93-8. [http://dx.doi.org/10.1007/s12298-009-0010-1] [PMID: 23572917]

[18] Mitra M, Gantait S, Mandal N. *Coleus forskohlii*: Advancements and prospects of *in vitro* biotechnology. Appl Microbiol Biotechnol 2020; 104(6): 2359-71. [http://dx.doi.org/10.1007/s00253-020-10377-6] [PMID: 31989223]

[19] McCown BH. Woody Plant Medium (WPM) : A mineral nutrient formulation for microculture for woody plant species. HortScience 1981; 16: 453.

[20] Murashige T, Skoog F. A revised medium for rapid growth and bio assays with tobacco tissue cultures. Physiol Plant 1962; 15(3): 473-97. [http://dx.doi.org/10.1111/j.1399-3054.1962.tb08052.x]

[21] Gamborg OL, Miller RA, Ojima K. Nutrient requirements of suspension cultures of soybean root cells. Exp Cell Res 1968; 50(1): 151-8. [http://dx.doi.org/10.1016/0014-4827(68)90403-5] [PMID: 5650857]

[22] Linsmaier EM, Skoog F. Organic growth factor requirements of tobacco tissue cultures. Physiol Plant 1965; 18(1): 100-27. [http://dx.doi.org/10.1111/j.1399-3054.1965.tb06874.x]

[23] Verma P, Mathur AK, Shanker K. Growth, alkaloid production, rol genes integration, bioreactor up-scaling and plant regeneration studies in hairy root lines of *Catharanthus roseus*. Plant Biosystems 2012; 146(sup1): 27-40.

[24] Sandhya M, Deepti L, Bhakti D, *et al.* Effect of growth regulator combination on in-vitro regeneration of *Catharanthus roseus*. Int J Life Sci 2016; 6: 1-4.

[25] Das A, Sarkar S, Bhattacharyya S, Gantait S. Biotechnological advancements in *Catharanthus roseus* (L.) G. Don. Appl Microbiol Biotechnol 2020; 104(11): 4811-35. [http://dx.doi.org/10.1007/s00253-020-10592-1] [PMID: 32303816]

[26] Begum F, Nageswara Rao SSS, Rao K, Prameela Devi Y, Giri A, Giri CC. Increased vincristine production from *Agrobacterium tumefaciens* C58 induced shooty teratomas of *Catharanthus roseus* G. Don. Nat Prod Res 2009; 23(11): 973-81. [http://dx.doi.org/10.1080/14786410802131153] [PMID: 19521912]

[27] Ravindra B, Neelambika T, Meti GS, Mulgund K, Nataraja S, Vijaya K. Synthesis of antimicrobial silver nanoparticles by callus cultures and *in vitro* derived plants of *Catharanthus roseus*. Res Pharm 2012; 2(6): 18-31.

[28] Pietrosiuk A, Furmanowa M, Łata B. *Catharanthus roseus* :Micropropagation and *in vitro* techniques. Phytochem Rev 2007; 6(2-3): 459-73. [http://dx.doi.org/10.1007/s11101-006-9049-6]

[29] Gantait S, Kundu S. Neoteric trends in tissue culture-mediated biotechnology of Indian ipecac [*Tylophora indica* (Burm. f.) Merrill]. 3 Biotech 2012; 7(3): 1-5.

[30] Kumar A, Prakash K, Sinha RK, Kumar N. *In vitro* plant propagation of *Catharanthus roseus* and assessment of genetic fidelity of micropropagated plants by RAPD marker assay. Appl Biochem Biotechnol 2013; 169(3): 894-900. [http://dx.doi.org/10.1007/s12010-012-0010-4] [PMID: 23292901]

[31] Mehta J, Upadhyay D, Paras P, Ansari R, Rathore S, Tiwari S. Multiple shoots regeneration of (anti-cancer plant) *Catharanthus roseus*: An important medicinal plant. Am J Pharm Tech Res 2013; 3: 785-93.

[32] Amiri S, Fotovat R, Tarinejad AR, Panahi B, Mohammadi SA. *In vitro* regeneration of periwinkle (*Catharanthus roseus* L.) and fidelity analysis of regenerated plants with ISSR Markers. J Plant Physiol Breeding 2019; 9(1): 129-35.

[33] Moreno PRH, van der Heijden R, Verpoorte R. Elicitor-mediated induction of isochorismate synthase and accumulation of 2,3-dihydroxy benzoic acid in *Catharanthus roseus* cell suspension and shoot cultures. Plant Cell Rep 1994; 14(14): 188-91. [http://dx.doi.org/10.1007/BF00233788] [PMID: 24192892]

[34] Mitra A, Khan B, Rawal S. Rapid *in vitro* multiplication of plants from mature nodal explants of *Catharanthus roseus.* Planta Med 1998; 64(4): 390. [http://dx.doi.org/10.1055/s-2006-957463] [PMID: 17253257]

[35] Mohammed F, Satyapal S, Tanwer BS, Moinuddin K, Anwar S. *In vitro* regeneration of multiplication shoots in *Catharanthus roseus* : An important medicinal plant. Adv Appl Sci Res 2011; 2(1): 208-313.

[36] Yuan YJ, Hu TT, Yang YM. Effects of auxins and cytokinins on formation of *Catharanthus roseus* G. Don multiple shoots. Plant Cell Tissue Organ Cult 1994; 37(2): 193-6. [http://dx.doi.org/10.1007/BF00043615]

[37] Zárate R, Memelink J, van der Heijden R, Verpoorte R. Genetic transformation via particle bombardment of *Catharanthus roseus* plants through adventitious organogenesis of buds. Biotechnol Lett 1999; 21(11): 997-1002. [http://dx.doi.org/10.1023/A:1005622317333]

[38] Hernández-Domínguez E, Campos-Tamayo F, Vázquez-Flota F. Vindoline synthesis in *in vitro* shoot cultures of *Catharanthus roseus.* Biotechnol Lett 2004; 26(8): 671-4. [http://dx.doi.org/10.1023/B:BILE.0000023028.21985.07] [PMID: 15200179]

[39] Swanberg A, Dai W. Plant regeneration of periwinkle (*Catharanthus roseus*) *via* organogenesis. HortScience 2008; 43(3): 832-6. [http://dx.doi.org/10.21273/HORTSCI.43.3.832]

[40] Lee ON, Ak G, Zengin G, *et al.* Phytochemical composition, antioxidant capacity, and enzyme inhibitory activity in callus, somaclonal variant, and normal green shoot tissues of *Catharanthus roseus* (L) G. Don. Molecules 2020; 25(21): 4945. [http://dx.doi.org/10.3390/molecules25214945] [PMID: 33114628]

[41] Vandana S, Ashwini K, Arun K, Sumit K. An efficient *in vitro* propagation protocol for *Catharanthus roseus* (L.). Res J Biotechnol 2020; 15: 1.

[42] Ramani S, Jayabaskaran C. Enhanced catharanthine and vindoline production in suspension cultures of *Catharanthus roseus* by ultraviolet-B light. J Mol Signal 2008; 3(1): 9. [http://dx.doi.org/10.1186/1750-2187-3-9] [PMID: 18439256]

[43] Hilliou F, Christou P, Leech MJ. Development of an efficient transformation system for *Catharanthus roseus* cell cultures using particle bombardment. Plant Sci 1999; 140(2): 179-88. [http://dx.doi.org/10.1016/S0168-9452(98)00225-8]

[44] Veerabathini S, Sarang S, Shalini S, Deepa Sankar P. Standardization of friable callus development in *Catharanthus roseus* (Linn.) G. Don. International Journal of Pharmacy and Pharmaceutical Sciences. 2015;7(3):111-113.

[45] Zhao J, Zhu WH, Hu Q. Effects of light and plant growth regulators on the biosynthesis of vindoline and other indole alkaloids in *Catharanthus roseus* callus cultures. Plant Growth Regul 2001; 33(1): 43-9.
[http://dx.doi.org/10.1023/A:1010722925013]

[46] Xu M, Dong J, Zhu M. Effect of nitric oxide on catharanthine production and growth of *Catharanthus roseus* suspension cells. Biotechnol Bioeng 2005; 89(3): 367-71.
[http://dx.doi.org/10.1002/bit.20334] [PMID: 15744842]

[47] Xu M, Dong J. Nitric oxide stimulates indole alkaloid production in *Catharanthus roseus* cell suspension cultures through a protein kinase-dependent signal pathway. Enzyme Microb Technol 2005; 37(1): 49-53.
[http://dx.doi.org/10.1016/j.enzmictec.2005.01.036]

[48] Datta A, Srivastava PS. Variation in vinblastine production by *Catharanthus roseus,* during *in vivo* and *in vitro* differentiation. Phytochemistry 1997; 46(1): 135-7.
[http://dx.doi.org/10.1016/S0031-9422(97)00165-9]

[49] Filippini R, Caniato R, Vecchia FD, *et al.* Somatic embryogenesis and indole alkaloid production in *Catharanthus roseus*. Plant Biosyst 2000; 134(2): 179-84.
[http://dx.doi.org/10.1080/11263500012331358444]

[50] Zhao J, Zhu WH, Hu Q. Promotion of indole alkaloid production in *Catharanthus roseus* cell cultures by rare earth elements. Biotechnol Lett 2000; 22(10): 825-8.
[http://dx.doi.org/10.1023/A:1005669615007]

[51] Zhao J, Zhu WH, Hu Q, Guo YQ. Improvement of indole alkaloid production in *Catharanthus roseus* cell cultures by osmotic shock. Biotechnol Lett 2000; 22(15): 1227-31.
[http://dx.doi.org/10.1023/A:1005653113794]

[52] Zhao J, Zhu WH, Hu Q, Guo YQ. Compact callus cluster suspension cultures of *Catharanthus roseus* with enhanced indole alkaloid biosynthesis. *In Vitro* Cell Dev Biol Plant 2001; 37(1): 68-72.
[http://dx.doi.org/10.1007/s11627-001-0013-2]

[53] Zhao J, Zhu WH, Hu Q, He XW. Enhanced indole alkaloid production in suspension compact callus clusters of *Catharanthus roseus:* impacts of plant growth regulators and sucrose. Plant Growth Regul 2001; 33(1): 33-41.
[http://dx.doi.org/10.1023/A:1010732308175]

[54] Junaid A, Mujib A, Sharma MP, Tang W. Growth regulators affect primary and secondary somatic embryogenesis in Madagaskar periwinkle (*Catharanthus roseus* (L.) G. Don) at morphological and biochemical levels. Plant Growth Regul 2007; 51(3): 271-81.
[http://dx.doi.org/10.1007/s10725-007-9171-5]

[55] Ataei-Azimi A, Hashemloian BD, Ebrahimzadeh H, Majd A. High *in vitro* production of ant-canceric indole alkaloids from periwinkle (*Catharanthus roseus*) tissue culture. Afr J Biotechnol 2008; 7(16).

[56] Shukla AK, Shasany AK, Verma RK, Gupta MM, Mathur AK, Khanuja SPS. Influence of cellular differentiation and elicitation on intermediate and late steps of terpenoid indole alkaloid biosynthesis in *Catharanthus roseus.* Protoplasma 2010; 242(1-4): 35-47.
[http://dx.doi.org/10.1007/s00709-010-0120-1] [PMID: 20217156]

[57] Singh R, Kharb P, Rani K. Rapid micropropagation and callus induction of *Catharanthus roseus in vitro* using different explants. World J Agric Sci 2011; 7(6): 699-704.

[58] Kim SW, In DS, Choi PS, Liu JR. Plant regeneration from immature zygotic embryo-derived embryogenic calluses and cell suspension cultures of *Catharanthus roseus.* Plant Cell Tissue Organ Cult 2004; 76(2): 131-5.
[http://dx.doi.org/10.1023/B:TICU.0000007254.51387.7f]

[59] Junaid A, Mujib A, Bhat MA, Sharma MP. Somatic embryo proliferation, maturation and germination in *Catharanthus roseus.* Plant Cell Tissue Organ Cult 2006; 84(3): 325-32.

[http://dx.doi.org/10.1007/s11240-005-9041-7]

[60] Ilah A, Mujib A, Junaid A, Samar F, Abdin MZ. Somatic embryogenesis and two embryo specific proteins (38 and 33 kD) in *Catharanthus roseus.* Biologia 2009; 64(2): 299-304. [http://dx.doi.org/10.2478/s11756-009-0031-9]

[61] Dhandapani M, Kim DH, Hong SB. Efficient plant regeneration *via* somatic embryogenesis and organogenesis from the explants of *Catharanthus roseus. In Vitro* Cell Dev Biol Plant 2008; 44(1): 18-25. [http://dx.doi.org/10.1007/s11627-007-9094-x]

[62] Junaid A, Mujib A, Bhat MA, Sharma MP, Šamaj J. Somatic embryogenesis and plant regeneration in *Catharanthus roseus.* Biol Plant 2007; 51(4): 641-6. [http://dx.doi.org/10.1007/s10535-007-0136-3]

[63] Aslam J, Mujib A, Fatima S, Sharma MP. Cultural conditions affect somatic embryogenesis in *Catharanthus roseus* L. (G.) Don. Plant Biotechnol Rep 2008; 2(3): 179-89. [http://dx.doi.org/10.1007/s11816-008-0060-9]

[64] Aslam J, Mujib A, Nasim SA, Sharma MP. Screening of vincristine yield in *ex vitro and in vitro* somatic embryos derived plantlets of *Catharanthus roseus* L. (G) Don. Sci Hortic 2009; 119(3): 325-9. [http://dx.doi.org/10.1016/j.scienta.2008.08.018]

[65] Aslam J, Khan SH, Siddiqui ZH, *et al. Catharanthus roseus* (L.) G. Don. An important drug: It's applications and production. Pharm Glob 2010; 4(12): 1-6.

[66] Aslam J, Mujib A, Fatima Z, Sharma MP. Variations in vinblastine production at different stages of somatic embryogenesis, embryo, and field-grown plantlets of *Catharanthus roseus* L. (G) Don, as revealed by HPLC. *In Vitro* Cell Dev Biol Plant 2010; 46(4): 348-53. [http://dx.doi.org/10.1007/s11627-010-9290-y]

[67] Yuan F, Wang Q, Pan Q, *et al.* An efficient somatic embryogenesis based plant regeneration from the hypocotyl of *Catharanthus roseus.* Afr J Biotechnol 2011; 10(66): 14786-95. [http://dx.doi.org/10.5897/AJB09.1186]

[68] Gulzar B, Mujib A, Rajam MV, *et al.* Shotgun label-free proteomic and biochemical study of somatic embryos (cotyledonary and maturation stage) in *Catharanthus roseus* (L.) G. Don. 3 Biotech 2021; 11(2): 1-5.

[69] Amiri S, Fotovat R, Panahi B, Tarinezhad A, Mohammadi SA. Review of abiotic and biotic elicitors' roles in secondary metabolites biosynthesis of periwinkle (*Catharanthus roseus* (Linn.) G. Don). Journal of Medicinal Plants. 2020; 19(74):1-24.

[70] Choi PS, Kim YD, Choi KM, Chung HJ, Choi DW, Liu JR. Plant regeneration from hairy-root cultures transformed by infection with *Agrobacterium rhizogenes* in *Catharanthus roseus.* Plant Cell Rep 2004; 22(11): 828-31. [http://dx.doi.org/10.1007/s00299-004-0765-3] [PMID: 14963692]

[71] Soumya V, Sowjanya A, Kiranmayi P. Evaluating the status of phytochemicals within *Catharanthus roseus* due to higher metal stress. Int J Phytoremediation 2021; 23(13): 1391-401. [http://dx.doi.org/10.1080/15226514.2021.1900063] [PMID: 33735592]

[72] Zhou ML, Zhu XM, Shao JR, Wu YM, Tang YX. An protocol for genetic transformation of *Catharanthus roseus* by *Agrobacterium rhizogenes* A4. Appl Biochem Biotechnol 2012; 166(7): 1674-84. [http://dx.doi.org/10.1007/s12010-012-9568-0] [PMID: 22328251]

[73] Traverse KK, Mortensen S, Trautman JG, Danison H, Rizvi NF, Lee-Parsons CW. Generation of Stable *Catharanthus roseus* Hairy Root Lines with *Agrobacterium rhizogenes.* Plant Secondary Metabolism Engineering. New York, NY: Humana 2022; pp. 129-44. [http://dx.doi.org/10.1007/978-1-0716-2185-1_11]

[74] Goklany S, Rizvi NF, Loring RH, Cram EJ, Lee-Parsons CWT. Jasmonate-dependent alkaloid

biosynthesis in *Catharanthus roseus* hairy root cultures is correlated with the relative expression of *Orca* and *Zct* transcription factors. Biotechnol Prog 2013; 29(6): 1367-76. [http://dx.doi.org/10.1002/btpr.1801] [PMID: 23970483]

[75] ten Hoopen HJG, Vinke JL, Moreno PRH, Verpoorte R, Heijnen JJ. Influence of temperature on growth and ajmalicine production by *Catharantus roseus* suspension cultures. Enzyme Microb Technol 2002; 30(1): 56-65. [http://dx.doi.org/10.1016/S0141-0229(01)00456-2]

[76] Davioud E, Kan C, Hamon J, Tempé J, Husson HP. Production of indole alkaloids by *in vitro* root cultures from *Catharanthus trichophyllus.* Phytochemistry 1989; 28(10): 2675-80. [http://dx.doi.org/10.1016/S0031-9422(00)98066-X]

[77] Verma P, Sharma A, Khan SA, Shanker K, Mathur AK. Over-expression of *Catharanthus roseus* tryptophan decarboxylase and strictosidine synthase in rol gene integrated transgenic cell suspensions of Vinca minor. Protoplasma 2015; 252(1): 373-81. [http://dx.doi.org/10.1007/s00709-014-0685-1] [PMID: 25106473]

CHAPTER 11

A Systematic Review on Micropropagation of Medicinal and Vulnerable Ashoka Tree [*Saraca asoca* (Roxb.) W.J.de Wilde]

Pradeep Bhat[1], **Sandeep R. Pai**[2], **Vinay Kumar Hegde**[3], **Poornananda Madhava Naik**[4] and **Vinayak Upadhya**[5,*]

[1] *ICMR-National Institute of Traditional Medicine, Nehru Nagar, Belagavi, Karnataka-590010, India*

[2] *Department of Botany, Rayat Shikshan Sanstha's Dada Patil Mahavidyalaya, Karjat-414402, Ahmednagar, Maharashtra, India*

[3] *Bhandimane Life Science Research Foundation, Sirsi-581401, Uttara Kannada, Karnataka, India*

[4] *Kanara E-vision Science and Commerce PU College, Chalageri-581145, Ranebennur, Karnataka, India*

[5] *Department of Forest Products and Utilization, College of Forestry, (University of Agricultural Sciences Dharwad), Sirsi-581401, Uttara Kannada, Karnataka, India*

Abstract: *Saraca asoca* (Family - Fabaceae) is well-known medicinal tree species used in codified and non-codified systems of traditional medicine. Tree parts, *viz*., bark, flower, leaf, root and fruit, are used to treat various disorders. A huge number of pharmaceutical products were prepared using bark as one of the major ingredients. Ashoka tree is categorized as vulnerable by the International Union for Conservation of Nature (IUCN) and endangered by the Conservation Assessment and Management Plan (CAMP) due to its overexploitation and limited distribution. Unsustainable utilization, deforestation and climate changes are the major threats to the existence of the species. Aiming at the conservation of the Ashoka tree, ample research works were performed to standardize the micropropagation techniques. The present chapter discusses the efforts made towards conservation and micropropagation studies on the Ashoka tree.

Keywords: Ayurveda, Conservation, Medicinal Plant, *Saraca asoca*, Tissue Culture, Vulnerable.

* **Corresponding author Vinayak Upadhya:** Department of Forest Products and Utilization, College of Forestry, (University of Agricultural Sciences Dharwad), Sirsi-581401, Uttara Kannada, Karnataka, India; Tel: 9449561591; E-mails: sirsivinayak@gmail.com and upadhyavs@uasd.in

T. Pullaiah (Ed.)

INTRODUCTION

Saraca asoca (Roxb.) W.J.de Wilde is an evergreen small-sized tree that belongs to the family Fabaceae [1]. It is popularly known as the Ashoka tree and other vernacular names of this tree in different languages of the Indian subcontinent are Sitaashoka (Sanskrit, Nepali, Hindi, and Gujarati), Aachange, Seeta ashokada mara, Eliyala, Kenkali mara, Kempuchinnada ele gida (Kannada), Asokamu (Telugu), Hemapushpam (Malayalam), Jasundi (Marathi), and Asogam (Tamil) [2]. It is native to the Indian subcontinent, and the scattered populations are found in the Indo-Malayan (from Pakistan to Malaysia) region. In India, it is mainly found in Eastern and Western Ghats, sub-Himalayan tracts from Uttar Pradesh to Eastern states of India [3 - 5]. Socio-culturally, it is a valued tree in India and elsewhere and occupies a privileged place in Hindu tradition. The term Ashoka means 'without sorrow' or 'the one which takes out the sorrow' [3]. The tree was mentioned in the ancient Indian treatise, the '*Ramayana*' [4].

Medicinal Uses

The tree is used to cure various disorders in codified (Ayurveda, Unani, Siddha) and non-codified folklore systems of traditional medicine. Bark and bark products are chiefly used to treat various gynecological disorders and other conditions (Table **1**) [4 - 9].

Table 1. Medicinal uses of *S. asoca* bark.

Gynecological Disorders	Other Medicinal Uses
Excessive bleeding, stress, gynecological disorders, irregular menses, premenstrual syndrome, ovarian cysts, fibroids, dysfunctional uterine bleeding, menstrual flow issues, uterine inflammation, menopause-related indications, menorrhagia, dysmenorrhea, leucorrhea, metrorrhagia, menopausal syndrome, postmenopausal syndrome, premenstrual tension, genitourinary diseases, pubertal and menopausal bleeding, spasmodic and lower back pain, stress and mood swings.	Fever, neurological disorders, snake bites, disease of the eye, wounds, skin diseases, including leprosy, anti-abortion agent, anemia, improved skin complexion, piles, burning sensation, tumors, dermatitis, cure indigestion, animal bite, hair tonic, biliousness, dyspepsia, dysentery, colic, piles, ulcers and pimples.

Folklore healers suggest bathing under the shade of the Ashoka tree for patients suffering from mental disorders. According to the local practitioners, mental peace or mental stability can be obtained by wearing lei (*Maala*) using root pieces of the tree. Further, the leaf, inflorescence, root, fruit, seed and whole plant parts were also reported to be used in the treatment of blood purification, dysentery, diabetes, menorrhagia, bleeding piles, dysentery, kidney stones, cough, and to prevent abortion, as cardiotonic and cooling agent [4, 6 - 8].

Phytoconstituents and Bioactivity

It is reported to contain a number of bioactive constituents in Ashoka tree such as catechin, leucocyanidin, epicatechin, procyanidine B-2, saracoside, β-sitosterol, lignin glycosides, procyanidin gallate, myoinositol, oleic, linoleic, palmitic and stearic acids, quercetin, kaempferol-3-O-P-D-glucoside, *etc*. These phytochemical constituents are responsible for antioxidant, antibacterial, antifungal, anticancer, antiulcer, analgesic, antiarthritic, anti-inflammatory, anti-nephrolithiatic, antidepressant, antidiabetic, hypolipidemic, larvicidal, antimutagenic, antimennorhagic, oxytocic, genoprotective and uterine tonic properties. It also acts cardioprotective, dermatoprotective and brain tonic agent [3 - 6, 8, 9].

Conservation Status

The bark of the tree is the main part harvested from the wild populations, and the market demand for bark is about 15,000 metric tons during 2007–2011. It is increasing over the time by crude drug market and pharmaceutical industries. Flowers and leaves were also reported to be marketed locally in India at a smaller scale [4]. Unsustainable harvesting of bark in large volumes from wild populations is one of the main reasons for the rapid depletion of the plant along with its sensitive niche. Habitat destruction, forest fire, encroachment, domestic animal grazing, developmental activities and changing climatic conditions are the other reasons for the rapid decrease in its natural habitat. Hence, the tree species is listed in the Rare, Endangered and Threatened (RET) category of vulnerable status by the International Union for Conservation of Nature (IUCN, 2022) [10] and endangered by Conservation Assessment and Management Plan (CAMP, 2001) [5, 11]. Further, fair to poor natural regeneration status was also reported for *S. asoca* [9]. Therefore, several efforts were made to develop *in vitro* tissue culture techniques to enhance the regeneration of the plant and ultimately to achieve the conservation strategies and action plans. The present chapter emphasizes the ample research works carried out on micropropagation studies of the endangered medicinal plant *S. asoca*.

Reproductive Biology

Ashoka tree is a habitat specific evergreen perennial tree that produces bright colored fragrant flowers in paniculate corymbose inflorescence during the month of December to May. The flowers change their color from light orange/yellow to scarlet from their initiation to wilting. The pollination is entomophilous, and the successful anthesis for cross pollination and pollen germination was reported in the early morning periods. The pods mature from May to July, however, variations in phenology were observed in the plants grown in different locations [12]. Seeds are recalcitrant, and seed germination studies showed physiological dormant stages with a non-orthodox storage behavior [13, 14]. Better seed germination percentage was reported in the seeds treated with 0.1 N sulphuric acid and *Vrikshayurveda* methods [13, 15]. A negative effect of abscisic acid (ABA) and a positive effect of gibberellic acid (GA3) were noted during the process of germination [16, 17].

Vegetative Propagation

Vegetative propagation in the Ashoka tree was studied by Smitha [18] through hardwood cuttings, air layering and grafting methods. The air layering process showed a 90% success rate, followed by whip and tongue grafting (87.67%). The least success was reported in hardwood cutting method (16.67%). Further, the effects of hormones, *viz.*, Indole-3-acetic acid (IAA), indole-3-butyric acid (IBA), and Naphthalene acetic acid (NAA), on root and leaf initiation, growth and survival success rates were also studied by different researchers [19, 20].

MICROPROPAGATION STUDIES

Surface Sterilization

Surface sterilization is one of the important steps in plant tissue culture to obtain contamination free tissues from the explants and regeneration of induced plantlets [21]. In *S. asoca*, several chemicals were used for surface sterilization by the researchers to obtain good results. All the reported studies used running water or distilled water for the initial cleaning of explants, and the time varies from three minutes to 30 minutes. Different chemical sterilants used for surface sterilization of *S. asoca* are given in Table **2**.

Table 2. Chemical sterilant used for surface sterilization of *S. asoca*.

S. No.	Name of the Sterilant	Explant Used	Exposure Period	References
1	NaOCl	Leaf, stem	2 min	[22]
	Ethanol		2 min	
	0.3% $HgCl_2$		2 min	
2	Bavistin	Shoot tip	120 min	[23]
	Detergent		20 min	
	Labolene (few drops)		10 min	
	70% Ethanol		1 min	
	0.1% $HgCl_2$		10 min	
3	0.1% $HgCl_2$	Tender leaf, tender stem, flower buds	4 min	[24]

(Table 2) cont.....

S. No.	Name of the Sterilant	Explant Used	Exposure Period	References
4	70% Ethanol	Nodal segment	Swabbed using cotton	[25]
	2% aqueous Cetrimide solution		30 min	
	0.2% Bavistin		30 min	
	0.5% Streptomycin		30 min	
5	2% aqueous Cetrimide	Pod	30 min	[26]
	0.2% aqueous Bavistin		30 min	
	0.5% aqueous Ambistryn-S		30 min	
	0.6% aqueous $HgCl_2$		30 min	
6	1% Bevistin	Shoot tip, nodal and internodal segment	-	[28]
	Chloramphenicol (400 ppm)			
	0.1% $HgCl_2$		5 min	
7	0.1% $HgCl_2$	Leaf	30 min	[28]
		Hypocotyls, epicotyls, coppiced shoots	9 min	
		Cotyledons, auxiliary bud	12 min	
8	Ethanol	Meristematic shoot tip, nodal segment, internodal segment, leaf bits, axillary bud, cotyledon, embryo, seed, anther, ovary, hypocotyl region	5 min	[29]
	0.1% $HgCl_2$		3 min	
9	0.1% Tween-20	Apical segment (bud), nodal segment	15 min	[30]
	0.1% Bavistin		15 min	
	0.5% Streptomycin		-	
	70% Ethanol		0.50 min	
	$HgCl_2$ (0.1, 0.15, 0.2%)		5, 10, 15 min	

Rout and Khare [21] studied the effect of different surface sterilants, *viz.*, mercuric chloride ($HgCl_2$), propanol and NaOCl with varied concentrations and exposure time on contamination and callus regeneration through leaf explants. Treatment with 0.1% $HgCl_2$ (exposure for 15 minutes) followed by 1% NaOCl for 2 minutes produced 93.33% aseptic culture, which significantly reduced the percentage of fungal contamination.

Vichitra [30] assessed the effect of three concentrations (0.1, 0.15 and 0.2%) of $HgCl_2$ at different exposure times (5, 10 and 15 minutes) on shoot segment (bud) and nodal explants. A profuse contamination was observed in the culture; hence, they carried out a bioassay experiment to check the resistance and susceptibility of identified fungus against different concentrations of fungicides *viz.* sectin, bayleton, indofil, bavistin, copper oxychloride and propiconazole. Aseptic culture up to 95% with bud break was found when sterilized with propicanazole. The effects of different concentrations of $HgCl_2$, exposure time and season of collection of explants were studied. The explants collected during the winter season treated with 0.15% $HgCl_2$ concentration for 15 minutes reported significant results with 87% aseptic conditions and a 71% survival rate. Poor performance of explants in the formation of aseptic cultures was found in the plant samples collected in the rainy season [30].

Callus Induction Studies

Waman *et al.* [28] reported the callus induction in the Ashoka tree for the first time. Young leaves, hypocotyls, epicotyls, coppiced shoots, cotyledons and axillary buds were used as explants and cultured on solidified full strength and half strength Murashige and Skoog medium (MS medium) containing agar (0.6%) with NAA (0.5 mg/L) + Benzyl adenine (BA, 2 mg/L) + sucrose (3%) as supplementary agents. Cultures were maintained at 25±1°C with a 16 h photoperiod. Half strength MS medium showed good white to brown colored calli in epicotyls (89.9%). The percentage of callus formation was varied from 13.3% to 83.3% in other explants. The mean callus initiation period was also observed from 1.5 to 9.4 weeks, and the young leaves cultured on MS media showed poor results.

Rout and Khare studied the importance and effect of surface sterilization on callus induction and growth. Maximum callus percentage was obtained in the MS medium after sterilization with 0.1% $HgCl_2$ and 1% NaOCl sterilizer [21]. Vignesh *et al.* [22] reported the callus production from leaf explants inoculated on MS medium supplemented with 6-Benzylaminopurine (BAP) and with 2,4-Dichlorophenoxyacetic acid (2,4-D). Significant callus fresh weight (42 mg) and 80% callus initiation were observed in the MS medium containing 0.3 mg/L BAP and 0.6 mg/L 2,4-D. Further, comparative studies on phytoconstituents present in leaves and calli were also estimated.

Standardization of callus induction was also studied by Mini and Sankaranarayanan using leaf, stem and flower bud as explants [24]. MS basal medium containing sucrose (3%), agar (0.8%) and different concentrations of 2,4-D and NAA (1-5 mg/L) were used to test the callus induction. The formation of 100% callus from leaf and shoot explants was reported, whereas it was poor in flower explants. Leaf explants were inoculated on MS media containing 2,4-D and NAA at 1:1 and 2:1 ratios. The optimum callus induction was observed in the cultures supplemented with 3 mg/L of 2,4-D. The combinations of NAA and 2,4-

D at 3 to 5 mg/L concentration induced root formation along with hard and white calli after two weeks of inoculation [24].

Callus induction using meristematic shoot tip, nodal segment, internodal segment, leaf bits, axillary bud, cotyledon, embryo, seed, anther, ovary, and hypocotyl region from explants was studied by Paranthaman *et al.* [29]. MS, B5 and Woody Plant Medium (WPM) supplemented with 2,4-D in different concentrations (0.5 to 4.0 mg/L) were used for the study. Among all, MS medium with 2,4-D (2 mg/L) showed maximum callus induction (81.91%), followed by B5 (66.47%) and WPM (65.60%). Maximum callus induction of 88% was observed when the ovary was used as an explant.

Organogenesis Studies

The effect of different hormones supplemented in three different basal media *viz.* WPM, Gamborg B5 medium and MS on shoot induction were tested by Sharma *et al.* [23]. Positive results were found in WPM, and it was accepted as the best medium for shoot induction when supplemented with BAP (6 mg/L), Kinetin (Kn, 1 mg/L), and NAA (1.5 mg/L). Subsequently, lesser results in shoot induction were detected in MS and Gamborg B5 medium. It was concluded that the growth hormones at lower concentrations did not support the shoot induction.

Nodal explants on four basal nutrient media *viz.* Gamborg B5, MS, Nitsch & Nitsch (NN) and WPM supplemented with five doses of BA (0, 2.2, 4.4, 8.8 and 17.8 μM) and their combination on induction and proliferation of shoots was studied by Shirin *et al.* [25]. The nitrogen source in the Gamborg B5 medium and five strengths of KNO_3 (0.25×, 0.5×, 1.0×, 1.25× and 1.5×) were tested to enhance the shoot number. Significant results with a maximum number of shoots (1.92) were reported with the supplementation of 0.25× strength of potassium nitrate (KNO_3) in Gamborg B5 medium. However, the changed strengths of KNO_3 failed to show a significant effect on the elongation of shoots. The study also reported the effect of quick dip and transfer of shoots after one, three, five and seven days of pulse treatment along with IBA (200 μM) in half strength MS liquid basal medium. Five days pulse treatment of shoots showed *in vitro* rooting rate of 37.5% and subsequently the shoots were shifted to half strength semisolid MS medium complemented with IBA (0.2 μM) + Phloroglucinol (3.96 μM). Further, hardened plantlets were shifted to a soil mixture in a shade house for successful regeneration [25].

Shirin *et al.* [26] studied the influence of B5 medium containing BA (0 and 2.5 μM) and Zinc (Zn at 0, 5, 10 and 20 μM concentration) on shoot regeneration using the embryonic axis of immature seeds as an explant. The study showed the formation of callus and shoot regeneration under 2.5 μM BA. However, it was found that Zn and its interaction with BA have no significant effect on the regeneration of shoots.

Subbu *et al.* [27] studied *in vitro* clonal propagation of *S. asoca* using shoot tip, nodal and internodal explants on MS medium supplemented with various concentrations of BAP (0.5% - 2.0%), Kn and 2,4-D along with the plant hormones *viz.* IAA and IBA (1.0 to 5.0 ml/L). Organogenesis and callus induction was observed in all the explants. Nodal explants cultured on 0.5 mg/L BAP showed maximum shoot organogenesis, whereas none of the explants produced shoot organogenesis in 2,4-D medium. IBA showed good *in vitro* rooting results compared to IAA and a 40% success rate of hardening was also achieved in the study.

In vitro micropropagation through bud proliferation, callogenesis, *in vitro* and *ex vitro* rooting methods were performed by Vichitra [30]. The nodal region of the plant was taken as explants and inoculated on five different media with different concentrations of BAP and plant growth regulators (2,4-D and Defol) at various concentrations. Among all, maximum sprouting (91%) was achieved in the Nitsch medium with 1.5 mg/L BAP concentration. In contrast, maximum shoot length (3.03 cm) was achieved in Gamborg's medium with BAP (1.0 mg/L). Significant callus production (60%) with soft, slightly green callus was developed in Gamborg's medium with BAP (0.5 mg/L) and 2,4-D (5.0 mg/L). Various combinations of plant growth regulators revealed significantly high (48%) callus differentiation with mean shoot number (2.92) and shoot length (3.41 cm) in BAP (1.0 mg/L) + Defol (0.5 mg/L). Gamborg's medium supplemented with 4.0 mg/L IBA showed the best results in *in vitro* rooting (58%) with 2.68 average root numbers and a maximum root length of 3.64 cm. Acclimatization of plantlets showed 53% survival in the field conditions. In the *ex vitro* rooting experiment, Gamborg's medium containing IBA (5000 ppm) showed maximum root induction (62%), root number (2.79) and root length (3.94 cm). Acclimatized *ex vitro* raised plants showed a 62% survival rate in the field.

Paranthaman and Usha Kuramari [31] reported standardization of tissue culture protocol in *S. asoca* collected from five locations in Tamil Nadu and Kerala states. Meristematic shoot tip, nodal segment, internodal segment, leaf bits, axillary bud, cotyledon, embryo, seed, anther, ovary, hypocotyl region were cultured on MS, B5 and WPM nutrient medium supplemented with different hormones *viz.*, BA, Zn, BAP, Kn and IAA for callus induction and organogenesis. Indirect organogenesis through the callus induction was reported significant compared to direct organogenesis. A maximum number of calluses were induced through ovary culture on MS medium supplemented with 2,4-D (2.0 mg/L). Multiple roots and shoot production were observed and further used for the establishment of hardening of plantlets.

An efficient *in vitro* clonal propagation method was developed by Ramasubbu and Chandra Prabhs [32] using shoot tip, nodal and inter nodal explants. The effect of BAP at 0.5 mg/L concentration induced 11.71±0.53 mean adventitious shoot numbers from the nodal explants. The highest (82%) frequency of shoot organogenesis was observed in nodal explants treated with 0.5 mg/L BAP. A significant number of calluses were formed on the media containing 2,4-D and the developed micro shoots were inoculated on MS medium supplemented with 4.0 mg/L IBA. Further, the rooting frequency of 40% shoot explants was hardened and acclimatized in the soil.

CONCLUSION

Saraca asoca, a high valued medicinal plant, contains abundant active constituents with curative properties. Potent medicinal uses and pharmacological activities have created a great demand from pharmaceutical industries. It has created destructive harvesting practices, and hence, its population has gradually declined in the wild. Attention to its conservation strategies and enhancing the production of active constituents are the needs of the time. These goals can be achieved through the combination of tissue culture and other advanced biological techniques. The present chapter indicates that micropropagation through tissue culture technique is a highly significant method for conserving highly medicinal and endangered Ashoka trees. However, these studies are of a pilot scale, and there is a huge scope for large-scale research to fulfill the needs of herbal industries and to conserve *S. asoca* in its natural habitat.

REFERENCES

[1] Anonymous *Saraca asoca* India biodiversity portal. Available from: https://indiabiodiversity.org /species/show/18020 (Accessed on June 2022).

[2] Anonymous *Seeta Ashoka* Flowers of India. Available from: https://www.flowersofindia.net/ catalog/slides/Sita%20Ashok.html (Accessed on May 2022).

[3] Hegde S. Studies on genetic and phytochemical variations of *Saraca asoca* (Roxb) De Wilde from Western Ghats. Belagavi: Ph.D Thesis, KLE University 2018.

[4] Singh S, Anantha Krishna TH, Kamalraj S, Kuriakose GC, Valayil JM, Jayabaskaran C. Phytomedicinal importance of *Saraca asoca* (Ashoka): An exciting past, an emerging present and a promising future. Curr Sci 2015; 109(10): 1790-801. [http://dx.doi.org/10.18520/cs/v109/i10/1790-1801]

[5] Patwardhan A, Pimputkar M, Mhaskar M, *et al.* Distribution and population status of threatened medicinal tree *Saraca asoca* (Roxb.) De Wilde from Sahyadri–Konkan ecological corridor. Curr Sci 2016; 111(9): 1500-6. [http://dx.doi.org/10.18520/cs/v111/i9/1500-1506]

[6] Pradhan P, Joseph L, Gupta V, *et al. Saraca asoca* (Ashoka): A review. J Chem Pharm Res 2009;

1(1): 62-71.

[7] Pandey A, Kumar A. Good field collection practices for medicinal plants (bark drugs). National seminar on Recent advances in research and development in medicinal and aromatic plants. A country scenario.

[8] Kulkarni RV. *Saraca asoca* (Ashoka): A review. World J Pharm Res 2018; 7(19): 536-44.

[9] Manna SS, Mishra SP. Diversity, population structure and regeneration of tree species in Lalgarh forest range of West Bengal, India. Intern J Bot Studies 2017; 2: 191-5.

[10] The IUCN Red List of Threatened species. Available from: https://www.iucnredlist.org/search?query=saraca%20asoca&searchType=species (Accessed on June 2022).

[11] CAMP. Conservation assessment and management plan for medicinal plants of Maharashtra state. FRLHT 2001.

[12] Smitha GR, Thondaiman V. Reproductive biology and breeding system of *Saraca asoca* (Roxb.) De Wilde: A vulnerable medicinal plant. Springerplus 2016; 5(1): 2025.
[http://dx.doi.org/10.1186/s40064-016-3709-9] [PMID: 27995002]

[13] Devan AS, Warier RR. *Saraca asoca* : Morphology and diversity across its natural distribution in India. Int J Complement Altern Med 2021; 14(6): 317-23.

[14] Jayasuriya KMGG, Wijetunga ASTB, Baskin JM, Baskin CC. Seed dormancy and storage behaviour in tropical Fabaceae: A study of 100 species from Sri Lanka. Seed Sci Res 2013; 23(4): 257-69.
[http://dx.doi.org/10.1017/S0960258513000214]

[15] Prachi S, Pramod K. Evaluation on the seed germination rate of Ashoka (*Saraca asoca* (Roxb.) de Wilde) with special reference to Vrikshayurveda. Intern J Ayurveda Pharma Res 2016; 4(10).

[16] Prajith TM, Anilkumar C, Ajith Kumar KG. Changes in abscisic acid levels in embryonic axis of *Saraca asoca* seeds during maturation and artificial dehydration. Indian J Plant Physiol 2017; 22(3): 354-7.
[http://dx.doi.org/10.1007/s40502-017-0315-y]

[17] Madhushree SI, Raviraja Shetty G, Souravi K, Rajasekharan PE, Ganapathi M, Ravi CS. Standardization of seed and vegetative propagation techniques in *Saraca asoca* (Roxb.) De Wilde: An endangered medicinal plant. Int J Curr Microbiol Appl Sci 2018; 7(4): 1327-35.
[http://dx.doi.org/10.20546/ijcmas.2018.704.148]

[18] Smitha GR. Vegetative propagation of Ashoka [*Saraca asoca* (Roxb.) De Wilde] : An endangered medicinal plant. Res Crops 2013; 14(1): 274-83.

[19] Dash GK, Senapati SK, Rout GR. Effect of auxins on adventitious root development from nodal cuttings of *Saraca asoka* (Roxb.) de Wilde and associated biochemical changes. J Hortic For 2011; 3(10): 320-6.

[20] Rout S, Khare N, Beura S. Vegetative propagation of Ashoka (*Saraca asoca* Roxb. De Wilde.) by stem cuttings. Pharma Innov 2018; 7(1): 489-588.

[21] Rout S, Khare N. Effect of various surface sterilant on contamination and callus regeneration of Ashoka (*Saraca asoca* Roxb. De Wilde) from leaf segment explant. Int J Curr Microbiol Appl Sci 2018; 7(7): 2027-33.
[http://dx.doi.org/10.20546/ijcmas.2018.707.239]

[22] Vignesh A, Selvakumar S, Vasanth K. Comparative LC-MS analysis of bioactive compounds, antioxidants and antibacterial activity from leaf and callus extracts of *Saraca asoca*. Phytomedicine Plus 2022; 2(1): 100167.
[http://dx.doi.org/10.1016/j.phyplu.2021.100167]

[23] Sharma V, Varma R, Varma A, Sharma J. Effect of various culture media on shoot initiations of *Saraca indica* L. endangered plant. Intern J Curr Res Life Sci 2020; 9(9): 3334-7.

[24] Mini ML, Sankaranarayanan R. Standardization of callus induction in *Saraca indica* auct, non Linn. J Chem Pharm Res 2013; 5(2): 250-2.

[25] Shirin F, Parihar NS, Shah SN. Effect of nutrient media and KNO_3 on *in vitro* plant regeneration in *Saraca asoca* (Roxb.) Willd. Am J Plant Sci 2015; 6(19): 3282-92. [http://dx.doi.org/10.4236/ajps.2015.619320]

[26] Shirin F, Parihar N, Rana PK, Ansari SA. *In vitro* shoot regeneration from embryonic axis of a multipurpose vulnerable Leguminous tree, *Saraca indica* L. Tree For Sci Biotechnol 2011; 5(1): 45-8.

[27] Subbu RR, Chandraprabha A, Sevugaperumal R. *In vitro* clonal propagation of vulnerable medicinal plant, *Saraca asoca* (Roxb.) De Wilde. Nat Prod Radiance 2008; 7(4): 338-41.

[28] Waman AA, Umesha K, Sathyanarayana BN. First report on callus induction in Ashoka (*Saraca indica*): an important medicinal plant. In: Prakash J, Ed. Proc. IVth IS on Acclimatization and establishment of micropropagated plants. 383-6.

[29] Paranthaman M, Kumari RU, Narayanan NL, Sivasubramaniam K. Morphological characterization and *in vitro* callus induction in [*Saraca asoca* (Roxb.) de Wilde.] a vulnerable medicinal tree. Journal of Tree Sciences 2017; 36(1): 103. [http://dx.doi.org/10.5958/2455-7129.2017.00014.0]

[30] Vichitra A. Micropropagation of *Saraca asoca* (Roxb) De Wilde in response to biochemical and seasonal variations. Dehradun: Ph.D. Thesis, Forest Research Institute University 2015.

[31] Paranthaman, M and R. Ushakumari. 2014. Tissue culture studies in Ashoka (*Saraca asoca* Roxb. de Wilde). Lambert Academic Publishing (Germany). ISBN: 978-3-659-31151-2.

[32] Ramasubbu R, Chandra Prabhs A. *In vitro* clonal propagation in *Saraca asoca* (Roxb.) De Wilde: A vulnerable medicinal plant. Plant Cell Biotechnol Mol Biol 2012; 13(3-4): 99-104.

CHAPTER 12

In Vitro Propagation of Yam as a Medicinal Plant

Jaindra Nath Tripathi[1,*], **Kannan Gandhi**[1] and **Leena Tripathi**[1]

[1] *International Institute of Tropical Agriculture (IITA), PO Box 30709-00100, Old Naivasha Road, Nairobi, Kenya*

Abstract: Yam (*Dioscorea* spp.), a tropical monocot flowering, perennial multi-species crop, belongs to the family Dioscoreaceae. It is a valuable source of medicines and food security crops in yam-growing regions in Asia, Africa, and southern American countries. More than 600 yam species are widely cultivated in tropical and subtropical countries and used as food and medication for various human diseases. It provides big starchy tuberous roots as a source of carbohydrates, protein, antioxidants, minerals, and vitamins. It is also high in vitamin C, B6, manganese, potassium, and antioxidant compounds, which nourish and protect against oxidative cell damage in the human body. In addition, they are rich in potent plant compounds, including anthocyanins, a color-producing chemical that helps to reduce blood pressure and inflammation and protect against cancer and diabetes. Exceptionally, yam is an excellent crop for food security and human health. Micropropagation of medicinal yam is essential for the large-scale multiplication and conservation of endangered species. So far, in micropropagation of medicinal yam spp., very few studies have been conducted. These studies used axillary buds, nodal cuttings, mature, immature leaves, and shoot tips as explants for micropropagation. Several tissue culture techniques are available for micropropagation of yam, especially direct and indirect organogenesis for *in-vitro* propagation for large-scale generation of plantlets.

Keywords: *Dioscorea* spp., Medicinal yams, Micropropagation, Organogenesis, Regeneration, Tissue culture, Yam basal medium.

INTRODUCTION

Origin, Cultivation Classification, Economic Value, Production

Yam (*Dioscorea* spp.), a tropical monocot flowering, perennial, and multi-species crop, belongs to the family Dioscoreaceae. It produces edible large-sized tubers for food as well as medicine. It possesses creeper plant-like stems with heart-shaped green leaves and white or green flowers and berry-like fruits. Yam is the third most important root and tuber crop next to cassava and sweet potato [1]. It is

* **Corresponding author Jaindra Nath Tripathi:** International Institute of Tropical Agriculture (IITA), PO Box 30709-00100, Old Naivasha Road, Nairobi, Kenya; Tel: +254 020 422 3666; E-mail: j.tripathi@cgiar.org

T. Pullaiah (Ed.)

a staple food as well as full of medicinal qualities, grown in several West African countries since 11,000 BC [2], mainly in Nigeria, Ghana, Côte d'Ivoire, Benin, and Togo, which are known as yam-belts in Africa that is responsible for more than 92% of the total yam production [3]. In the region, yam cultivation covers over 8.1 million hectares, with a total annual production of over 67 million tons [4]. Ghana and Nigeria alone account for 77% of the product of yam. The crop also contributes much more protein to the human diet than the more widely grown cassava and even more than meat protein [5].

Medicinal Importance of Yam

More than 600 yam varieties are available globally in the genus *Dioscorea,* and only twelve yam varieties are extensively disseminated in the growing region, especially in Africa, Asia, Oceania, and South America. Other yam varieties are grown as wild plants in nature. These wild plant varieties are bitter and full of medicinal compounds. Waris *et al.* reported that the tuber, leaves, and stem of *D. deltoidea* are used to treat jaundice [6]. Tubers of *D. dumetorum* are used as a birth control agent for controlling the human population [7]. In 2018, Mustafa *et al.* [8] reported that the tuber and leaves of *D. belophylla* are used to treat malaria, jaundice, and dysentery. It contains exceptional medicinal properties enriched with alkaloids and steroids [9, 10]. Plant-based medicine has been used for centuries as an alternative medicine for various human diseases, mainly menopausal symptoms, diabetes, rheumatoid arthritis, and muscular cramps [11]. The tubers of wild yams contain a chemical compound called diosgenin, which can produce various essential hormones in our body, especially estrogen [12]. Traditional healers in West African countries use wild yam tubers as an alternative to hormone replacement therapy during menopausal conditions of older women. Various formulations like tablets, capsules, powders, tinctures, and creams are commercially available. Yam tubers also contain a chemical known as dioscoretine, which regulates blood sugar levels to the optimal range in the animal model study [13]. It needs to confirm further research and validation on human research. It is also helpful in chronic joint pain, rheumatoid, and several muscular cramp-related disorders [14]. Some species contain vitamin C and B6, manganese, and potassium [15]. The African yam (*Dioscorea* spp.) contains thiocyanate, which can potentially protect against sickle cell anemia [16]. Tubers of certain wild species of *Dioscorea*, such as *D. nipponica*, were found to contain diosgenin, a steroid sapogenin extracted and used for the commercial synthesis of cortisone, pregnenolone, progesterone, and other steroid products [17]. Wild yam tubers have been reported to be a preventive or therapeutic medicine against several ailments, including arthritis, cancer, diabetes, gastrointestinal disorders, high cholesterol, and inflammation in Memorial Sloan-Kettering Cancer Center [18]. The Chinese medicinal yams are known for their high therapeutic value to human health, which can be used to treat chronic diarrhea, chronic enteritis, spleen malfunctions, lung infections, gastric diseases, diabetes, nocturnal emission, enuresis, and underlying embolism [19, 20]. Yams have also been used as healthy food and herbal medicinal ingredients in traditional Chinese medicine [21]. In humans, Yam extracts showed significant antioxidant activity and modified serum lipid levels [22]. Yam flour was reported to protect rats from chemical-induced toxicity [23]. Several previous researchers report the medicinal properties and uses of different species of medicinal yams. Pillai *et al.* reported that the starch content of *D. esculenta* makes it viable for therapeutic purposes [24]. In 2021, Parida and Sarangi reported that tubers of *D. glabra*, *D. puber*, and *D. wallichii* have many medicinal uses. In traditional Chinese medicine, a decoction of pieces of yam is also a popular method of consumption [25].

D. alata, also known as purple yam, looks peculiar and contains high nutritional content. The flesh of this yam is purple and has a potato texture when cooked. Many people like its sweet and nutty flavor. It can be cooked in a variety of ways. Apart from their taste, purple yams are also a rich source of antioxidants, vitamins, and minerals. It is also high in vitamin C and can increase antioxidant levels by up to 35%, protecting against oxidative cell damage. In addition, they are rich in potent plant compounds and antioxidants, including anthocyanins, which give them their lively color. Studies have shown that anthocyanins may help reduce blood pressure and inflammation and protect against cancer and type 2 diabetes [26, 27].

Yam as a Staple Food

Yam is a tuber-producing crop that serves as a valuable food source in tropical and sub-tropical countries across Africa, Southeast Asia, South America, the Caribbean, and the Pacific islands [28]. Yams are used not only as fresh vegetables but also as processed foods like chips, dry roasted slices, flakes, flours, fried in oils, grilled, baked, pounded paste (fufu), and barbecued. It can be cooked with rice, plantain, beans, sweet potato, lamb, chicken, and butternut as squash soup [29]. The significant challenges in yam production can be categorized into several biotic and abiotic factors, including lack of clean, disease-free planting material, pests, diseases, decreasing soil fertility, and yield potential [30, 31]. Climate change has a significant impact on crop phenology, like the formation of authentic medicinal materials. Several studies reported [32 - 34] that authenticity is the core symbol of quality restorative materials, especially in Chinese medicinal yams. The specific locations of medicinal yams have been a manifestation of their prominent geographical characteristics, which are closely related to their demands for unique climate, soil, and other ecological conditions [35].

Yam Seed System Macro-propagation

The unavailability of quality seeds is mainly due to the traditional methods of seed yam production in West Africa. Many farmers hold 25% of the harvested yam and use it as planting material for the next crop where the number of seed yams required is significant; when there is an expansion in farm size, the proportion of planting materials may be consistently higher [36]. The first traditional method is milking harvests, where physiologically immature tubers at 60% to 70% maturity are harvested in the growing season without destroying the feeding roots system and sold at the market [37]. The parental plant redevelops small new yam tubers used as seeds for the following planting season before total senescence. The second traditional method of seed production uses small whole tubers from varieties that produce multiple tubers per stand or by sorting small tubers from a ware crop. Still, the chances of infection with pests and diseases are high in the field. The third traditional method involves cutting mature ware tubers into small portions (100–250 g) [38]. These three traditional methods are slow, with a multiplication ratio of about 1:6, compared to some cereals, which multiply at 1:200 [39]; as such, these methods cannot supply seed in sufficient quantity and quality needed for natural growth of the yam sector throughout the yam belt of West Africa.

Micropropagation

In-vitro micropropagation and direct organogenesis are the main biotechnological techniques that could be adopted to obtain healthy seedlings, increasing the potential for cultivation [40, 41], especially in commercially valuable species, such as *D. alata* [42]. Most of the yam accessions were multiplied by nodal vine cuttings through micropropagation.

Sterilization of the Nodal Cuttings

Nodal cuttings were collected from farmers' fields or glasshouse plants for sterilization. Briefly, nodal cuttings were washed in tap water and surface sterilized in 5% sodium hypochlorite (NaOCl) containing two drops of Tween 20 for 15-20 minutes, with continuous agitation, and immersed in 70% ethanol for 1 min. Further, these nodal cuttings were washed thrice with distilled water and transferred to a culture medium. This sterilization technique can be used with any of the yam varieties.

In-vitro Multiplication Through Direct Organogenesis

Direct organogenesis is the micropropagation process from complete plants that can be generated using various explants like axillary buds, leaves, nodal cuttings, and stems Figs. (**1a & 1b**). For the direct yam organogenesis, nodal cuttings or axillary bud explants were sub-cultured and multiplied in Yam Basal Media (YBM) according to the *in-vitro* method reported [43, 44]. The YBM culture medium contains Murashige and Skoog medium (MS basal salts with MS vitamins) [45], sucrose (2%) 6-benzyl amino purine (BAP) (0.05 mg/L), 1-Naphthaleneacetic acid (NAA) (0.02 mg/L), ascorbic acid (25 mg/L), and gelrite (2.4 g/L). The pH of the culture medium can be adjusted to 5.7 and then autoclaved at 121°C for 20 min. Yam cultures are maintained in a growth room at 25 ± 2 °C and light/dark cycle (16h/8h), and photoperiod is provided by cool-white, fluorescent tubes (30 µmol/m^2, sec). This YBM basal medium was successfully tested in 27 yam varieties for micropropagation at the International Institute of Tropical Agriculture (IITA), Kenya. All the yam accessions provide multiple well-rooted shoots with 100% efficiency within 4 to 6 weeks. This culture medium can be applied to any of the yam cultures, either medicinal or food cultivars, for *in-vitro* multiplication and maintenance with slight modifications of the concentrations of the plant growth regulators, especially BAP and NAA. All the plant tissue culture chemicals and reagents were bought from DUCHEFA BIOCHEMIE BV, Haarlem, Netherlands.

Fig. (1). *In-vitro* micropropagation of Yam (*D. rotundata*). a) Regeneration of small shoots with branching, b) Small plantlets with leaves and roots, c) Well developed plants with several leaves in the humid chamber, d) Fully developed potted plants.

Victoria *et al.* reported an efficient protocol for *in-vitro* multiplication of the medicinal yam, *D. bulbifera* L., also known as bitter yam, native to Africa, Asia, and Australia, from nodal and axillary bud explants. The MS medium supplemented with adenine sulfate (15.0 mg/l) and Kinetin (Kn) (2.0 mg/l) was proven to be an excellent medium for shoot regeneration (100%) from nodal explants. Multiple shoots (15-20 shoots/explant) were generated using the same media within 15-20 days. MS basal medium supplemented with NAA (1.0 mg /L) was used for optimal root development [46] (Table **1**).

Table 1. Various types of culture media used for the regeneration of shoots, roots and calli in *Dioscorea* spp.

Species	Explant	Callus Induction	Shoot Induction	Rooting	References
D. oppositifolia and *D. pentaphylla*	Nodal explants	-	MS + BAP (8.8 μM) + Activated charcoal (0.3%)	MS + NAA (2.67 μM)	[49]
D. bulbifera	Nodal and shoot explants	-	MS + Adenine sulphate (15 mg/L) + Kn (2 mg/L)	MS + NAA (1 mg/L)	[46]
D. alata L.	Nodal segments	-	MS + Kn (1.5 mg/L) + IAA (2 mg/L)	MS + IAA (2.5 mg/L)	[51]
D. rotundata	Immature leaf lobes & Axillary bud	MS + Picloram (12 mg/L) + casein hydrolysate (600 mg/L) + Proline (1 g/L)	-	-	[44]
D. rotundata	Stem internode & Root segment	MS + 2,4-D (0.5 mg/L) or MS + Picloram (3 mg/L)	-	-	[44]
D. rotundata	Nodal segments	-	MS + BAP (1.5 mg/L) + IAA (0.5 mg/L)	MS + NAA (0.5 mg/L)	[52]
D. alata and *D. rotundata*	Vine cuttings	-	MS + Kn (0.5 mg/L) + Cysteine (20 mg/L) and MS + BAP (0.05 mg/L) + NAA (0.02 mg/L)	-	[53]

The in-vitro propagation system developed by Lakshmisita *et al.* for medicinal yam, *D. floribunda,* is indigenous to central Mexico to northern central America and grown commercially for diosgenin by MS basal medium supplemented with NAA and Kn [47].

D. opposita, Guangfeng medicinal yam plantlets, rapid micropropagation *in-vitro* technique was studied by Yin *et al.* [48]. They sterilized yam nodal stem cuttings with alcohol (70%) for 20 – 30 seconds, followed by sterile water three times, then washed with mercuric chloride (0.1%) for 10 min, and later washed with sterile water three times. For micropropagation culture medium for stems and axillary buds was MS media supplemented with 6-BAP (2 mg/L) and NAA (0.1 mg/L). The rooting culture medium contained MS-supplemented NAA (0.5 mg/L), which provided proper roots in the nodal cutting.

Poornima and Ravi optimized micropropagation of wild medicinal yam *D. oppositifolia* and *D. pentaphylla* using MS basal media supplemented with 6-BAP (1mg/L) and activated charcoal (0.3%) [49].

Using stem cuttings, Chen *et al.* optimized for rapid *in-vitro* micropropagation of *D. zingiberensis* Wright. Half MS medium was supplemented with 20.0 g/L sucrose and 8.0 g/L agar as basal medium. Lateral buds on nodal cuttings grew into shoots within 20 days after culture on basal medium supplemented with 6-BAP (1 mg/L) and NAA (0.2 mg/L). Rooted plantlets were generated on MS medium with IBA for three weeks [50].

Das *et al.* reported the medium for *in-vitro* multiplication of *D. alata* using nodal explants. The best multiplication was recorded on MS medium with Kn (1.5mg/L) and Indole-3-acetic acid (IAA) (2mg/L) for shoot induction and MS medium with IAA (2mg/L) for rooting [51].

Taha *et al.* mentioned that *D. rotundata in vitro* multiplication was done using nodal segments with MS medium supplemented with BAP (1.5mg/L) and IAA (0.5mg/L), and the rooting was performed with NAA (0.5 mg/L) [52]. Bomer *et al.* optimized a protocol for rapid shoot multiplication for *D. alata* and *D. rotundata* using vine cuttings. The shoot induction was achieved on MS medium with Kn (0.5mg/L) and cysteine (20mg/L) for *D. alata* and MS with BAP (0.05mg/L) and NAA (0.02mg/L) for *D. rotundata* [53].

Indirect Organogenesis Through Embryogenic Calli

In vitro, the production of *Dioscorea* spp. through somatic embryogenesis or indirect organogenesis, and the effect of activated charcoal and various additives on the generation of the primary somatic embryos (SE) and secondary somatic embryos (SSE) were thoroughly investigated [44]. Various explants (leaves, nodes, internodes, roots, and axillary buds) were investigated to develop friable embryogenic calli. The explants were cultured on an MS medium supplemented with various concentrations of auxins, especially 2,4-dichloro phenoxy acetic acid (2,4-D), NAA, and picloram. The cultures were incubated in the complete dark at temperatures ranging from 25 ± 2 °C. The axillary bud explants cultured on MS medium supplemented with picloram (0.5mg/l) favored the production of a large number of calli. The optimal proliferation of calli was achieved on MS medium supplemented with picloram (0.5 mg/L), casein hydrolysate (600 mg/L), and proline (1 g/L). Fully developed eight weeks old calli obtained on MS medium supplemented with picloram (0.5 mg/L) in combination with casein hydrolysate (600 mg/l) and proline (1 g/l) were transferred on MS basal medium augmented with activated charcoal (1%) for the induction of indirect somatic embryos. The calli were incubated for six weeks at 25 ± 2 °C with 16 h daylight and an 8 hr dark cycle for the proliferation of calli to develop complete shoots.

Chen *et al.* reported that shoots proliferation through the mature leaf's explants of *D. zingiberensis*. The 80% of leaf explants developed calli on the culture media containing 6-BAP (1-5 mg/L) and NAA (1-2 mg/L) in 60 days, and calli generated adventitious buds within 50 days. All the adventitious buds developed roots on the medium with indole butyric acid (IBA) in 20 days; the regenerated plantlets were planted in the soil with a high success rate of survival [54].

The embryogenic calli were cultured to MS basal medium supplemented with BAP (0.4 mg/l) to induce embryos. The lower concentration of BAP works exceptionally to induce the germination of embryos. Well-developed germinated embryos were separated from the clump of calli and transferred separately in Yam basal media (MS basal salts with MS vitamins), sucrose (2%), BAP (0.05 mg/L), NAA (0.02 mg/L), ascorbic acid (25 mg/L), and gelrite (2.4 g/L) for proper shoot and root development.

Weaning and Potting

Well-rooted plantlets were transferred to a sterile poly pot (2.5 cm) filled with peat moss+ vermiculite (1:1) (Fig. **1c**). Plants were placed in a growth room at 28°C with relative humidity up to 70-80%. For acclimatization, plantlets were kept in a humid chamber for three weeks. Plants were irrigated with ¼ MS solution in three days intervals up to 3 weeks, and well-developed shoots were shifted to larger pots (20 cm diameter) containing a mixture of soil and vermiculite (1:1). These plants were ready to plant in the soil after four weeks. The adapted plants were then transferred to pots containing soil and compost in a ratio of 1:1 and kept in the screen house for further growth and observation (Fig. **1d**).

Temporary Immersion Bioreactor System (TIBS)

A temporary immersion bioreactor system (TIBS) is to scale up the micro-propagation of the plantlets to generate the certified seed [55]. The clean planting materials constitute stocks for the rapid multiplication of superior yam varieties, ensuring that the virus is not passed on to subsequent generations. Plantlets from TIBS are of superior quality than those from conventional tissue culture (CTC), as they are more vigorous and resilient to post-flask acclimatization due to more efficient process control. Large batches are handled more easily for scale-up propagation with lower risks of mix-ups. The propagation ratio in TIBS was five to six per plantlet every 8–10 weeks compared to three to four every 12–16 weeks with CTC. In addition, the rate of subculturing in TIBS was 100 cuttings per person per hour, double that of CTC, reducing the cost of labor as well, and the use of liquid without agar/geltrite reduces medium cost by 50%. This is an excellent procedure for mass production of medicinal yam accessions at commercial scaling up of *in-vitro* plantlets.

CONCLUSION

Yam, a versatile crop, plays a crucial role in ensuring food security and has the potential to revolutionize the medicinal industry. The *in-vitro* multiplication technique can provide an excellent way to produce a large number of clean planting materials to enhance the process of medicinal uses commercially. There are several yam varieties available to explore more potential to develop plant-based medicines that can cure several human diseases in the future, and it has enormous potential to research this medicinal crop.

REFERENCES

[1] Fu Y, Chen S, Huang PY, Li YJ. Application of bubble separation for quantitative analysis of choline in *Dioscorea* (yam) tubers. J Agric Food Chem 2005; 53(7): 2392-8. [http://dx.doi.org/10.1021/jf048501h] [PMID: 15796568]

[2] Coursey DG. The origins and domestication of yams in Africa. Gastronomy: the anthropology of food and food habits 1975; 187-212.

[3] Maroya N, Balogun M, Aighewi B, Mignouna DB, Kumar PL, *et al.* Transforming yam seed systems in West Africa.Root. Tuber and Banana Food System Innovations Springer 2022. [http://dx.doi.org/10.1007/978-3-030-92022-7_14]

[4] FAOSTAT. Food and Agriculture Organization of the United Nations. 2020. Available from: https://www.fao.org/faostat/en/#home

[5] FAOSTAT. Food and Agriculture Statistics. 2021. Available from: https://www.fao.org/food-agriculture-statistics/en/

[6] Waris R, Tripathi S, Shukla AC, Agnihotri P. An overview of the genus *Dioscorea* L. (Dioscoreaceae) in India. Plant Sci Today 2021; 8(1): 72-8. [http://dx.doi.org/10.14719/pst.2021.8.1.878]

[7] Mishra S, Kumar S. *Dioscorea dumetorum* (Kunth) T. Durand & H. Schinz.: A new addition to the flora of India. Species 2021; 22(69): 84-8.

[8] Mustafa A, Ahmad A, Tantray AH, Parry PA. Ethnopharmacological potential and medicinal uses of miracle herb *Dioscorea* spp. J Ayurvedic herb Med 2018; 4(2): 79-85.

[9] Bantilan C. Health and nutrition benefits of yams. New York: Healthline Media, Inc. 2019.

[10] Mignouna HD, Abang MM, Asiedu R. Genomics of yams, a common source of food and medicine in the tropics. Genomics of Tropical Crop Plants 2008; 549-70. [http://dx.doi.org/10.1007/978-0-387-71219-2_23]

[11] Komesaroff PA, Black CVS, Cable V, Sudhir K. Effects of wild yam extract on menopausal symptoms, lipids and sex hormones in healthy menopausal women. Climacteric 2001; 4(2): 144-50. [http://dx.doi.org/10.1080/cmt.4.2.144.150] [PMID: 11428178]

[12] Hill NR, Fatoba ST, Oke JL, *et al.* Global prevalence of chronic kidney disease : A systematic review and meta-analysis. PLoS One 2016; 11(7): e0158765. [http://dx.doi.org/10.1371/journal.pone.0158765] [PMID: 27383068]

[13] Malviya N, Jain S, Malviya S. Antidiabetic potential of medicinal plants. Acta Pol Pharm 2010; 67(2): 113-8. [PMID: 20369787]

[14] Mukesh R, Namita P. Medicinal plants with antidiabetic potential : A review. Am-Eurasian J Agric Environ Sci 2013; 13(1): 81-94.

[15] Bhandari MR, Kasai T. Kawabata. Nutritional evaluation of yam (*Dioscore*a spp.) in Nepal. Food Chem 2003; 82: 619-23. [http://dx.doi.org/10.1016/S0308-8146(03)00019-0]

[16] Agbai O. Anti-sickling effect of dietary thiocyanate in prophylactic control of sickle cell anemia. J Natl Med Assoc 1986; 78(11): 1053-6. [PMID: 3795284]

[17] Marker RE, Sterols CV. The preparation of testosterone and related compounds from sarsasapogenin and diosgenin. J Am Chem Soc 1940; 62(9): 2543-7. [http://dx.doi.org/10.1021/ja01866a077]

[18] Wireko I, Béland D. The challenge of healthcare accessibility in sub-Saharan Africa: The role of ideas and culture. Eur J Int Manag 2013; 7(2): 171-86. [http://dx.doi.org/10.1504/EJIM.2013.052852]

[19] Cheng J, Li D. Advances in research on the function and active ingredients of Chinese yam. Northwest Pharma J 2010; 25(5): 398-400.

[20] Wang Q, Wyman DA, Xu JF, *et al.* Petrogenesis of Cretaceous adakitic and shoshonitic igneous rocks in the Luzong area, Anhui Province (eastern China): Implications for geodynamics and Cu–Au mineralization. Lithos 2006; 89(3-4): 424-46. [http://dx.doi.org/10.1016/j.lithos.2005.12.010]

[21] Liu SY, Wang JY, Shyu YT, Song LM. Studies on yams (*Dioscorea* spp.) in Taiwan. J Chin Med 1995; 6: 111-26.

[22] Araghiniknam M, Chung S, Nelson-White T, Eskelson C, Watson RR. Antioxidant activity of *Dioscorea* and *dehydroepiandrosterone* (DHEA) in older humans. Life Sci 1996; 59(11): PL147-57. [http://dx.doi.org/10.1016/0024-3205(96)00396-7] [PMID: 8795709]

[23] Farombi EO, Nwankwo JO, Emerole GO. Possible modulatory effect of browned yam flour diet on chemically-induced toxicity in the rat. Food Chem Toxicol 1997; 35(10-11): 975-9. [http://dx.doi.org/10.1016/S0278-6915(97)87266-3] [PMID: 9463531]

[24] Pillai S, Netravali IA, Cariappa A, Mattoo H. Siglecs and immune regulation. Annu Rev Immunol 2012; 30(1): 357-92. [http://dx.doi.org/10.1146/annurev-immunol-020711-075018] [PMID: 22224769]

[25] Xu GJ, Xu LS. Species systematization and quality evaluation of commonly used Chinese traditional drugs. Fuzhou: Fujian Sci Tech Press 1997; II.

[26] Zhu JK. Abiotic stress signaling and responses in plants. Cell 2016; 167(2): 313-24. [http://dx.doi.org/10.1016/j.cell.2016.08.029] [PMID: 27716505]

[27] Lin CJ, Chen TL, Tseng YY, *et al.* Honokiol induces autophagic cell death in malignant glioma through reactive oxygen species-mediated regulation of the p53/PI3K/Akt/mTOR signaling pathway. Toxicol Appl Pharmacol 2016; 304: 59-69. [http://dx.doi.org/10.1016/j.taap.2016.05.018] [PMID: 27236003]

[28] Okonkwo SN. The botany of the yam plant and its exploitation in enhanced productivity of the crop. Adv Yam Res 1985; 3-25.

[29] Umar AG, Nwafor MS, Likita S, Adoko S. The Indigenous yam storage technology and post-harvest losses in Nigeria. Intern J food and Agri Res 2016; 5(2): 113-39.

[30] Amusa AN, Adegbite AA, Muhammed S, Baiyewu RA. Yam diseases and its management in Nigeria. Afr J Biotechnol 2003; 2(12): 497-502. [http://dx.doi.org/10.5897/AJB2003.000-1099]

[31] Adegbite AA, Adesiyan SO. Root extracts of plants to control root-knot nematode on edible soybean. J Veg Sci 2006; 12(2): 5-12. [http://dx.doi.org/10.1300/J484v12n02_02]

[32] Guo H, Xu M, Hu Q. Changes in near-surface wind speed in China: 1969-2005. Int J Climatol 2011; 31(3): 349-58. [http://dx.doi.org/10.1002/joc.2091]

[33] Xu J, Grumbine RE, Shrestha A, *et al.* The melting Himalayas: Cascading effects of climate change on water, biodiversity, and livelihoods. Conserv Biol 2009; 23(3): 520-30. [http://dx.doi.org/10.1111/j.1523-1739.2009.01237.x] [PMID: 22748090]

[34] Li L, Zhang M, Bhandari B. Influence of drying methods on some physicochemical, functional and pasting properties of Chinese yam flour. Lebensm Wiss Technol 2019; 111(111): 182-9. [http://dx.doi.org/10.1016/j.lwt.2019.05.034]

[35] Wang S, Bai Y, Shen C, *et al.* Auxin-related gene families in abiotic stress response in *Sorghum bicolor*. Funct Integr Genomics 2010; 10(4): 533-46. [http://dx.doi.org/10.1007/s10142-010-0174-3] [PMID: 20499123]

[36] Katung PD, Idem NU A, Showemimo FA. Tuber and fiber crops of Nigeria: Principles of production and utilization. Ade Commercial Press 2006; 22: p. 239.

[37] Aighewi BA, Asiedu R, Maroya N, Balogun M. Improved propagation methods to raise the productivity of yam (*Dioscorea rotundata* Poir.). Food Secur 2015; 7(4): 823-34. [http://dx.doi.org/10.1007/s12571-015-0481-6]

[38] Aighewi BA, Akoroda MO, Asiedu R. Seed yam (*Dioscorea rotundata* Poir) production, storage, and quality in selected yam zones of Nigeria. Afr J Root Tuber Crops 2002; 5(1): 20-3.

[39] Mbanaso ENA, Egesi CN, Okogbenin E, Ubalua AO, Nkere CK. Plant biotechnology for genetic improvement of root and tuber crops.Root and Tuber Crops Research for Food Security and Empowerment. Umudike, Nigeria: National Root Crops Research Institute 2011; pp. 45-64.

[40] Alizadeh S, Mantell SH, MariaViana A. *In vitro* shoot culture and microtuber induction in the steroid yam *Dioscorea composita* Hemsl. Plant Cell Tissue Organ Cult 1998; 53(2): 107-12. [http://dx.doi.org/10.1023/A:1006036324474]

[41] Royero M, Vargas TE, Oropeza M. Micropropagación y organogénesis de *Dioscorea alata* (ñame). Interciencia 2007; 32(4): 247-52.

[42] Balogun MO, Fawole I, Ng SYC, Ng NQ, Shiwachi H, Kikuno H. Interaction among cultural factors in microtuberization of white yam (*Dioscorea rotundata*). Trop Sci 2006; 46(1): 55-9. [http://dx.doi.org/10.1002/ts.61]

[43] Nyaboga E, Tripathi JN, Manoharan R, Tripathi L. Agrobacterium-mediated genetic transformation of yam (*Dioscorea rotundata*): An important tool for functional study of genes and crop improvement. Front Plant Sci 2014; 15(5): 463.

[44] Manoharan R, Tripathi JN, Tripathi L. Plant regeneration from axillary bud derived callus in white yam (*Dioscorea rotundata*). Plant Cell Tissue Organ Cult 2016; 126(3): 481-97. [http://dx.doi.org/10.1007/s11240-016-1017-2]

[45] Murashige T, Skoog F. A revised medium for rapid growth and bioassays with tobacco tissue cultures. Physiol Plant 1962; 15(3): 473-97. [http://dx.doi.org/10.1111/j.1399-3054.1962.tb08052.x]

[46] Victoria PK, Batra P, Dhillon S, Chowdhury VK. *In vitro* micropropagation of medicinal air yam (*Dioscorea bulbifera*). Res Crops 2011; 12: 226-9.

[47] Sita GL, Bammi RK, Randhawa GS. Clonal propagation of *Dioscorea floribunda* by tissue culture. J Hortic Sci 1976; 51(4): 551-4. [http://dx.doi.org/10.1080/00221589.1976.11514725]

[48] Yin MH, Xu ZJ, Zhang SQ, Lv SJ, Zeng YH, *et al.* Study on rapid micropropagation *in vitro* technique of Guangfeng medicinal yam (*Dioscorea opposita*) Zhong Yao Cai. Chinese 2015; 38(11): 2245-9. [PMID: 27356371]

[49] Poornima GN, Ravishankar RV. *In vitro* propagation of wild yams, *Dioscorea oppositifolia* (Linn) and *Dioscorea pentaphylla* (Linn). Afr J Biotechnol 2007; 6(20): 2348-52. [http://dx.doi.org/10.5897/AJB2007.000-2368]

[50] Chen Y, Fan J, Yi F, Luo Z, Fu Y. Rapid clonal propagation of *Dioscorea zingiberensis*. Plant Cell Tissue Organ Cult 2003; 73(1): 75-80. [http://dx.doi.org/10.1023/A:1022683824635]

[51] Supriya D, Manabendra DC, Pranab BM. Micropropagation of *Dioscorea alata* L. through nodal segments. Afr J Biotechnol 2013; 12(47): 6611-7. [http://dx.doi.org/10.5897/AJB2013.12191]

[52] Taha SS, Abdelaziz ME. *In vitro* propagation of yam *via* nodal segment culture. Bios Res 2017; 14(4): 1217-22.

[53] Bömer M, Rathnayake AI, Visendi P, *et al.* Tissue culture and next-generation sequencing: A combined approach for detecting yam (*Dioscorea* spp.) viruses. Physiol Mol Plant Pathol 2019; 105: 54-66. [http://dx.doi.org/10.1016/j.pmpp.2018.06.003] [PMID: 31007374]

[54] Chen YQ, Fan JY, Yi F, Fu YS, Luo ZX. Studies on plantlet regeneration from the mature leaves of *Dioscorea zingiberensis*. Zhongguo Zhong Yao Za Zhi 2004; 29(2): 129-32. [PMID: 15719676]

[55] Balogun MS, Huang Y, Qiu W, Yang H, Ji H, Tong Y. Updates on the development of nanostructured transition metal nitrides for electrochemical energy storage and water splitting. Mater Today 2017; 20(8): 425-51. [http://dx.doi.org/10.1016/j.mattod.2017.03.019]

CHAPTER 13

Micropropagation of the Medicinal Plant 'Sarpagandha' [*Rauvolfia serpentina* (L.) Benth. *ex* Kurz] and its Applications in Human Welfare

Suproteem Mukherjee[1], **Diptesh Biswas**[1] and **Biswajit Ghosh**[1,*]

[1] *Plant Biotechnology Laboratory, Post Graduate Department of Botany, Ramakrishna Mission Vivekananda Centenary College, Rahara, Kolkata -700118, India*

Abstract: *Rauvolfia serpentina* (L). Benth. ex Kurz., commonly known as Sarpagandha (Indian snakewood), of the family Apocynaceae, is a medicinally important woody shrub. Since ancient times, the root of this shrub has been used for treating numerous diseases, especially hypertension, mental agitation and cardiovascular diseases. In addition to eighty different alkaloids, all well-known for their pharmaceutical properties, the plant also contains reserpine, recognized as the world's first antihypertensive drug. Thus, the demand for this plant has only grown in the pharmaceutical industry. However, overexploitation and abysmal traditional propagation methods have endangered this valuable species' natural vegetation, creating an unpleasant gap between the demand and availability. In this scenario, the *in vitro* micropropagation technique comes as an alternative strategy to help replenish this threatened shrub's natural vegetation loss and commercial needs. Furthermore, the beneficial features of the plant tissue culture technique by providing genetically uniformed disease-free true-to-type plant propagation within a short time, and conserving elite variety plantlets makes this technique an inevitable tool for the rapid production of economically important plants in the 21st century. Therefore, this chapter focuses on the different *in vitro* plant tissue culture techniques applied to regenerate *R. serpentina* plants. In addition, the roles of various physical and chemical factors that could affect the regeneration rate, geographical distribution, bioactive compounds and their bioactivity have also been discussed. The comprehensive data could be helpful for further studies on this valuable plant.

Keywords: Bioactive compounds, Micropropagation, Organogenesis, Pharmacology, *Rauvolfia serpentina*, Somatic Embryogenesis.

[*] **Corresponding author Biswajit Ghosh:** Plant Biotechnology Laboratory, Post Graduate Department of Botany, Ramakrishna Mission Vivekananda Centenary College, Rahara, Kolkata -700118, India; Tel: 91-9432113696; E-mail: ghosh_b2000@yahoo.co.in

T. Pullaiah (Ed.)

INTRODUCTION

Rauvolfia serpentina (L). Benth. ex Kurz., commonly known as Indian snakewood or Sarpagandha, is considered to be one of the most valuable plants, as it shows a wide range of medicinal properties and contains the world's first antihypertensive drug, reserpine [1]. In literature, like Ayurveda, Siddha, and Unani, it has been found that the roots of *R. serpentina* have been used to cure diseases like high blood pressure, anxiety, insomnia, epilepsy and several central nervous system diseases [2]. *Rauvolfia serpentina* originated in tropical and sub-tropical climatic regions of South-East Asia, and grows up to the elevation of 1300–1400 m. It is indigenous to moist deciduous forests of the Himalayas and Indian peninsula and occurs in India, Bangladesh, Bhutan, Nepal, Pakistan, Sri Lanka, China, Myanmar, Indonesia, Malaysia and Vietnam [3 - 9].

R. serpentina contains an array of bioactive compounds [7, 10]. Among these, the alkaloids are chiefly the reasons for the major bioactivities of *R. serpentina.* The roots and root bark of *R. serpentina* are potent sources of more than 30 indole NAA alkaloids (0.7–2.4%), the most crucial one being reserpine [6, 10, 11]. Reserpine is the most effectively utilized medicinal phytocompound isolated from *R. serpentina* and is a natural tranquilizer. It is a sympathomimetic agent that acts on the sympathetic nervous system, controls cardiac contractions and heart rate, and lowers blood pressure during hypertension. Even in minimal oral doses, reserpine demonstrates its antihypertensive actions by acting as a depressant on the central and peripheral nervous systems [11, 12].

Nevertheless, the overexploitation of this shrub for pharmaceutical utilization has threatened its natural vegetation in some parts of the Southern Western Ghats and North-East regions of India [13]. Besides overexploitation, other issues that caused the decline of *R. serpentina* vegetation at a high rate are its nominal seed germination rate because of poorly viable seeds and a significantly lower rate of vegetative propagation through cuttings [14]. These issues have been marked as severe drawbacks for large-scale production, and eventually, *in vitro* propagation became the solution for these issues [3].

The optimum production of *R. serpentina* can be achieved using several formulations containing various combinations of cytokinin, auxin and additives. *In vitro* propagation of *R. serpentina* has been in attention since the late 20th century [15]. In the 21st century, newly adopted biotechnological approaches made this technique more competent for industrial and conservational purposes [16].

MICROPROPAGATION OF *RAUVOLFIA SERPENTINA*

Factors Affecting Micropropagation

For the *in vitro* regeneration of *R. serpentina* through micropropagation, shoot tips, leaves, nodal and internodal pieces, roots, and embryos have been chosen by several experimenters as explants (Fig. **1**, Table **1**). Many reports on *R. serpentina* suggest that a temperature of 24–25 ± 1–2° C can be the best for plant growth [4, 6]. Besides the temperature, a light intensity of 3000 lux and a 16h photoperiod were found to be optimum for *in vitro* regeneration [10]. The *in vitro* culture experiments have maintained a relative humidity of 50–70% for proper growth of tissues [3]. As the *in vitro* propagation is conducted in a heterotrophic condition, it is evident that the supplemented amount of carbon has played a pivotal role in the *R. serpentina* organ development process. Commonly 3.0% sucrose has been widely used in the case of *R. serpentina* micropropagation. However, some investigators found that half strength of sucrose or a range of 0.03-3.0% sucrose can also be an alternative (Table **1**). In *R. serpentina*, many authors have used MS (Murashige and Skoog, 1962) medium for its relatively high content of nitrate-ammonium salts (Table **2**). Woody plant media, another widely used nutrient medium for woody shrubs, is also used in the *R. serpentina* culture [36].

Fig. (1). Micropropagation of Rauvolfia serpentina; (**a–b**) initiation and growth of shoot primordia, (**c–d**) multiplication of shoots, (**e**) *In vitro* rooting, (**f**) a rooted plantlet, (**g**) acclimatization of rooted plantlet, (**h**) *ex vitro* plant.

Table 1. Direct organogenesis experiments reported from *R. serpentina* explants.

S. No.	Explant	Basal Medium	Carbohydrate Source	PGR Combination and Concentration	Additives	Response	References
1.	Shoot tip	MS	0.03% sucrose	2.5 mg/l BAP + 0.1 mg/l NAA	NM	Multiple shoot	[17]
				0.2 mg/l IBA + 0.2 mg/l NAA		Rooting	
2.	Shoot apices, Nodal cuttings	MS	3.0% sucrose	2.0 mg/l BAP + 1.5 mg/l BAP + 0.5 mg/l NAA	NM	Shoot multiplication	[18]
				2.0 mg/l NAA and 1.5 mg/l BAP		Rooting	
3.	Nodal explants	MS	2.5% sucrose	1.0 mg/l BAP + 0.1 mg/l IBA	NM	Shoot multiplication	[19]
				0.5 mg/l IBA		Rooting	
4.	Nodal explants	MS	3.0% sucrose	10.0 µM BAP + 0.5 µM IAA	50 µM ascorbic acid, 25 µM each of AdS, arginine and citric acid	Shoot multiplication	[20]
				50.0 µM IBA + 50.0 µM NOA	NM	Rooting	
5.	Shoot apex	MS	3.0% sucrose	4.0 mg/l BAP + 0.5 mg/l NAA	NM	Shoot multiplication	[21]
6.	Shoot apices	MS	3.0% sucrose	4.4 µM BAP + 0.54 µM NAA	NM	Shoot multiplication	[22]
7.	Node	MS	NM	1.0 mg/l BAP	NM	Shoot multiplication	[23]
8.	Nodal explant	MS	3.0% sucrose	0.5 mg/l BAP + 0.5 mg/l 2,4-D	NM	Shoot multiplication	[24]
				2.0 mg/l IBA		Rooting	
9.	Nodal explant	MS	3.0% sucrose	1.0 mg/l NAA + 2.0 mg/l BAP	NM	Shoot proliferation	[25]

(Table 1) cont.....

S. No.	Explant	Basal Medium	Carbohydrate Source	PGR Combination and Concentration	Additives	Response	References
10.	Shoot apex, single node	MS	3.0% sucrose	4.4 µM BAP + 0.54 µM NAA	NM	Shoot regeneration, Shoot multiplication	[12]
				5.71 µM IAA + 4.14 µM IBA		Rooting	
11.	Shoot	MS	3.0 % sucrose	0.5 mg/l BAP	NM	Shoot multiplication	[26]
12.	Leaf, nodal segment	MS	1.5 % sucrose	2.0 mg/l BAP + 0.1mg/l IAA or 2.0 mg/l BAP + 0.2 mg/l NAA	NM	Shoot multiplication	[27]
				PGR free		Rooting	
13.	Nodal explant	MS	3.0% sucrose	1.0 mg/l BAP + 0.125 mg/l IBA	NM	Shoot multiplication	[28]
14.	*In vitro* generated shoot apices, nodal segment	MS	3.0% sucrose	2.0 mg/l BAP + 0.5 mg/l BAP + 0.5 mg/l NAA	NM	Shoot multiplication	[29]
				1.0 ppm NAA		Rooting	
15.	Nodal segment	MS	3.0% sucrose	10.0 mg/l BAP + 0.1 mg/l NAA	NM	Shoot multiplication	[30]
				1.0 mg/l NAA		Rooting	
16.	Nodal segment	MS	3.0% sucrose	1.0 µM BA + 1.0 µM NAA	NM	Axillary shoot bud development, shoot multiplication	[31]
				50.0 µM NAA		Rooting	
17.	Shoot apex, nodal segment	MS	3.0% sucrose	1.0 mg/l BAP + 0.1 mg/l Kn or 1.0 mg/l BAP + 0.1 mg/l Kn + 0.1 mg/l GA_3	NM	Shoot multiplication	[32]
				1.0 mg/l IBA + 1.0 mg/l NAA		Rooting	
18.	Apical bud, nodal explant	MS	3.0% sucrose	2.22 µM BAP + 2.32 µM Kn + 0.54 µM NAA	NM	Shoot multiplication	[33]

(Table 1) cont.....

S. No.	Explant	Basal Medium	Carbohydrate Source	PGR Combination and Concentration	Additives	Response	References
19.	Nodal explant	MS	3.0% sucrose	17.74 μM BAP + 32.57 μM AdS	NM	Shoot multiplication	[34]
20.	Young shoot	WPM	3.0% sucrose	5.0 μM BA + 1.0 μM NAA	NM	Shoot multiplication	[35]
				1.0 μM NAA		Rooting	
21.	Nodal explant	WPM	3.0% sucrose	7.5 μM BA + 2.5 μM NAA	NM	Shoot regeneration	[36]
				150 μM IBA		Rooting	
22.	Shoot	MS	3.0% sucrose	5.0 μM BAP + 1.0 μM NAA	NM	Shoot multiplication	[37]
23.	Shoot	MS	3.0% sucrose	2.5 mg/l BAP	NM	Shoot multiplication	[38]
				0.5 mg/l NAA		Rooting	
24.	Leaf	MS	NM	1.5 mg/l BAP + 1.5 mg/l Kn	NM	Shoot regeneration	[39]
				0.5 mg/l IBA		Rooting	
25.	Shoot	MS	3.0% sucrose	2.0 mg/l BAP + 0.25 mg/l NAA	NM	Shoot multiplication	[10]
		½ MS	2.0% sucrose	0.5 mg/l IBA		Rooting	
26.	Shoot tips	MS	3.0% sucrose	2.5 mg/l BAP + 0.1 mg/l NAA	0.5 mg/l thiamine-HCl + 1.0 mg/l pyridoxine HCl + 0.5 mg/l nicotinic acid	Shoot proliferation	[9]
		½ MS	NM	0.4 mg/l NAA + 0.1 mg/l IAA	NM	Rooting	
27.	Internodes	MS	NM	2.0 mg/l BAP	NM	Direct shoot regeneration	[40]
				0.2 mg/l IBA		Root induction	
28.	Shoot apices	MS	3.0% sucrose	0.5 mg/l BAP + 0.5 mg/l IAA	NM	Shoot regeneration	[41]
				3.0 mg/l BAP + 3.0 mg/l IAA		Shoot elongation	
				3.0 mg/l IBA		Rooting	

(Table 1) cont.....

S. No.	Explant	Basal Medium	Carbohydrate Source	PGR Combination and Concentration	Additives	Response	References
29.	Shoot	MS	3.0% sucrose	0.2 mg/l IAA + 0.2 mg/l BAP + 0.2 mg/l Kn	NM	Shoot regeneration	[42]
				0.5 mg/l IBA		Rooting	
30.	Nodal segment	MS	3.0% sucrose	0.2-0.5 mg/l TDZ or 0.2-0.5 mg/l TDZ + 0.5 mg/l NAA	NM	Shoot proliferation	[43]
				0.1 mg/l IBA		Rooting	
31.	Nodal segments	MS	3.0% sucrose	10.0 µM BAP + 0.5 µM NAA	20 µM CuSO4 or 25 µM ZnSO4	Shoot Multiplication	[44]
		½ MS		0.5 µM IBA	NM	Rooting	
32.	Nodal segments	MS	3.0% sucrose	50.0 µM TDZ	NM	Shoot multiplication	[45]
				0.5 µM IBA		Rooting	
33.	Nodal segments	MS	3.0% sucrose	1.5 mg/l BAP + 0.5 mg/l NAA	5% $^{v}/_{v}$ Coconut water	Shoot multiplication	[46]
		½ MS	NM	1.0 mg/l NAA + 0.5 mg/l IBA	NM	Rooting	
34.	Nodes	MS	NM	2.5 mg/l Kn + 0.4 mg/l IAA	NM	Shoot multiplication	[47]
				2.0 mg/l IAA + 0.6 mg/l Kn		Rooting	
35.	Shoot tips	MS	NM	3 mg/l BAP + 0.5 mg/l NAA	NM	Shoot multiplication	[48]
				0.5 mg/l IAA or IBA		Rooting	
36.	Apical and nodal buds	MS	3.0% sucrose	1.0 mg/l BAP + 0.125 mg/l IBA	NM	Shoot induction	[49]
37.	Shoot apices	Liquid MS	3.0% sucrose	2.0 mg/l BAP + 2.0 mg/l IBA + 1.0 mg/l Kn	NM	Shoot multiplication	[11]
				3.0 mg/l IBA + 1.0 mg/l Kn		Shoot growth	

(Table 1) cont.....

S. No.	Explant	Basal Medium	Carbohydrate Source	PGR Combination and Concentration	Additives	Response	References
38.	Shoot tips	MS	3.0% sucrose	1.5 mg/l BAP	NM	Shoot multiplication	[50]
				1.0 mg/l IAA		Rooting	
39.	Nodal explants	MS	NM	0.8 μM TDZ	NM	Shoot regeneration	[5]
				1.0 μM IBA		Rooting	
40.	Leaf and nodes	MS	NM	2.0 mg/l BAP + 0.5 mg/l IAA + 0.2 mg/l NAA	NM	Shoot multiplication	[51]
				0.2 mg/l IBA		Rooting	
41.	Shoot tips, nodes	MS	NM	2.5 mg/l BAP	NM	Shoot regeneration	[52]
				1.0 mg/l NAA		Rooting	

Table 2. Indirect organogenesis experiments reported from *R. serpentina* explants.

S. No.	Explant	Basal Medium	Carbohydrate Source	PGR Combination and Concentration	Response	References
1.	Nodal segment	MS	5.0% sucrose	2.0 mg/l BAP + 1.0 mg/l 2,4-D	Callus induction	[24]
				1.0 mg/l BAP + 1.5 mg/l AdS	Shoot formation	
				2.0 mg/l IBA	Root formation	
2.	Nodal segment	MS	3.0% sucrose	3.0 mg/l 2,4-D + 2.0 mg/l BAP	Callus formation	[25]
				4.0 mg/l TDZ	Shoot formation	
			4.0% sucrose	3.0 mg/l IBA	Root formation	
3.	Root	MS	3.0% sucrose	10.0 mg/l IBA + 10.0 mg/l NAA	Callus formation	[26]

(Table 2) cont.....

S. No.	Explant	Basal Medium	Carbohydrate Source	PGR Combination and Concentration	Response	References
4.	Leaf, Nodal segment	MS	1.5% sucrose	1.5 mg/l BAP + 0.5 mg/l IAA + 2.5 mg/l 2,4-D	White granular callus	[27]
				0.1 mg/l Picloram + 1.0 mg/l Kn	Loose friable callus	
				3.0 mg/l BAP + 1.0 mg/l NAA + 1.0 mg/l Kn	Shoot formation	
				PGR-free	Root formation	
5.	Shoot, leaf	MS	3.0% sucrose	0.125 mg/l 2,4-D + 1.0 mg/l BAP and 0.125 mg/l IBA + 1.5 mg/l BAP	Callus	[28]
6.	Leaves	MS	3.0% sucrose	2.0 mg/l NAA + 2.0 mg/l Kn	Callus induction	[53]
7.	Node, internode, shoot apex, leaf	MS	3.0% sucrose	1.0 ppm NAA + 2.0 ppm 2,4-D	Callus induction	[29]
				2.0 ppm BAP	Shoot formation	
				1.0 ppm NAA	Root formation	
8.	Nodal explant	MS	3.0% sucrose	0.5 mg/l BAP + 2.0 mg/l NAA	Callus induction	[54]
				2.0 mg/l BA + 0.2 mg/l NAA	Adventituous shoot formation	
		½ MS	3.0% sucrose	1.0 mg/l IBA + 1.0 mg/l IAA	Root formation	
9.	Node, internode	MS	3.0% sucrose	1.0 mg/l 2,4-D + 1.0 mg/l Kn	Callus induction	[55]
				3.0 mg/l 2,4-D+ 10 % coconut milk 1.0 mg/l IAA	Shoot regeneration	
10.	Leaf	MS	3.0% sucrose	1.0 mg/l BAP + 0.5 mg/l IAA	Callus formation	[56]
				2.5 mg/l BAP + 0.4 mg/l IAA with 1.0 ppm GA_3	Shoot regeneration	
				2.5 mg/l BAP + 0.5 mg/l IAA + 0.5 mg/l NAA	Rooting	

(Table 2) cont.....

S. No.	Explant	Basal Medium	Carbohydrate Source	PGR Combination and Concentration	Response	References
11.	Leaf, nodal segments	MS	3.0% sucrose	2.0 mg/l 2,4-D + 1.0 mg/l BAP	Callus	[32]
				1.0 mg/l BAP + 0.1 mg/l Kn + 0.1 mg/l GA_3	Shoot formation	
				1.0 mg/l IBA + 1.0 mg/l IAA	Rooting	
12.	Young leaves, stem	MS	3.0% sucrose	2.0 mg/l 2,4-D + 1.0 mg/l BAP	Callus induction	[57]
				2.0 mg/l BAP + 0.5 mg/l NAA	Shoot formation	
		½ MS		0.2 mg/l IBA + 0.2 mg/l NAA	Rooting	
13.	Leaf	MS	NM	Combination of BAP and NAA	Callus induction	[34]
				22.19 µM BAP + 8.64 µM AdS	Shoot regeneration	
14.	Leaf, stem	MS	3.0% sucrose	2.5 mg/l 2,4-D	Callus induction	[38]
				0.2 NAA mg/l + 1.5 mg/l BA	Shoot regeneration	
				0.5 mg/l NAA	Rooting	
15.	Leaves	MS	NM	NM	Callus induction	[39]
				3.0 mg/l BAP + 2.0 IAA	Indirect shoot regeneration	
				0.5 IBA	Rooting	
16.	Shoot apices	MS	3.0% sucrose	0.5 mg/l IAA + 0.5 mg/l NAA	Callus induction, shoot regeneration	[41]
				3.0 mg/l BAP + 3.0 mg/l IAA	Shoot elongation	
				3.0 mg/l IBA	Rooting	
17.	Shoot	MS	3.0% sucrose	0.5 mg/l BAP + 1.0 mg/l Kn + 0.5 mg/l IAA	Callus induction	[42]
				0.5 mg/l BAP + 0.5 mg/l Kn + 1.0 mg/l IAA	Shoot regeneration	
				0.5 mg/l IBA	Rooting	

(Table 2) cont.....

S. No.	Explant	Basal Medium	Carbohydrate Source	PGR Combination and Concentration	Response	References
18.	Nodal segments, leaf discs	MS	3.0% sucrose	2.0 mg/l 2,4-D/NAA or 2.0 mg/l 2,4-D/NAA with 0.5 mg/l TDZ	Callus initiation and plantlet regeneration	[43]
				0.1 mg/l IBA	Rooting	
19.	Leaf and internode	MS	3.0% sucrose	1.5 mg/l 2,4-D	Callus induction	[46]
		½ MS	NM	1.0 mg/l NAA + 0.5 mg/l IBA	Rooting	
20.	Mature cotyledons and embryos	MS	NM	2.0 mg/l 2,4-D + 0.5 mg/l BAP	Callus induction	[58]
21.	Leaf	½ MS	3.0% sucrose	5.0 mg/l 2,4-D + 2.0 mg/l NAA	Callus production	[59]
		MS		6.0 BAP + 2.0 GA_3	Shoot regeneration and elongation	
		½ MS		3.0 mg/l IBA	Rooting	
22.	Node, leaf	MS	3.0% sucrose	2.0 mg/l 2,4-D + 0.5 mg/l BAP	Callus induction	[60]
				0.5 mg/l 2,4-D + 0.25 mg/l BAP	Callus proliferation	
23.	Inflorescence segments	MS	2.0% sucrose	0.5, 1.0 mg/l NAA	Callus induction	[8]
				1.0 mg/l BAP	Plantlet formation	
24.	Leaves	MS	NM	2.0 mg/l BAP	Callus induction and indirect shoot multiplication	[51]
				0.2 mg/l IBA	Rooting	
25.	Seed	MS	3.0% sucrose	Without PGR	*In vitro* seed germination	[29]
				1.0 mg/l BAP + 0.5 mg/l NAA	Callus induction, Somatic embryogenesis	
				0.5 mg/l BAP + 0.1 mg/l NAA	Shoot regeneration	
				1.0 mg/l NAA + 0.5 mg/l BAP	Rooting	

(Table 2) cont.....

S. No.	Explant	Basal Medium	Carbohydrate Source	PGR Combination and Concentration	Response	References
26.	Leaf	MS	NM	2.0 mg/l 2,4-D + 0.5 mg/l BAP	Callus induction	[52]
				0.5 mg/l 2,4-D + 0.25 mg/l BAP	Callus proliferation	
				2.0 mg/l BAP + 0.5 mg/l NAA	Shoot regeneration	
				1.0 mg/l NAA	Rooting	
27.	Root segment	MS	3.0% sucrose	4.0 mg/l BAP + 4.0 mg/l NAA	Friable callus, whole plantlet, aerial part formation	[61]
		½ MS		0.2 mg/l IBA	Root formation	
28.	Leaf disk	MS	NM	2.0 mg/l 2,4-D + 1.0 mg/l BAP	Callus induction	[62]
				4.5 mg/l BAP + 0.5 mg/l NAA	Shoot regeneration	
				1.0 mg/l NAA + 0.1 mg/l BAP	Rooting	
29.	Leaves	MS	3.0% sucrose	1.5 mg/l 2,4-D	Callus induction	[15]
				1 mg/l TDZ and 0.2 mg/l NAA	Indirect shoot regeneration	
30.	Leaves and stem internodal segments	MS	3.0% sucrose	2.0 mg/l 2,4-D and 1.0 mg/l BAP	Callus induction	[63]
31.	Leaves	MS	NM	5.0 mg/l BAP	Callus induction	[64]
32.	Shoot apices	MS	NM	4.0 µM BAP and 4.0 µM NAA	Callus induction and shoot regeneration	[7]
		½ MS		12.0 µM NAA	Rooting	

Organogenesis

Direct Organogenesis

It is well established that a precise level of cytokinin can play a critical role in direct shoot organogenesis in *R. serpentina*. Different concentrations of 6-benzylaminopurine (BAP) have often been used for shoot regeneration, shoot multiplication, and multiple shoot proliferation [22, 26, 40, 50, 52]. It is also found that cytokinin combinations like 17.74 µM BAP and 32.57 µM adenine sulphate (AdS) can also be a promising alternative for shoot induction and multiple shoot regeneration (Table **1**). BAP with α-naphthalene acetic acid (NAA) has been found as a widely used cytokinin-auxin combination for shoot multiplication and multiple shoot proliferation of *R. serpentina* [9, 10, 12, 17, 21, 27, 30, 31, 46, 48, 65]. Combined with indole-3-acetic acid (IAA) or indole-3 butyric acid (IBA), BAP was also used for shoot regeneration and multiplication [19, 20, 27, 41, 51]. Kinetin (Kn), the first artificial cytokinin, was also used in the direct organogenesis of *R. serpentina* (Table **1**). Kn has been very effective for shoot formation and multiple shoot proliferation when combined with BAP or auxins (NAA, IAA, IBA) [11, 33, 39, 42, 47]. Mondal *et al.* [33] also found that different BAP, Kn and NAA combinations resulted *in vitro* flowering besides shoot regeneration (Table **1**). Another novel cytokinin, thidiazuron (TDZ), is used in plant tissue culture for its potential to protect chlorophyll from degradation in isolated leaves [66]. Applying a low amount (0.8 μM) of TDZ could be convenient for shoot regeneration [5]. TDZ was also effective for shoot multiplication and shoot proliferation in *R. serpentina* [43, 44].

Indirect Organogenesis

Various combinations of cytokinin and auxin have been applied to achieve the optimum callus growth in *R. serpentina*. In most experiments, nodal segments and leaves were used as explants for callus formation (Table **2**). Auxins, combined with cytokinin or individually, have been found to be effective in indirect organogenesis in the Indian snake plant. Among the different auxins, 2,4-dichlorophenoxy acetic acid (2,4-D) was found more effective in callus induction. 1.5 mg/l 2,4-D alone has been effective in callus induction from leaf and internodal explants [15, 46]. Several reports also suggest that 2,4-D, combined with cytokinins like BAP, was very effective for callus proliferation (Table **2**). Bhatt *et al.* [28] used 0.125 mg/l 2,4-D with 1.0 mg/l BAP to induce callus successfully. However, applying a high amount of 2,4-D (4–5 mg/l) combined with NAA was also effective for friable callus formation [59, 61]. Pant and Joshi [29] used that particular auxin combination but in a low concentration of 1.0 ppm NAA and 2.0 ppm 2,4-D and successfully obtained healthy callus. Yahya *et al.* [26] found that 10 mg/l IBA and 10 mg/l NAA were optimum for callus formation in *R. serpentina*. In some investigations, BAP combined with NAA was found to be the desired combination for callus formation [7, 34, 54]. With a novel approach, Khan *et al.* [51] found that only BAP could be enough for callus induction. Besides BAP, Kn was also found to be an alternative cytokinin for callus induction (Table **2**). In their work, Bhadra *et al.* [27] got loose friable callus from the nodal segment by introducing 0.1 mg/l picloram with 1.0 mg/l Kn. Combining the same amount of NAA and Kn resulted in profuse callus induction

[53]. Kn combined with 2,4-D also has shown remarkable callus induction from nodal parts of the plant [55].

In the last two decades, a substantial objective of the findings was to identify the optimum combination of PGRs for organogenic regeneration of the callus. Several reports suggest that cytokinins are eventually used for adventitious shoot regeneration from callus. Meanwhile, auxin has been used for root regeneration. Cytokinins like TDZ, BAP, and Kn are often used for shoot regeneration from the callus (Table **2**). Sometimes, BAP is used singly in a low concentration of 2 ppm to regenerate shoot [8, 29], whereas some investigators got a better result when BAP is applied with adenine sulfate [24, 34]. Gibberellic acid (GA_3) combined with BAP has also been used for shoot regeneration and shoot elongation from the callus [59, 56]. Adopting a different approach, Aryal and Joshi [55] used 10% (w/w) coconut milk as an additive with 3.00 mg/l 2,4-D in their experiment and observed healthy shoot bud regeneration from node-derived callus. In many studies, NAA and BAP have often been used for shoot regeneration from callus. Several workers used Kn combined with the NAA and BAP to optimize the growth of regenerated shoots from callus (Table **2**). TDZ, another novel cytokinin, also has been applied in various concentrations ranging from 2–4 µM to regenerate shoots [25]. Furthermore, TDZ has been effective when applied with NAA for indirect shoot regeneration [15].

Somatic Embryogenesis

In *R. serpentina*, Zafar *et al.* [60] employed direct embryogenesis from root explant. They used liquid MS basal medium supplemented with 1.0 mg/l NAA and 0.5 mg/l BAP to induce somatic embryos. As a result, they obtained somatic embryos on the basal and terminal portions of the root segment. Mature bipolar embryos were further subcultured in liquid MS medium supplemented with 1.25 mg/l BAP and 0.25 mg/l GA_3 to get better somatic embryo germination results (Table **3**). Several researchers carried out indirect embryogenesis from the embryogenic callus. Uikey *et al.* used 2,4-D to initiate somatic embryogenesis, whereas Pant and Joshi [21] found NAA and BAP combination more favourable.

Contrary to previous works, Banu *et al.* observed somatic embryo formation in MS medium supplemented with 1.0 mg/l BAP and 0.5 mg/l Kn. For embryo germination from indirect embryogenesis, NAA is used at a concentration range from 0.5–1.0 mg/l plantlet regeneration. On the other hand, the application of cytokinins like TDZ and BAP ranged from 0.2–1.0 mg/l for germination and plantlet regeneration. Roots from regenerated embryos were induced at various concentrations of IBA and NAA with BAP (Table **3**).

Table 3. Somatic embryogenesis experiments reported from *R. serpentina* explants.

S. No.	Explant	Basal Medium	Carbohydrate Source	PGR Combination and Concentration	Response	References
1.	Callus	MS	NM	0.5 mg/l 2,4-D + 1.0 mg/l 2,4-D	Somatic embryogenesis	[43]
	Somatic embryo		2.0 % sucrose	0.5 mg/l BAP, TDZ, NAA	Shoot proliferation	
	Shoot		1.5 % sucrose	0.1-1.0 mg/l IBA	Rooting	
2.	Callus	MS	3.0 % sucrose	1.0 mg/l BAP + 0.5 mg/l NAA	Somatic embryogenesis	[6]
	Somatic embryo			0.5 mg/l BAP + 0.1 mg/l NAA	Shoot regeneration	
	Shoot			1.0 mg/l NAA + 0.5 mg/l BAP	Rooting	
3.	Root	Liquid MS	NM	1.0 mg/l NAA + 0.1 mg/l BAP	Somatic embryogenesis	[60]
	Somatic embryo	MS		1.5 mg/l BAP + 0.25 GA_3	Somatic embryo germination	
4.	Callus	MS	3.0 % sucrose	1.0 mg/l BAP + 0.5 mg/l Kn	Somatic embryo formation	[61]

In vitro Root Generation

Root induction is a significant step for *in vitro* plantlet regeneration (Fig. **1**). Especially for *R. serpentina*, it is undoubtedly very crucial as the main bioactive compound accumulates in the root. Exogenous auxins can be used either alone or combined with other PGRs. In plant tissue culture, IBA, IAA and NAA are predominantly used as plant growth regulators in healthy root formation (Tables **1-3**). IBA has frequently been used in experiments to induce roots from regenerated shoots in *in vitro* culture [41, 59]. Some findings have found that a meager amount of IBA (1 ppm) can also be effective for healthy root induction [29]. Sometimes, a combination of NAA and IBA is found suitable for rooting, while some reports suggest that a combination of NAA with BAP is a better alternative (Tables **1-3**). According to some reports, IBA incorporated with β-Naphthoxyacetic acid (NAA) can induce roots [20]. A report observed that PGR-free media could also initiate root [27].

Ex vitro Establishment

Several researchers used different conditions and substances to acclimate the *in vitro* grown *R. serpentina* plantlets. Applying a mixture of soil, sand and leaf manure during acclimatization results in great survival rates ranging from 85–95% [10, 17, 20, 31]. Different reports also suggest that when combined with vermicompost, the soil and sand mixture results in notable survival frequency [59]. Some findings revealed that autoclaved soil and Soilrite™ could be effective as an acclimatization substance. However, some reports suggest that garden soil with compost mixture shows a 100% survival rate [51].

CONCLUSION

Rauvolfia serpentina, a potent medicinal plant with great pharmaceutical values, is facing a severe threat to its natural vegetation due to inadvertent consequences of overexploitation and poor seed viability. On the other hand, vegetative propagation through cuttings is a deleterious process as it can be harmful to the health of mother plants. Keeping in mind these facts and considering the medicinal importance of this shrub, the micropropagation method can be a better alternative for germplasm conservation through mass propagation and the production of bioactive secondary metabolites. Furthermore, the plant tissue culture technique offers genetically uniformed, disease-free plantlets within a relatively short period of time. Several variables, such as physical factors like light intensity, temperature, relative humidity, and chemical elements, have affected the micropropagation attributes and production rate of *R. serpentina* plantlets. Furthermore, different constituents of nutrition media, sugar concentrations, and the versatile concentration and combinations of plant growth regulators have also impacted *in vitro* plant propagation of *R. serpentina*. In this chapter, different approaches to *in vitro* plant tissue culture of *R. serpentina* that are adopted in the 21st century have been summarized and discussed, which could be helpful for both research purposes and further commercial applications.

ACKNOWLEDGEMENTS

All the authors are thankful to Swami Kamalasthananda, Principal, Ramakrishna Mission Vivekananda Centenary College, Rahara, Kolkata, India, for the facilities provided during the present study. DB acknowledges the Human Resource Development Group, Council of Scientifc & Industrial Research (CSIR-HRDG) for providing CSIR−Senior Research Fellowship. SM also acknowledge the West Bengal State Government for providing SVMCM fellowship during the study.

REFERENCES

[1] Srivastava B, Sharma VC, Singh R, Pant P, Jadhav AD. Substitution of roots with small branches of *Rauwolfia serpentina* for therapeutic uses-a phytochemical approach. Ayushdhara 2015; 2(6): 373-8.

[2] Poonam P, Mishra S. Physiological, biochemical and modern biotechnological approach to improvement of *Rauwolfia serpentina.* IOSR J Pharm Biol Sci 2013; 6(2): 73-8. [http://dx.doi.org/10.9790/3008-0627378]

[3] Mukherjee E, Gantait S, Kundu S, Sarkar S, Bhattacharyya S. Biotechnological interventions on the genus *Rauvolfia*: recent trends and imminent prospects. Appl Microbiol Biotechnol 2019; 103(18): 7325-54.
[http://dx.doi.org/10.1007/s00253-019-10035-6] [PMID: 31363825]

[4] Kumari R, Rathi B, Rani A, Bhatnagar S. *Rauvolfia serpentina* L. Benth. ex Kurz.: Phytochemical, pharmacological and therapeutic aspects. Int J Pharm Sci Rev Res 2013; 23(2): 348-55.

[5] Hussain SA, Ahmad N, Anis M, Alatar AA, Faisal M. Role of thidiazuron in modulation of shoot multiplication rate in micropropagation of *Rauvolfia* species. Thidiazuron: From Urea Derivative to Plant Growth Regulator. Singapore: Springer 2018; pp. 429-38.
[http://dx.doi.org/10.1007/978-981-10-8004-3_24]

[6] Pant KK, Joshi SD. *In-vitro* somatic embryogenesis and organogenesis from the node induced calli of *Rauvolfia serpentina* Benth. ex Kurz. Nepal J Agric Sci 2018; 17: 167-73.

[7] Varnika V, Sharma R, Singh A, Shalini S, Sharma N. Micropropagation and screening of phytocompounds present among *in vitro* raised and wild plants of *Rauvolfia serpentina.* Walailak J Sci Technol 2020; 17(11): 1177-93.
[http://dx.doi.org/10.48048/wjst.2020.6492]

[8] Kaur S. *In vitro* callus induction in inflorescence segments of medicinally important endangered plant *Rauwolfia serpentina* (L.) Benth. ex Kurz : A step towards *ex situ* conservation. Annals of Plant Sciences 2018; 7(2): 1986.
[http://dx.doi.org/10.21746/aps.2018.7.2.1]

[9] T S, G SR, D J. Standardization of protocol for *in vitro* propagation of an endangered medicinal plant *Rauwolfia serpentina* Benth. J Med Plants Res 2013; 7(29): 2150-3.
[http://dx.doi.org/10.5897/JMPR11.066]

[10] Senapati SK, Lahere N, Tiwary BN. Improved *In vitro* clonal propagation of *Rauwolfia serpentina* L. Benth–An endangered medicinal plant. Plant Biosyst 2014; 148(5): 885-.

[11] Kad A, Pundir A, Sharma S, Sood H. Development of suspension cultures and *ex-vitro* rooting in *Rauwolfia serpentina* for rapid and largescale multiplication. Int J Innov Res SciEng 2017; 3(1): 135-43.

[12] Rajasekharan PE, Ambika SR, Ganeshan S. *In vitro* regeneration and slow growth studies on *Rauvolfia serpentina.* IUP J Biotech 2007; 1(1): 63-7.

[13] Kumar V, Mehra B, Daniel S, Jan N. Effect of organic manures and irrigation schedule on growth and yield of Sarpagandha under teak based Agroforestry system. Pharma Innovation Journal 2019; 8(6): 86-8.

[14] Andreev I, Adnof D, Spiridonova K, Kunakh V. Long-term stability of two *Rauwolfia serpentina* cell strains. Catrina Int J Environ Sci 2007; 2(2): 133-6.

[15] Mukherjee E, Sarkar S, Bhattacharyya S, Gantait S. Ameliorated reserpine production *viaIn vitro* direct and indirect regeneration system in *Rauvolfia serpentina* (L) Benth ex Kurz 2020; 10(7): 1-4.

[16] Cardoso JC, Sheng Gerald LT, Teixeira da Silva JA. Micropropagation in the twenty-first century. Plant cell culture protocols 2018.
[http://dx.doi.org/10.1007/978-1-4939-8594-4_2]

[17] Ahmed S, Amin MN, Anjum A, Haque ME. *In vitro* antibacterial activity of *Rauvolfia serpentina* and its tissue culture. Niger J Nat Prod Med 2002; 6(1): 45-9.
[http://dx.doi.org/10.4314/njnpm.v6i1.11693]

[18] Sehrawat AR, Uppal S, Chowdhury JB. Establishment of plantlets and evaluation of differentiated roots for alkaloids in *Rauwolfia serpentha.* J Plant Biochem Biotechnol 2002; 11(2): 105-8.
[http://dx.doi.org/10.1007/BF03263144]

[19] Jain V, Singh D, Saraf S, Saraf S. *In-vitro* micropropagation of *Rauwolfia serpentina* through multiple

shoot generation. Anc Sci Life 2003; 23(1): 44-9.
[PMID: 22557112]

[20] Kataria V, Shekhawat NS. Cloning of *Rauvolfia serpentina*: An endangered medicinal plant. J Sustain For 2005; 20(1): 53-65.
[http://dx.doi.org/10.1300/J091v20n01_04]

[21] Baksha R, Jahan MAA, Khatun R, Munshi JL. *In vitro* rapid clonal propagation of *Rauvolfia serpentina* (Linn.) Benth. Bangladesh J Sci Ind Res 1970; 42(1): 37-44.
[http://dx.doi.org/10.3329/bjsir.v42i1.353]

[22] Goel MK, Kukreja AK, Khanuja SP. Cost-effective approaches for *in vitro* mass propagation of *Rauwolfia serpentina* Benth. ex Kurz. Asian J Plant Sci 2007; 6(6): 957-61.
[http://dx.doi.org/10.3923/ajps.2007.957.961]

[23] Goel MK, Kukreja AK, Singh AK, Khanuja SPS. *In vitro* plant growth promoting activity of phyllocladane diterpenoids isolated from *Callicarpa macrophylla* Vahl. in shoot cultures of *Rauwolfia serpentina.* Nat Prod Commun 2007; 2(8): 1934578X0700200.
[http://dx.doi.org/10.1177/1934578X0700200802]

[24] Ilahi I, Rahim F, Jabeen M. Enhanced clonal propagation and alkaloid biosynthesis in cultures of *Rauwolfia.* Pak J Plant Sci 2007; 13(1): 45-56.

[25] Pandey VP, Kudakasseril J, Cherian E, Patani G. Comparison of two methods for *in vitro* propagation of *Rauwolfia serpentina* from nodal explants. INDIAN DRUGS 2007; 44(7): 514-9.
[http://dx.doi.org/10.53879/id.44.07.p0514]

[26] Yahya AF, Hyun J, Lee J, Jung M. Effect of explant types, auxin concentration and light condition on *in vitro* root production and alkaloid content of *Rauvolfia serpentina* (L.) Benth. ex Kurz. J Korean For Soc 2007; 96(2): 178-82.

[27] Bhadra SK, Bhowmik TK, Singh P. *In vitro* micropropagation of *Rauvolfia serpentina* (L.) Benth through induction of direct and indirect organogenesis. Chittagong University. J Biol Sci 2008.

[28] Bhatt R, Arif M, Gaur AK, Rao PB. *Rauwolfia serpentina*: Protocol optimisation for *in vitro* propagation. Afr J Biotechnol 2008; 7(23): 4265-8.

[29] Pant KK, Joshi SD. Rapid Multiplication of *Rauvolfia serpentina* Benth. ex. Kurz through tissue culture. ScientWorldJ 2008; 6(6): 58-62.

[30] Goel MK, Mehrotra S, Kukreja AK, Shanker K, Khanuja SP. *In vitro* propagation of *Rauwolfia serpentina* using liquid medium, assessment of genetic fidelity of micropropagated plants, and simultaneous quantitation of reserpine, ajmaline, and ajmalicine In: Protocols for *In vitro* Cultures and Secondary Metabolite. Access J Medicinal Aromat. Humana Press 2009; pp. 17-33.
[http://dx.doi.org/10.1007/978-1-60327-287-2_2]

[31] Mishra Y, Usmani G, Mandal AK. *In vitro* cloning of *Rauvolfia serpentina* (L.) Benth. var. CIM-Sheel and evaluation of its field performance. J Biol Res 2010; 13: 85.

[32] Bahuguna RN, Joshi R, Singh G, Shukla A, Gupta R, Bains G. Micropropagation and total alkaloid extraction of Indian snake root (*Rauwolfia serpentina*). Indian J Agric Sci 2011; 81(12): 1124-9.

[33] Mondal S, Silva JT, Ghosh PD. *In vitro* flowering in *Rauvolfia serpentina* (L.) Benth. ex. Kurz. Intern J Plant Dev Biol 2011; 5(1): 75-7.

[34] Saravanan S, Sarvesan R, Vinod MS. Identification of DNA elements involved in somaclonal variants of *Rauvolfia serpentina* (L.) arising from indirect organogenesis as evaluated by ISSR analysis. Indian J Sci Technol 2011; 4(10): 1241-5.
[http://dx.doi.org/10.17485/ijst/2011/v4i10.19]

[35] Abdurahman AA, Alwathnani AH. High frequency shoot regeneration and plant establishment of *Rauvolfia serpentina*: An endangered medicinal plant. J Med Plants Res 2012; 6(17): 3324-9.
[http://dx.doi.org/10.5897/JMPR12.111]

[36] Faisal M, Alatar AA, Ahmad N, Anis M, Hegazy AK. Assessment of genetic fidelity in *Rauvolfia serpentina* plantlets grown from synthetic (encapsulated) seeds following *in vitro* storage at 4 °C. Molecules 2012; 17(5): 5050-61. [http://dx.doi.org/10.3390/molecules17055050] [PMID: 22555295]

[37] George S, Geetha SP, Balachandran I. *In vitro* multiplication of *Rauvolfia serpentina* (L.) Benth. ex Kurz-An endangered medicinal plant. MedPlants-Int J Phytomed Rel Indus 2012; 4(1): 45-8.

[38] Mallick SR, Jena RC, Samal KC. Rapid *in vitro* multiplication of an endangered medicinal plant sarpgandha (*Rauwolfia serpentina*). Am J Plant Sci 2012; 3(4): 437-42. [http://dx.doi.org/10.4236/ajps.2012.34053]

[39] Rani A, Kumar M, Kumar S. Rapid *in vitro* propagation of Serpgandha (*Rauvolfia serpentina* Linn.) Benth through young leaves. Adv Appl Res 2013; 5(2): 131-4.

[40] Hardjo PH, Rijanto A, Yuantara L. Micropropagation and antibacterial activity of *Rauvolfia serpentina* (L.) Benth. ex Kurz. Joint Symposium on Frontier Research in Biodiversity and Agricultural Resources.

[41] Rani A, Kumar M, Kumar S. Effect of growth regulators on micropropagation of *Rauvolfia serpentina* (L.) Benth. J Appl Nat Sci 2014; 6(2): 507-11. [http://dx.doi.org/10.31018/jans.v6i2.490]

[42] Rani A, Kumar M, Kumar S. Micropropagation of serpgandha (*Rauvolfia serpentina* L. Benth): An endangered medicinal plant. Appl Biol Res 2014; 16(1): 114-8. [http://dx.doi.org/10.5958/0974-4517.2014.00058.5]

[43] Uikey DS, Tripathi MK, Tiwari G, *et al. In vitro* plant regeneration *via* organogenesis in. Plant Cell Biotechnol Mol Biol 2014; 15: 136-49. [*Rauvolfia serpentina* (L.) Benth.].

[44] Ahmad N, Alatar AA, Faisal M, *et al.* Effect of copper and zinc on the *in vitro* regeneration of *Rauvolfia serpentina.* Biol Plant 2015; 59(1): 11-7. [http://dx.doi.org/10.1007/s10535-014-0479-5]

[45] Alatar AA. Thidiazuron induced efficient *in vitro* multiplication and *ex vitro* conservation of *Rauvolfia serpentina* : A potent antihypertensive drug producing plant. Biotechnol Biotechnol Equip 2015; 29(3): 489-97. [http://dx.doi.org/10.1080/13102818.2015.1017535]

[46] Kumari A, Kumar S, Anam A, Ahmad S, Naseem M. *In vitro* cloning of an endangered medicinal plant, *Rauwolfia serpentina* (L.). J Plant Dev Sci 2015; 7: 555-61.

[47] Murab T, Chandurkar P, Choudhary A, Tripathi N. *In vitro* propagation of *Rauwolfia serpentina* (Linn.) Benth: an attempt to save an endangered medicinal plant. Intern J Pharmaceut Res Bio-Sci 2015; 4(3): 398-408.

[48] Rana SK, Sehrawat AR, Chowdhury VK. Assessment of clonal fidelity in micropropagated plantlets of *Rauwolfia serpentina* Benth. Ex. Kurz. MedPlants-Int J Phytomed Rel Indus 2015; 7(4): 258-63. [http://dx.doi.org/10.5958/0975-6892.2015.00038.6]

[49] Koul PM. *In vitro* shoot proliferation from apical and nodal explants of *Rauwolfia serpentina* (L.) Benth-An endangered medicinal plant. World J Pharm Res 2016; 5(11): 746-55.

[50] Prakasha D, Ramya G, Nidagundi R. *In vitro* micropropgation of Sarpagandha (*Rauvolfia serpentina*). Bioscan 2017; 12(3): 1391-5.

[51] Khan S, Banu TA, Akter S, *et al. In vitro* regeneration protocol of *Rauvolfia serpentina* L. Bangladesh J Sci Ind Res 2018; 53(2): 133-8. [http://dx.doi.org/10.3329/bjsir.v53i2.36674]

[52] Zafar N, Mujib A, Ali M, *et al.* Genome size analysis of field grown and tissue culture regenerated *Rauvolfia serpentina* (L) by flow cytometry: Histology and scanning electron microscopic study for *in vitro* morphogenesis. Ind Crops Prod 2019; 128: 545-55.

[http://dx.doi.org/10.1016/j.indcrop.2018.11.049]

[53] Nurcahyani N, Solichatun S, Anggarwulan E. The reserpine production and callus growth of Indian snake root (*Rauvolfia serpentina* (L.) Benth. ex Kurz) culture by addition of Cu^{2+}. Biodiversitas 2008; 9(3): 177-9.
[http://dx.doi.org/10.13057/biodiv/d090305]

[54] Salma U, Rahman MSM, Islam S, *et al.* The influence of different hormone concentration and combination on callus induction and regeneration of *Rauwolfia serpentina* L. Benth. Pak J Biol Sci 2008; 11(12): 1638-41.
[http://dx.doi.org/10.3923/pjbs.2008.1638.1641] [PMID: 18819656]

[55] Aryal S, Joshi SD. Callus induction and plant regeneration in *Rauvolfia Serpentina* (L.) Benth ex. Kurz. J Nat Hist Mus 2009; 24: 82-8.
[http://dx.doi.org/10.3126/jnhm.v24i1.2245]

[56] Singh P, Singh A, Shukla AK, Singh L, Pande V, Nailwal TK. Somatic embryogenesis and *in vitro* regeneration of an endangered medicinal plant sarpagandha (*Rauvolfia serpentina* L.). Life Sci J 2009; 6(2): 57-62.

[57] Panwar GS, Attitalla IH, Guru SK. An efficient *in vitro* clonal propagation and estimation of reserpine content in different plant parts of *Rauwolfia serpentina* L. American-Eurasian J Scientific Res 2011; 6(4): 217-22.

[58] Uikey DS, Tripathi MK, Tiwari G, Patel RP, Ahuja A. Embryogenic cell suspension culture induction and plantlet regeneration of *Rauvolfia serpentina* (L.) Benth.: Influence of different plant growth regulator concentrations and combinations. Med Plant 2016; 8: 153-62.

[59] Gantait SS, Dutta K, Majumder J. *In vitro* regeneration and conservation of an endangered medicinal plant sarpagandha (*Rauvolfia serpentina* L.). J Hortic Sci 2017; 12(1): 71-7.
[http://dx.doi.org/10.24154/jhs.v12i1.74]

[60] Zafar N, Mujib A, Ali M, Tonk D, Gulzar B. Aluminum chloride elicitation (amendment) improves callus biomass growth and reserpine yield in *Rauvolfia serpentina* leaf callus. Plant Cell Tissue Organ Cult 2017; 130(2): 357-68.
[http://dx.doi.org/10.1007/s11240-017-1230-7]

[61] Banu TA, Khan S, Goswami B, Afrin S, Habib A, Akter S. Indirect organogenesis and somatic embryogenesis for regeneration of *Rauvolfia serpentina* L. from root explants. Bangladesh J Bot 2020; 49(4): 1021-7.
[http://dx.doi.org/10.3329/bjb.v49i4.52534]

[62] Mahadik SM, Sawardekar SV, Kelkar VG, Gokhale NB. *In vitro* regeneration technique in *Rauwolfia serpentina* and quantification of reserpine. Int J Chem Stud 2020; 8(6): 520-5.
[http://dx.doi.org/10.22271/chemi.2020.v8.i6h.10827]

[63] Singh PS, Patni B. Elucidating the effect of salinity stress in enhancing the phenolic acid content in *Rauwolfia Serpentina in vitro.* J Stress Physiol Biochem 2020; 16(3): 103-10.

[64] Sinha NK, Kumar G. *Rauwolfia serpentina*: Protocol optimization for callus induction from leaf explants. Int J Innovative Sci Res Tech 2020; 5(10): 289-92.

[65] Rajasekharan PE, Ambika SR, Ganeshan S. *In vitro* Regeneration and Conservation of *Rauvolfia serpentina* (L.) Benth. ex Kurz. a critically endangered medicinal plant species. ICFAI J Biotech 2017.

[66] Pai SR, Desai NS. Effect of TDZ on various plant cultures. Thidiazuron: From urea derivative to plant growth regulator. Singapore: Springer 2018; pp. 439-54.
[http://dx.doi.org/10.1007/978-981-10-8004-3_25]

CHAPTER 14

Micropropagation of *Wrightia* Species

S. Asha[1,*]

[1] *Department of Biotechnology, VFSTR (Deemed to be University), Vadlamudi, Guntur Dt., A.P., India*

Abstract: The genus *Wrightia* belongs to the Apocyanaceae family and encompasses 32 species. This genus has many pharmacological properties and is used for many of the human ailments in the traditional systems of medicine. It also has commercial importance for its timber, dye, *etc.* Due to its commercial importance, some of the species of this genus, like *Wrightia tinctoria* and *W. arborea* are overexploited and have become endangered. There is a need to conserve these species. One of the techniques to conserve plants and multiplication is micropropagation. In this chapter, regeneration studies that include collection, sterilization, shoot and root generation, and acclimatization of *Wrightia tinctoria* and *W. arborea* are described.

Keywords: Acclimatization, Auxins, Cytokinins, Explant, *in vitro* Seedling, Microrpopagation, Nodes, Root Regeneration, Shoot Regeneration, Sterilization. *Wrightia.*

INTRODUCTION

The genus of *Wrightia* belongs to the Apocyanaceae family. This genus is distributed throughout the world as shrubs or small trees. It has many pharmacological properties. Dao [1] identified 32 species and out of which a few species of *Wrightia* are *W. tinctoria, W. arborea, W. coccinea, W. mollissima* and *W. pubescens* [2]. Through the literature survey, it was found that micropropagation studies were carried out for four species, *i.e.*, *W. tinctoria, W. arborea* (Syn.: *W. tomentosa*), *W. religiosa* and *W. sirikitae.* In this chapter, micropropagation studies of *W. tinctoria* and *W. Arborea* are reviewed.

MICROPROPOGATION IN *W. TINCTORIA* ROXB.

Wrightia tinctoria Roxb. is a small to medium-sized deciduous tree that produces milky white latex. It is a flowering plant that can be found in Australia, India, My-

* **Corresponding author S. Asha:** Department of Biotechnology, VFSTR (Deemed to be University), Vadlamudi, Guntur Dt., A.P., India; Tel: 9440757505; E-mail: sai848@gmail.com

T. Pullaiah (Ed.)

anmar, Nepal, Timor, and Vietnam. It is referred to as the Pala indigo plant or dyer's oleander, Sweet indrajao, Dudhalo, Dudhi, Mitha-indrajau, Karayaja, Kala Kuda and Ankudu chettu. It inhabits both dry and moist regions within its habitat. It has anti-helminthic, antidiarrheal, anti-psoriatic, diuretic, anticancer, antiulcer, analgesic, and antioxidant properties [3].

The herb has historically been used to cure conditions such as psoriasis, eczema, scabies, jaundice, leukaemia, gynaecological diseases, toothache, headaches, dandruff, and diarrhoea. The white soft wood is used for furniture, toys, matchboxes, miniature boxes, turnery, and carving.

W. tinctoria has a short seed viability period and low germination rate and cutting vigour. Due to overexploitation and a lack of quick natural regeneration, the population of this plant has significantly decreased. There is a need for widespread propagation and protection of this precious softwood plant because of these factors, as well as the challenges the toy manufacturing sector faces. One method for the speedy regeneration of propagules that is available and well-established is micropropagation [4]. It provides a quick way to produce clonal planting stock for afforestation, the production of woody biomass, and the preservation of rare and elite genotypes.

As this plant has medicinal and economic value, many scientists like Purohit and Kukda [4, 5], Kairamkonda *et al.* [6], Aftab *et al.* [7], Arulanandam *et al.* [8], Mridula and Nair [9] and Priya *et al.* [10] have worked on the regeneration of this plant using different explants like nodal segments from *in vitro* seedling and adult tree internodes, leaf segments, hypocotyls *etc.*

Materials and Methods

All the required chemicals like mercuric chloride, ethyl alcohol, Tween 20, Labolene, activated charcoal, sucrose, ascorbic acid, agar, bavistin, soilrite, *etc.*, and hormones such as IAA (Indole acetic acid), NAA (Naphthaleneacetic acid), BAP (6-Benzylaminopurine), 2,4-D (2,4-Dichlorophenoxyacetic acid), IBA (Indole butyric acid), Kn (Kinetin), GA (Gibberellic acid), TDZ (Thidiazuron), Phloroglucinol, 2,4,5-T (2,4,5-Trichlorophenoxyacetic acid), *etc.*, were procured from Fischer Scientific, Sigma Aldrich, Thermo Scientific, Karnataka Explosives and High Media, India respectively.

Media Composition

MS medium [12]:MS salts - 100 mg L^{-1} myoinositol, 2 mg L^{-1} thiamine-HCl, 0.5 mg L^{-1} pyridoxine-HCl, 0.5 mg L^{-1} nicotinic acid) containing 3% (w/v) sucrose and supplemented with different concentrations of required auxins or cytokinins.

The pH of the media was adjusted to 5.8 either with 0.1 N NaOH or 0.1 N HCl before adding 0.8% (w/v) agar-agar prior to autoclaving. The medium was sterilized at 121°C under 15psi in an autoclave for 15-20 min.

Culture Conditions

The cultural conditions for *in vitro* seed germination for the first five days are dark at 72 ^{0} humidity and 25^{0} C temp and later 8 h dark and 16 h light regime.

On the other hand, cultural conditions for shoot and root regeneration are 29±2^{0} C during the day and 25±2^{0} C during the night with 2 photoperiods of 10±2 hours and 12±2 hours and 72% humidity.

Explants, Sterilization, Media, Inoculation and Callus Induction

Purohit and Kukda [5] developed an *in vitro* method for the propagation of *W. tinctoria* using cotyledonary node segments. They used both cotyledonary nodes and hypocotyls from 21 days old *in vitro* seedlings as explants. Among these, 2.0-2.5 cm cotyledonary node segments were found most suitable. They gathered ripe, dry *W. tinctoria* follicles from a superior tree in the Kevda woodland area near Udaipur. The seed surface was sterilised for 5 minutes with 0.2% mercuric chloride, after which the seed was thoroughly washed with sterile distilled water and aseptically inoculated with 0.8% water agar for germination.

Purohit and Kukda [4] described an efficient and reproducible *in vitro* clonal multiplication protocol for *W. tinctoria* adult trees. They obtained *in vitro* multiple shoots from nodal segments of more than 30-year-old trees that have axillary branches. The tree was selected based on quality wood and marked, from which shoots are harvested throughout the year. Sterile distilled water with a few drops of Tween-20 was used to sterilize the explants. Surface sterilised nodal segments measuring 2.5–3 cm long and 0.5–0.8 cm thick were rinsed in sterile distilled water after being exposed to 0.1% (w/v) $HgCl_2$ for five minutes. Explants were inoculated vertically on MS [11], B5 [12], woody plant medium [13], SH [14], and White [15] media.

Kairamkonda *et al.* [6] attempted to multiply *W. tinctoria in vitro* using zygotic embryo cultures. Seeds were removed from the fruit and cleaned for 2 hours under running water. 0.1% (w/v) $HgCl_2$ was used to surface sterilize the seeds for 2–3 minutes, followed by 3–4 rinses in sterile distilled water and soaked for 24 h in sterile distilled water. Zygotic embryos were extracted and inoculated on MS medium (MS salts - 100 mg/L myoinositol, 2 mg/L thiamine-HCl, 0.5 mg/L pyridoxine-HCl, 0.5 mg/L nicotinic acid) containing 3% (w/v) sucrose and supplemented with various concentrations (0.5- 2.5 mg/L) of IAA. All IAA concentrations resulted in the germination of zygotic embryos. Early germination (8 days) was observed with 2.0 mg/L IAA (2 weeks), when compared to controls, that were healthy and normal.

Aftab *et al.* [7] attempted to regenerate *W. tinctoria* plantlets through axillary shoot proliferation, callus culture, and direct adventitious regeneration. They used various types of explants like nodal shoot segments, apical shoot segments, leaf segment, and internode segments from *in vitro* shoot cultures for *in vitro* studies. Leaf and shoot segments were used as explants to obtain callus for adventitious regeneration and somatic embryogenesis. Callus induction was carried out using leaf and shoot segments on MS medium with additives + auxins (IAA, NAA, IBA, 2, 4-D, and 2, 4,5-T 0.5 - 1.0mg/L) either alone or with Kn (0.25mg/L). The leaf segment was proved better for callus induction than the shoot segment on MS with auxin. Out of the various auxins used, MS medium containing additives + 2,4-D (1.0 mg/L) proved the best to induce healthy callus and further multiplication, followed by NAA (1 mg/L) + Kn (0.25 mg/L).

Arulanandam *et al.* [8] used both nodal and internodal segments from the mother plant *W. tinctoria.* They took internodal explants from healthy plants and carefully rinsed with running tap water for 10 minutes before being treated with Bavistin (2%) and Tween 20 for 10-15 minutes to get rid of bacterial and fungal spores as well as surface dust particles. Explants were cleaned three times in the Laminar Air Flow chamber with double-distilled water following surface sterilisation with 0.1% $HgCl_2$ for 3 minutes. In order to generate calli and shoots, nodal segments of about 1.0 cm were produced aseptically and implanted vertically in MS medium mixed with 0.6% (w/v) agar, 3% sucrose, and various combinations and concentrations of NAA (0.2-1.0 mg/L) and BAP (0.5-3.0 mg/L). Among various concentrations of growth regulators, such as NAA (0.2-1.0 mg/L) and BAP (0.5-3.0 mg/L), and utilised for shoot initiation and callus induction, MS medium enriched with NAA (0.8 mg/L) and BAP (2.5 mg/L) produced the highest percent of white friable calli.

Mridula and Nair [9] reported a protocol for *in vitro* rapid production and acclimatization of *W. tinctoria.* They have used a single field grown order to minimize variations caused by genetic variability and statistical errors in data analysis [16]. Two weeks after chopping off one major branch, nodal areas formed from the order's recent growth bursts were used as explants [5]. The explants were treated with 0.3% (w/v) carbondazim fungicide for 15 minutes after being rinsed in 2% polysorbitol detergent—Labolene. After cleaning the explants, surface sterilisation with 0.1% mercuric chloride was carried out for 4-5 minutes.

Auxins and cytokinins were added to the MS medium, which contains 3% sucrose and 0.75% agar. In 250 ml culture flasks with 50 ml of sterilised media, five to seven explants were grown. They were then incubated at 25±2 °C with 40 mol m^{-2} s^{-1} irradiance for an 8-hour photoperiod and 55± 5% relative humidity. For initial explant establishment for 10 days, full strength MS media was amended with 2 μM BAP and NAA at 2, 4, or 6 μM. For subcultures, activated charcoal (AC) was added to the full strength MS media amended with 2 μM BAP and 6 μM NAA, at a rate of 0.05 to 0.10% to enable simultaneous shooting and rooting for 14 days.

Priya *et al.* [10] selected nodal segments from *in vitro* grown seedlings of WTR. One gram seeds were weighed, cleaned thoroughly under running tap water for 15 min, and washed with Tween 20 for 2min. Seeds were continuously cleaned with sterile water to completely remove the foam. The sterile seeds were placed in an LAF, treated with 0.1% $HgCl_2$, and then washed with sterile water, 70% ethanol (60 sec.), and sterile water (thrice). The seeds are then air dried, put on sterile Whatman no. 1 filter paper, MS media, and incubated for seed germination in a dark environment at a humidity of 72% and a temperature of 25 degrees.

To enhance the percentage of seed germination, overcome dormancy, and reduce the time of seed germination, physical and chemical treatments like hot water, 20% sulphuric acid, GA_3, and 20% sulphuric acid + GA_3 are applied to seeds at various intervals and concentrations. Following treatment, the seeds are washed with distilled water three times, then exposed to 70% ethanol for 60 seconds, followed by sterile water washes three times, inoculated on MS basal media and initially incubated for the first five days under dark at 72 ^{0}humidity and 25^0 C temp and later exposed to 8h dark and 16 h light regime to determine the percentage of germination. Among the tested treatments at different time intervals and concentrations, the seeds treated with 20% sulphuric acid for 5 minutes implanted on MS basal media + GA_3 (2.5 mg/ml) achieved 95% germination on the fourth day of seed inoculation. The number of days for seed germination after this treatment fell from 15 to 4 days. Within 15 days of germination, the plantlet grew up to 5 cm.

Shoot Regeneration

In order to promote multiple shoot proliferation, Purohit and Kukda [5] removed hypocotyls, epicotyledonary nodes, or cotyledonary nodes from 21-day old *in vitro* grown seedlings and inoculated them on MS basal medium that had been altered with various concentrations of cytokinins (0.5-8.0 $mgdm^{-3}$ Kn and BAP) and NAA (0.1-1.0 mg dm^{-3}) individually and in combination. To optimize the salt requirement by proliferating explants, B5, White's and SH media are screened. They divided clusters of shoots into groups during each subculture and put them onto the media for continued proliferation. They subcultured regenerated shoots again on the same or modified medium every three weeks. They observed different responses of cotyledonary nodes to various cytokinins inoculated on MS medium. The implanted cotyledonary nodes did not develop any shoots in the absence of growth regulators. Shoots subcultured on medium with the ideal concentration of Kn demonstrated growth in length up to 4.8 cm but no further multiplication. However, when taken alone, BAP (5.0 mg dm^{-3}) resulted in 7–8 shoots per node in just 21 days.

In comparison to shoots obtained on Kn, shoots on BAP fortified media were shorter.

A shoot length of 1.2 cm on average was recorded at the optimum BAP concentration. On BAP containing media, the regenerated shoots' leaves were smaller and yellow. Any alteration in the medium's BAP concentration resulted in a significant reduction in the number of shoots or suppression of shoot length. At any concentration, the medium's combined cytokinins (BAP and Kn) were unable to increase shoot length or proliferation.

By assuming that BAP at 5.0 $mgdm^{-3}$ is ideal for maximum shoot proliferation, several amounts of NAA (0.1-1.0 mg dm^{-3}) were put in the medium with BAP. With a small increase in shoot length from 0.01 mg dm^{-3} NAA paired with 5.0 mg dm^{-3} of BAP, the cultures' overall growth improved. More NAA was present, which led to callusing. The results showed that the best growth conditions for the cultures' overall development, shoot proliferation, and shoot length were 0.01 $mgdm^{-3}$ NAA and 5.0 mg dm^{-3} BAP. With this particular media combination, induction of 8 shoots/explants with an average shoot length of 1.4 cm was noted in 21 days. Within 84 days of inoculation, every subculture of regenerated shoots grown on the same medium saw a threefold multiplication rate that produced an average of 230 shoots per node with an average length of 2.3 cm (involving 3 subcultures of 21 d each). When BAP concentration was reduced from 5.0 to 1.0 $mgdm^{-3}$ during culturing, shoot length was enhanced without affecting the rate of multiplication.

Purohit and Kukda [4] have used 0.5-5mg/L BAP and Kn individually and in combination for shoot proliferation from nodal explants. They took a minimum of 20 replicates for each experiment, repeated thrice, and recorded data every 3W. For every three weeks, on fresh medium, *in vitro* produced shoots were sub cultured. Maximum bud break response was achieved from nodal segments of young lateral branches collected during March-June that yielded 91% aseptic cultures. To study the impact of season on bud break, new cultures of nodal segments generated from lateral branches were grown on MS + 2 mg/L BAP at regular intervals throughout the year. The best response in terms of shoot vigour, number of shoots, and frequency of bud break came from cultures started between March and June. Shoots along with mother explants were sub-cultured on MS+BAP (0.5-1 mg/L) every 3 W. A three-fold multiplication was observed on MS supplemented with BAP (1mg/L) every 3 W, where shoots are longer than initial cultures with BAP (2mg/L).

Aftab *et al.* [7] have found nodal shoot segment is better for shoot initiation than other explants. Multiple shoots and adventitious shoots were induced on MS medium with additives viz; ascorbic acid [50 mg/L], citric acid [25 mg/L], cysteine [25 mg/L] + IAA 0.1mg/L + cytokinin (Kn, BAP and TDZ) 0.5 to 2.5 mg/L either alone or in combination with nodal shoot segment and apical shoot segment. For further shoot multiplication, MS medium with additives and various cytokinins viz; BAP and TDZ were used.

For adventitious shoot induction, MS medium with NAA (1.0 mg/L) and various cytokinins, *i.e.*, Kn and BAP (0.5 to 2.5 mg/L) and TDZ (0.1 to 0.25 mg/L) were used. They found that MS medium with additives + IAA (0.1mg/L) + BAP (1.0 mg/L) was the best for shoot induction (3.5 shoots per explant average and shoot length 5.00 cm) and further multiplication in 4 weeks period. However, they didn't observe adventitious shoots on any of the growth hormone treatments used.

Arulanandam *et al.* [8] have subcultured nodal segment-derived proliferated shoots to induce additional multiple shoots. The maximum number of shoots per nodal explant was 7.62±1.6 among the various concentrations of growth regulators such as NAA (0.2-1.0 mg/L) and BAP (0.5-3.0 mg/L) and used for shoot initiation. The maximum percentage (80%) of shoot initiation was obtained on MS medium with BAP (1.5 mg/L) + NAA (0.4 mg/L). It was discovered that BAP alone in the medium is insufficient to elicit multiple shoot initiation, proving the need for auxin and cytokinin combinations.

Mridula and Nair [9] reported the highest shoot and root numbers, as well as the longest roots from the explants, inoculated on MS media supplemented with 6 µM NAA and 0.05% AC. The longest shoot length and highest chlorophyll A content were observed in plants cultured on MS medium supplemented with 6 µM NAA and 0.1% AC. The most *in vitro* leaves were produced by explants seeded on MS media supplemented with 4 M NAA and 0.1% AC. Maximum leaf fall was seen in shoots grown on MS medium supplemented with 4 M NAA and 0.05% AC.AC had a favourable and stimulatory effect on the roots of shoots and lowered basal callus.

For shoot regeneration, Priya *et al.* [10] utilised MS media enriched with phloroglucinol (0.5-5.5 mg/L), Kn (0.5-5.5 mg/L), BAP (0.5-5.5 mg/L), and NAA

(0.1 mg/L), inoculated with 4-5 cotyledonary nodal explants (1.5 cm) removed from 15-day-old *in vitro* seedling. They incubated the inoculated bottles for shoot regeneration under 16 h light, 8 h dark, 25 ^{0}C temperature, and 72% humidity for shoot regeneration. The newly prepared medium with NAA (0.1 mg/L) + BAP (1.5 mg/L) was subcultured every three weeks with the regenerated multiple shoots, and data was collected every two weeks. The cluster of shoots was divided into smaller groups and moved to a medium during each subculture. The number of shoots and shoot length per explant were then counted. After 30 days of incubation, the greatest number of shoots per explant supplemented with individual hormones such as Kn (3.5 mg/L), phloroglucinol (2.5 mg/L), and BAP (1.5 mg/L) is seven with shoot lengths of 8 cm, three with shoot lengths of 5.4 cm, and twelve with shoot lengths of 3.7 cm. The tested combination of NAA (0.1 mg/L) and BAP (1.5 mg/L) resulted in 14 shoots that measured 5.7 cm in length.

Root Regeneration

The elongated shoots were carefully removed by Purohit and Kukda [5] and transferred with the cut end slightly inserted in the medium aseptically for rooting. They employed individual doses of NAA, IAA, IBA, 2,4,5-T, and 2,4-D at varying amounts (0.1 - 2.0 mg dm^{-3}). Excised shoots received a 5-minute dip treatment with 50-2000 mgdm^{-3} of pre-autoclaved IBA for root induction. After receiving an IBA treatment, they inoculated shoots into an MS medium containing various salts (basal,3/4,1/2,1/4, and 1/8) and decreased sucrose levels (1.0%). Each experiment was carried out three times, with 20 repetitions of each treatment. By using a dip treatment with pre-autoclaved IBA (500 mgdm^{-3} for 5 min) implanted upon 1/4thMS salts, they were able to achieve rooting of differentiated shoots and saw 80% rooting within 8–10 days of shoot inoculation.

Purohit and Kukda [4] tried shoot rooting under *in vitro* and *in vivo* conditions. They treated the excised shoots with pre-autoclaved IBA (50-1000mg/L) for 5-15 min in the *in vivo* method and directly transferred to a potting mixture, *i.e.*, autoclaved garden soil and soilrite in 1:1 (v/v) and inorganic nutrients of MS media (pH 5.0) were irrigated.

Different concentrations (0.5-10 mg/L) of IAA, NAA, and IBA are individually added in full half and quarter strength of MS in the *in vitro* method. They also tested the two-step root induction method as auxin didn't favour rooting but resulted in callusing. They used 50-1000 mg/L pre-autoclaved IBA to dip the cut ends of shoots for 5-15 min and cultured on 1/3,1/2, and 1/4th MS and maintained culture for 5-7 days under dark at 30±2^0 C and 60-70% RH initially.

They observed partial success of *in vivo* rooting and 20% of shoot rooting by 100 mg/L IBA pulse treatment for 10 min. 0.5mg/L IBA resulted in two short roots

(1.4 cm) within 3W in 20% shoots with intense callusing. IBA pulse treatment resulted in improved rooting and minimization of callus.

Maximum root induction (40%) was obtained when cultivated on MS semisolid medium following a 10-minute treatment with 100 mg/L IBA. By reducing the concentration of MS salts to 1/4th and adding charcoal (200 mg/L), the rooting percentage was enhanced to 68%. In this medium, over 3 roots/shoot with 3.8 cm length from the basal end within 10-12 d were developed without callus and browning.

Aftab *et al.* [7] used MS 1/4thbasal salts medium with different auxins (NAA, IBA, IAA 0.5-2.0 mg/L) either alone or in mixture to obtain high frequency rooting from *in vitro* shoots. *In vitro* shoots induced rooting on MS 1/4th medium with IBA (1.0 mg/L) within 3 to 4 weeks period at 25± 2°C temperature.

Arulanandam *et al.* [8] excised individual shoots from regenerated multiple shoots and placed them onto MS media with varying IBA concentrations (0.2-2.0 mg/L) for root induction. They achieved rooting of 60% on MS medium supplemented with IBA (0.8 mg/L) in 12 days, and the maximum number of roots per shoot was 4.78±0.23. However, induction of basal callus instead of roots by IBA (2.0 mg/L) alone was observed.

Mridula and Nair [9] reported the highest shoot and root number and longest roots when the explants were inoculated on MS media supplemented with 6 µM NAA and 0.05% AC. Plants cultured on MS amended with 6 µM NAA and 0.1% AC exhibited the greatest shoot length and chlorophyll A content. The explants inoculated on MS media fortified with 4 µM NAA and 0.1% AC resulted in a maximum number of in *vitro* leaves. Maximum leaf fall was seen in the shoots grown on MS medium amended with 0.05% AC and 4 µM NAA. AC showed a positive and stimulatory effect in the rooting of shoots and reduced basal callus formation.

Extracted well-developed shoots with 4-5 leaves were placed into a 1/4th MS supplemented with varying concentrations of IBA (0.5-5.5 mg/L), NAA (0.5-5.5 mg/L), and IAA (0.5-5.5 mg/L), and the length and number of roots on each explant were counted after 28 days [10]. They observed the initiation of three roots per shoot with a maximum root length of 6.4 cm, three roots per shoot with a maximum root length of 5.8 cm, and nine roots with a maximum root length of 6.9 cm with respective hormone concentrations of 2.5 mg/L NAA, 3 mg/L IAA, and 3.5 mg/L IBA among tested concentrations of hormones. 99% of the shoots began to initiate roots without callus formation within 21 days following the shoot's inoculation.

Acclimatization

Purohit and Kukda [5] transferred the plantlets to culture bottles having a soil-soilrite mixture (1:1) and fortified with $1/4^{th}$MS salts after 15 days of rooting. After 10d of partial hardening in the bottles, plantlets were transferred to pots having soil-soilrite (1:1) mixture and kept under polythene covered chambers and maintained a relative humidity of greater than 70%, which resulted in 60% success for transplantation. They concluded that acclimatization is the most critical factor for achieving success in pot transfer of the regenerated plantlets. The progressive transition of plants from medium to culture bottles without sucrose and with low salt concentration causes stress in the plants and forces them to become mostly autotrophic; this process is helpful in obtaining greater acclimatisation success.

Purohit and Kukda [4] transferred 12W old rooted cultures after *in vitro* hardening or directly to soil. Plantlets with roots were transferred to culture bottles $1/4^{th}$ filled with soilrite and irrigated with 40 ml of $1/4^{th}$ MS and incubated in a culture room for 20-25 d followed by transferring to green house (25-28 ^{0} C and 35% RH) and kept for 15 d for *in vitro* hardening. After 15 d, they removed the caps and covered with inverted caps. After another 15 days, the lids were taken off and plantlets were allowed to develop in bottles with tap water irrigation. Polybags soil, FYM and soilrite are used for transferring the hardened plants and kept for 4-5 months, followed by nursery shade shifting and exposure to an open environment.

When plantlets were placed directly into potting mixes and grown in a natural environment, failure to survive was seen. However, *in vitro* hardened plantlets were first transferred to polybags, then shifted to the nursery, which resulted in 80% survival. During *in vitro* hardening under greenhouse, elongation of shoots, turning of leaves to greener and expansion of leaves and extensive root system was observed. Of the 2000 plantlets, more than 1600 successfully underwent hardening, and more than 200 were moved to naturally growing circumstances.

Kairamkonda *et al.* [6] transferred *in vitro* germinated plants obtained from zygotic embryo cultures to plastic pots containing sterilised vermiculite: garden soil after being washed with sterile distilled water to remove the agar residue (1:1). Each plastic pot was covered with a polythene bag to keep the relative humidity (RH) at 80 to 90% and acclimated for four weeks in the culture chamber. Then, the plants were placed in plastic pots filled with garden soil and housed in a growth chamber at a temperature of 25 to 27 °C for four weeks after removing the polythene bags. After 4 weeks, plants were transferred to earthenware pots filled with garden soil and kept in shade settings in the research field and observations were made periodically.

After hardening, plantlets were sent to the research field (after 6 weeks of germination) and 75% of the plants survived. All morphological characteristics of the plants created through zygotic embryo culture were found to be normal and similar to their parents, except for the flowering time. Among the fifty plants developed through zygotic embryo culture, only one among them was found to be an early flowering somaclonal variant. It was isolated and maintained separately in the research field, and it blossomed after one month of planting.

Before sending the plantlets into the field, Arulanandam *et al.* [8] maintained the *in vitro* produced plantlets in small polycups with sterile garden soil and sand (3:1), covered with un-perforated poly bags, and hardened for four weeks in a mist room.

Mridula and Nair [9] transferred the shoots with roots directly to mini pots containing a potting mixture (sand and vermiculite 1:1) after two weeks of subculture. After a 3 minutes dip in Bavistin at 0.1, 0.5, or 1%, they watered the plantlets and covered with polypropylene bags. Acclimatization was ensured for seven days with the maximum relative humidity, during which the plants were not irrigated. Temperature was maintained between 29 and 25 ^{0}C during the day and night, with two photoperiods of 12 and 10 hours. They didn't observe any leaf fall during the acclimatisation period, and the *in vitro* leaves were retained throughout the acclimatisation and subsequent growth stages. Highest survival percentage (86.8 ± 2.1) was obtained with longer photoperiods and 0.5% fungicide treatment than shorter photoperiods.

Priya *et al.* [10] hardened the plantlets after 28 days of rooting, under different stages and conditions using red sterile soil, coco peat and sand in a 1:1:1 ratio and 1/4th MS broth in initial stages and vermicompost and red soil in a 1:1 ratio before shifting to the field. The survival percentage varied from stage to stage, and ultimately, it was 55% after 8 weeks of shifting to the field.

MICROPROPAGATION *WRIGHTIA ARBOREA* (DENSST.) MABB.

Wrightia arborea (Densst.) Mabb. (Syn. *Wrightia tomentosa* Roem. & Schult.) is commonly called Wooly Dyeing, Rosebay, jaundice curative tree, bitter indrajao, and dyer's oleander. It is mainly found in the Himalayas at 600m altitude and all over warm parts of India and in Myanmar, Sri Lanka, and Thailand. It is a deciduous tree that grows around 20 meters long with small branches. It is categorized as an endangered species [17]. In Africa, it is occasionally cultivated as an ornamental.

In traditional medicine, *Wrightia arborea* has been used for curing cancer [18], antipyretic, hemostatic [19], amoebic dysentery, menstrual irregularities, and renal abnormalities [20]. Bark, roots, and stems from this plant in India are used for treating snake bite [21] and scorpion venom envenoming [22]; leaf extracts for hypertension treatment [23]. Devi *et al.* [24] reported that this plant cures diseases of the skin, thirst, pittam and vatam, dysentery, pile, psoriasis, eczema, worm infestation, antidysenteric, astringent, stomachic, venereal diseases, anthelmintic, antipyretic, leprosy, and diarrhea.

It is a tree with many uses, valuable timber for industry and therapeutic value. It is extensively used in the toy manufacturing sector for its ivory-like wood. The species has become endangered due to overexploitation, and there is a need to conserve this species and develop regeneration protocols.

Explants, Sterilization and Media Inoculation

Purohit *et al.* [25] described the micropropagation protocol of an adult tree *Wrightia arborea*. They selected more than 30-year-old trees of *W. arborea* marked for the quality of wood and which was a source of explants. Throughout the year from these plants, shoots were harvested. Sterilized distilled water with a few drops of Tween 20 was used for explants through washing.2.0-2.5 cm long and 0.2-0.5 cm thick nodal shoot segments were surface sterilized for 5 min with 0.1% (w/v) mercuric chloride and washed with sterilized water. The explants were inserted vertically on different culture media [MS (11), SH (14), WP (13)].

Purohit *et al.* [26] harvested mature and dried follicles from 30-year-old trees of *W. arborea*, which was recognised as elite by the State Forest Department for the wood quality. Explants like hypocotyls, epicotyledonary and cotyledonary nodes were excised from *in vitro* grown seedlings (15 days old). They have vertically positioned explants on MS [11], SH [14], WP [13], and B5 [12].

Vyas and Purohit [27] examined the development and multiplication of shoots of *W. arborea* in an environment with regulated CO_2 levels. They have collected mature and dried *W. arborea* follicles from an identified tree 30 years old. They inoculated the seeds for germination aseptically on water agar (0.8%) after being surface sterilised with 0.2% mercuric chloride for 5 minutes. They were then extensively rinsed with autoclaved distilled water 4-5 times.

Purohit *et al.* [28] developed a reproducible and highly efficient micropropagation protocol for *W. arborea* with sexually adult material. They chose above 30-year-old trees, identified for the wood quality as the source of explant and collected in the April-June period. They cut one main fork of the donor tree, and any juvenile shoots that emerged around the ends of the cuts were collected and used for explantation. Every fortnight, newly flushed shoots were harvested and the subsequent flushes were referred to as first (F1), second (F2), third (F3), fourth (F4), and fifth (F5). Nodal shoot segments measuring 1.5, 3.0, and 4.5 cm in length and 0.2 to 0.5 cm in diameter were examined and placed horizontally in two distinct orientations, one by resting flat on the medium (H1) and by inserting one side of a node with an axillary bud into the medium and leaving the other exposed to air (H2) on the medium. The explant was positioned vertically in three different ways, with the node totally submerged in the media (V1), on top of the medium (V2), and one centimetre above the medium (V3).

Surface-sterilization of explants was carried out for 5 min with 0.1% (w/v) mercuric chloride, and thorough washing with sterilised distilled water with a few drops of Tween-20. Surface sterilized explants were inoculated on MS media amended with 2 mg/L BAP. Explants were also imbibed with TDZ (50-10,000 nM), Kn (0.5-5.0 mg/L), and GA_3 (1.0-2.0 mg/L) at various doses on MS medium. Proliferated shoots from F5 flush nodes were further subcultured on MS media with various dosages of TDZ (0.1,1.0, and 10.0 M) or BAP (2 mg/L).

According to Nagalakshmi *et al.* [29], *W. arborea* can be preserved *via* direct organogenesis from the hypocotyl. They collected mature follicles of *W. arborea* from Bailruti of Kurnool district of Andhra Pradesh, India, in February. Seeds extracted from the shade-dried pods were treated serially with a few drops of Tween- 20 for five minutes and 70% alcohol for three minutes, followed by thorough washing with distilled water, and soaked for 4-6 hours. Soaked seeds were aseptically surface sterilized with 4.5% (v/v) of 30% H_2O_2 (v/v) for ten minutes, followed by thorough rinsing with sterilized distilled water in Laminar air flow chamber and inoculated in 0.57% agar medium maintained at a pH 7.

Hypocotyls were excised from 30 days old seedlings and inoculated in a vertical orientation on MS medium supplemented with an additive adenine sulphate ($AdSO_4$, 1mg/L) along with growth hormone auxins (IAA, IBA, NAA) and, cytokinins (BA, Kn) at concentrations of 0.5mg/L, 1mg/L, 1.5mg/L. The hypocotyl segments were categorized into three different regions, *i.e.*, the segment closer to the apical bud (upper portion), the middle segment, and the segment closer to the root (bottom portion) were implanted vertically (one explant per tube) on the culture media and maintained under controlled conditions. After 45 days, the percentage of caulogenetic response, number of shoots, and shoot length were measured. They discovered that hypocotyl segments' capacity to create shoot buds differed depending on where they were located on the seedling. In MS medium enriched with BAP (0.5 mg/L) and $AdSO_4$ (1 mg/L), hypocotyl segments taken from closer to the root were very sensitive and produced a maximum of 12 shoots per explant.

The impact of polyamines on *W. arborea*'s *in vitro* growth, shoot multiplication, and rooting has been studied by Joshi *et al.* [30]. In the months of February and March, they gathered mature *W. arborea* follicles from a wide range of trees growing at Kevda-Ki-Nal and Loira. In dehiscing pods, the seeds were separated from the floss. The same tree's seeds were gathered, cleaned, weighed, and preserved after drying. The seeds were thoroughly washed with sterile distilled water after being thoroughly rinsed in 90% alcohol for 30 seconds, 0.1% mercuric chloride for 10 minutes, and 90% alcohol for 30 seconds. For soaking, surface sterilised seeds were left in sterile water for up to an hour. Lastly, 0.8% of water agar was put with seeds. After 48 hours of incubation in a dark, warm environment, radicles began to emerge.

Shoot Regeneration

Different concentrations of BAP and Kn (0.5-5.0 mg/L each, singly, and in combination) have been employed by Purohit *et al.* [25] for the regulated proliferation of shoots from nodal explants. After the ideal culture conditions for the induction of the best explant shoots were established, the *in vitro* produced shoots were subcultured on fresh medium every three weeks.

The results revealed the induction of multiple shoots *in vitro* from nodal shoot segments through forced axillary branching. Culture initiation was strongly influenced by the nature of explants and the season. 90% of the nodal segments taken from juvenile lateral branches on MS media supplemented with 2.0 mg/L BAP produced an average of 4 shoots per node in 3 weeks. After culture establishment and the emergence of shoot buds, a number of shoots, including the mother explant, were transplanted to medium with a lower (1.0 mg/L) concentration of BAP. Every three-week subculture resulted in a threefold increase in the number of shoots. *In vitro* grown shoot nodal segments were also used to start a fresh culture cycle. Without losing their vigour, the shoots could continue to grow for at least 24 months.

Purohit *et al.* [26] used various doses of BAP and Kn (0.5–8.0 mg/L) alone and in combination. Cotyledonary nodes were discovered to be superior to epicotyledonary nodes among the explants studied. Hypocotyl explants failed to sprout a shoot on any of the media.

Bud break and the quantity of shoots produced from explants were most common on MS media, followed by B5, SH, and WP media. 98% of cultures displayed bud break within three weeks on MS medium containing 5.0 mg/L BAP, and had more than eight shoots per explant. Kn at various concentrations (0.5 to 8.0 mg/L) either alone or in conjunction with various concentrations of BAP (0.5 to 8.0 mg/L) were examined. 5.0 mg/L BAP caused the greatest amount of bud break (98%) and shoot yield per explant of any concentration. However, compared to BAP used alone, shoot growth was better when Kn was administered alone.

The shoots were subcultured every three weeks on a medium containing the same or lower concentrations of BAP after their initial proliferation on a medium containing 5.0 mg/L BAP, together with their mother explant. A three-fold multiplication rate per three weeks was attained on a medium containing 1.0 mg/L BAP. However, shoots grown at this concentration were longer than those obtained when cultures were started on a medium containing 5.0 mg/L BAP. A lower BAP concentration (1.0 mg/L) sustained a 3-fold rate of multiplication and longer shoots, which was shown to be ideal.

98% of the nodal segments on MS with a 5.0 mg/L BAP supplement produced an average of eight shoots per node in three weeks. After culture establishment and the appearance of shoot buds, a cluster of shoots from the mother explant was transferred to medium with a lowered (1 mg/L) concentration of BAP in order to obtain a three-fold rate of multiplication during each subculture of three weeks. *In vitro* grown shoot nodal segments were also employed to start fresh culture cycles.

On standard multiplication medium (SM), which is made up of MS basal medium supplemented with 2.0 mg/L BA, 3.0% sucrose, and gelled with 0.8% agar for axillary shoot proliferation [26], Vyas and Purohit [27] cultured cotyledonary leaves from three-week-old *in vitro* grown seedlings aseptically. Proliferating shoots were repeatedly subcultured every three weeks with new SM media. Different CO_2 concentrations (0.0, 0.6, 10.0, and 40.0 g m^{-3}) were administered in acrylic boxes, each with a capacity of 7500 cm and an air-tight cover on the top as per the method of Solarova *et al.* [31].

Clusters of five shoots were cultivated on 40 ml of SM medium that included sucrose as well as SM media devoid of sucrose (SCSM and SFSM, respectively). Shoot SCSMs were added to each CO_2 chamber, each of which was then covered with a lid and taped shut (Miracle, 5.0 cm wide). After 63 days and two subcultures of three weeks each, shoot clusters from *W. arborea* cotyledonary nodes inoculated on SCSM medium multiplied at a rate of around 4.0 times in the growing room's ambient air and generated more than 20 shoots per explant. On this medium, multiplication resulted in an average of roughly 250 leaves per cluster with an area larger than 19.0 mm. Different reactions were seen with

SCSM in shoot cultures grown under regulated CO_2. The cultures grew almost as well without CO_2 as they did with ambient air.

Purohit *et al.* [28] cultured explants on shoot multiplication media containing phloroglucinol (50, 100, and 250 mg/L) and maintained under regulated conditions. After the establishment of the best conditions for shoot induction, the *in vitro* grown shoots were subcultured on a new medium every 3 weeks.

The results showed that induced axillary branching *in vitro* could induce numerous shoots from nodal shoot segments. The best explants for getting more than 7 shoots per node were discovered to be those from the fifth flush (F5). Subcultures that followed multiplied at a slightly higher than 2.5-fold rate. The rate of multiplication was sped up to three times faster per subculture when phloroglucinol (100 mg/L) was added to the multiplication medium containing BAP (2 mg/L). The multiplication medium could also be used with 10 mM TDZ alone and yield a comparable response. *In vitro* grown shoot nodal segments were also used to start a fresh culture cycle. Without losing their vigour, the shoots may be replicated for at least 24 months.

Nagalakshmi *et al.* [29] have maintained the thin elongated shoots for one passage in a shoot multiplication medium with activated charcoal (AC) of different concentrations (25 mg/L, 50 mg/L, 75 mg/L and 100 mg/L) before root induction. It was found that BAP at 0.5 mg/L was the best concentration for increased shoot bud initiation, while IBA was quite effective in shoot elongation. A low concentration of cytokinins in combination with $AdSO_4$ favoured multiple shoot induction and increased the number of shoot buds per regenerating explant. The best shoot induction medium (SIM) was MS supplemented with BAP (0.5 mg/L) and $AdSO_4$ (1mg/ L). IBA at 1 mg/L concentration, when added to SIM, exhibited a profound effect on elongation and multiplication, and this combination was used as a shoot multiplication medium (SMM) where the frequency of lateral branches was more in number from the basal node of the stem than from the upper nodes.

Joshi *et al.* [32] determined the stomatal characteristics, relative water loss during various phases of micropropagation, and strategies for successfully hardening plantlets for their mass production. They induced shoots from cotyledonary nodes on MS media containing 22.2M BAP and repeatedly subcultured every three weeks with decreasing concentrations of BAP (8.9 M) in the medium.

In order to lower the cost of micropropagation, Joshi *et al.* [33] looked into the effects of gelling agents, carbon sources, vessel types, and liquid culture systems during the rooting phase in *W. arborea.* Studies have shown that a medium comprising sugar cubes and crude agar significantly improves the rate of shoot multiplication. It was established that vented vessels aided in culture growth and the liquid culture system during the rooting phase. The findings indicated a significant potential for lowering the price of *in vitro* plants.

By keeping the cotyledonary nodes from seedlings vertically on MS media enriched with 5 mg/L BAP and varied concentrations (0.1,0.5, and 1.0 mM) of put, spd, and spm for axillary bud growth, Joshi *et al.* [30] assessed the outcomes after 21 days.

Different concentrations (0.1,0.5, and 1.0 mM) of put, spd, spm, L-arg, and 0.1 m and 1.0 m D-arg were kept for multiplication, and proliferating groups of shoots were cultivated on MS media with 2 mg/L BAP. Proliferating shoots were then subcultured for up to three passages on the same medium without being separated. The findings demonstrated that the addition of polyamines at concentrations ranging from 0.1 mM to 1.0 mM in MS medium in combination with 5.0 mg/L BAP increased axillary shoot bud proliferation and rate of multiplication. The maximum rate of bud proliferation (9.61 shoot buds/explant) was observed on 1.0 mM spd, followed by 1.0 mM spm and 0.5 mM spm, which created 9.59 and 9.39 shoot buds/explant, respectively. This was in contrast to 5.0 mg/L BAP (3.80 shoot buds/explant), which served as the control. At all polyamine concentrations, the shoot length greatly outperformed control. Explants treated with 0.5 mM spm showed the maximum shoot length (2.52 cm), which was noticeably longer than the control (1.48 cm).Significantly different concentrations of polyamines added affected the rate of multiplication. On 0.5 mM spm, the highest rate of shoot multiplication was observed (3.93-fold), as opposed to 2.98-fold in control. When polyamines were added to the multiplication medium in contrast to the control, there was no appreciable increase in shoot length, with the exception of by spd at 1.0 mM concentration.

Root Regeneration

Purohit *et al.* [25] tried to root shoots both *in vitro* and *in vivo*. The excised branches were immediately moved to a potting mix comprised of autoclaved garden soil and soilrite in a 1:1 ratio (v/v) and irrigated with the inorganic nutrients of MS medium after being pulse-treated with pre-autoclaved IBA (50-1000 mg/L) for 5-15 minutes (pH 5.0). Auxins (IAA, NAA, and IBA) in varying doses (0.5-10.0 mg/L) were added one at a time to full, half, and quarter strength MS media for *in vitro* rooting. A two-step root induction procedure was used because these auxins did not encourage roots and resulted in unfavourable callusing. After being dipped in various concentrations of pre-autoclaved IBA solution (50–1000 mg/L) for varying amounts of time (5–15 min), the cut ends of shoots were implanted on three-quarters, half, and quarter strength MS salt medium. The cultures were first incubated for 5-7 days at 30 ± 2°C and 60–70%

relative humidity in the dark or with the lowest portion of the culture jars wrapped in black paper. 70% of the shoots were able to develop roots when their lower ends were submerged in pre-autoclaved IBA solution (100 mg/L) for 15 minutes before being implanted on a modified MS medium (major salts dropped to 14 strength).

Purohit *et al.* [26] tried to root shoots both *in vitro* and *in vivo*. After that, they were planted in MS basal medium and watered with inorganic nutrients in a potting mixture made of autoclaved garden soil and soilrite in a 1:1 ratio (v/v) (pH 5.0). *In vivo* rooting was only partially successful. All other treatments resulted in shoots that turned pale within 5-7 days before ultimately dying, with the exception of a pulse treatment of 500 mg/L IBA for 5 minutes, which formed roots in 30% of the shoots.

Auxins (2,4-D, 2,4,5-T, IAA, NAA, and IBA) were introduced at varying quantities (0.5-10.0 mg/L) to the full, three-fourth, half, and quarter strength MS mediums for *in vitro* rooting. As the auxins that were added to the media did not stimulate roots or cause callusing, they tried a two-step root induction approach. After being implanted on full, three-fourths, half, and quarter strength MS media, the shoots' cut ends were subjected to a range of pre-autoclaved IBA solution concentrations (50-200 mg/L) for variable periods of time (5-15 min). The cultures were first incubated for 5-7 days at 30 ± 2°C and 60-70% relative humidity in the lowest section of culture jars or in complete darkness.

IBA exhibited inductive qualities when compared to all other auxins tested for *in vitro* rooting. There was extensive callusing with all auxin kinds and concentrations, including IBA. However, there was rooting with all IBA concentrations between 0.1 and 2.0 mg/L within a week. With 0.5 mg/L IBA, the highest rooting response (45%) was attained. To enhance the rooting response, the MS medium's salt content was decreased to 3/4, 1/2, and 1/4. A significant improvement in rooting (70% of shoots rooted) with minimal callusing was noted on $1/4^{th}$ MS salt media containing 0.5 mg/L IBA.

IBA dip treatment enhanced roots and reduced callusing in excised shoots. 80% rooting was achieved when shoots were implanted on $1/4^{th}$ MS media after receiving 500 mg/L IBA treatment for 5 minutes. Over three roots/shoots with an average length of 4.0 cm were observed throughout the rooting process, which took 6-7 days. For all other IBA treatment concentrations and durations, there was callusing, root induction was postponed, and root length and number were decreased. When the lower ends of the shoots were dipped in pre-autoclaved IBA solution (500 mg/L for 5 min), followed by their implantation on modified MS medium (major salts lowered to 1/4thstrength), 80% of the shoots were effectively rooted.

The shoots that had undergone three, six, and nine passes in the multiplication media were employed for rooting by Purohit *et al.* [28]. The cut ends of 2-to 3-cm shoots were removed, dipped in IBA solution at various concentrations for varying periods of time (5–15 min), and then placed on a standard rooting medium composed of agar, sucrose (1%) and quarter strength MS salts (0.6%). Additionally, the conventional rooting medium was tested with activated charcoal (50, 100, 200, and 400 mg/L). The culture pots were first maintained for 5-7 days at 32°C with 60-70% relative humidity, covered in black paper or in complete darkness. When treated with pre-autoclaved IBA solution (100 mg/L) for 10 minutes and implanted on modified MS medium (major salts decreased to ¼th strength and 400 mg/L activated charcoal), rooting was seen in more than 60% of the shoots obtained after the sixth subculture.

Nagalakshmi [29] transferred the healthy, sturdy shoots to quarter-strength MS medium with charcoal (50mg/L) devoid of growth regulators for rooting that resulted in 75% rooted shoots within 20- 25 days under a 12 hours photoperiod.

By giving cut ends a 10-minute pulse treatment with pre-autoclaved IBA solution (22.2 M), Joshi *et al.* [32] were able to root extended shoots (2–3 cm) *in vitro*. Following that, they were implanted on 1/4th MS salt medium.

For root induction, Joshi [30] has used extended shoots measuring more than 1.5 cm that were acquired from replicating cultures. In rooting media, IBA-treated excised shoots were inserted. For 10 minutes, shoot bases were submerged in 200 mg/L of pre-autoclaved aqueous IBA solution. Tak's [34] recommended 1/4th strength MS salts with sucrose (1%) and agar (0.6%) were regarded as the standard for rooting medium (SR Medium). To examine their impact on root induction, polyamines, put, spd, and spm were introduced to the rooting media at doses of 0.1, 0.5, and 1.0 mM. According to the findings, adding polyamines to the rooting media had no appreciable positive impact on rooting percentage, number and length.

Hardening

Purohit *et al.* [26] have either directly transplanted or *in vitro* hardened rooted shoots from 12-week-old cultures to the soil. The plantlets were moved from the rooting medium into 400 ml screw cap glass bottles for *in vitro* hardening. These bottles were one-fourth filled with soilrite and were irrigated with a 40 ml inorganic salt solution. Plantlets were permitted to stay in the bottles for an additional two days after the bottle caps were taken off after 15 days. The plantlets received the same irrigation throughout this time as before. Plants that had been *in vitro* toughened were placed in plastic containers. Pots were placed right away in greenhouse polythene tents (1 m x 1 m x 1 m). In polythene tents, humidity was kept at 70-80% with routine water spraying. The hardened plants were exposed to the outside environment after 15 days. The medium and soilrite had both undergone a 15-minute steam sterilisation process at a pressure of 1.06 kg cm^{-2}. Out of 500 plantlets, 411 (or around 82%) were successfully toughened under water-misted polythene tents before being planted in the ground.

Purohit *et al.* [26] have transplanted rooted shoots from 12-week-old cultures either immediately or after *in vitro* hardening to potting soil and soilrite in black plastic pots. *In vitro* toughened plants were placed in plastic containers. The humidity was controlled between seventy and eighty percent in the greenhouse while pots were enclosed under polythene tents. After 15 days, the hardened plants were placed in the outside environment. Out of 600 plantlets, 500(83%) have been successfully hardened before being misted in polythene tents and planted in the ground.

Prior to *ex vitro* exposure, Purohit *et al.* [28] hardened the rooted shoots from 3-week-old cultures. They have tried three techniques to harden the material. In the first way, 25 pots of soilrite were put horizontally in pickle bottles that were 30 cm long and could hold individual plantlets (W1). The second method involved putting individual plantlets in glass troughs (30 cm in diameter) that were covered in polythene sheets after they had been transplanted into netted pots (W2). Thirdly, 400 ml screw cap glass bottles containing soilrite were autoclaved and then irrigated with 40 ml of an inorganic salt solution (major salts of MS medium reduced to $1/4^{th}$ strength, pH 5.0) (W3). Each bottle contained four plantlets that were kept in the culture chamber for 30 days.

Plantlets were placed one at a time into plastic pots that were 10 cm high and covered with polythene bags after being hardened using the W1 and W2 methods for 30 days. By perforating the polythene cover, opening it for one hour each day, and then removing it entirely, the humidity was gradually reduced. Plantlets that had been *in vitro* toughened (W3) were stored in closed bottles until they came in contact with the caps (nearly after 30 days). The caps were first loosened, then the mist house (with an RH of 70–85%) was used to finally open them. After a month, the plants were moved into pots and housed in a greenhouse with an evaporative cooling system that maintained an 80–40% humidity gradient.

Plants that had been successfully rooted were hardened *in vitro* in glass bottles filled with soil rite and irrigated with a $1/4^{th}$ strength MS salt solution (pH 5.0). *In vitro* hardening of more than 5,000 plantlets was accomplished, and they were then moved to a greenhouse for acclimatisation. The plants had a survival rate of more than 95% during the hardening process.

Nagalakshmi *et al.* [29] hardened the rooted plantlets after removal from the culture bottles after 25 days, washed gently under running tap water, placed for 5 minutes in Bavistin 0.25% (fungicide) and transferred to trays containing sterilized 'vermiculite' for 45 days and kept in poly tunnels in a glasshouse set at 18-25 °C; 80 ± 85% relative humidity. After 45 days, shoots were transferred to polybags containing potting mixture with 1:2:2 of soil, sand, and farmyard manure for further planting in the field.

Joshi *et al.* [32] generated more than 10,000 plantlets and exposed 50 plants each to one of three alternative hardening methods like transferring the rooted plants directly into pen pots; covering the plants in pots with polyethylene bags for 15 days, after which the bags were gradually removed and left open for the following 15 days; and 3) hardening rooted plants *in vitro* in culture bottles containing autoclaved soilrite moistened with 1/4th MS nutrients. After 30 days, the caps were first partially opened, then they were removed after another 30 days. For the following 15 days, plantlets were permitted to stay in open bottles. All of these procedures were carried out in a hardening unit under controlled conditions and in three months, the plants were prepared for field translocation.

The findings showed that plantlets moved immediately into pots in the open air and could not endure for more than a week. Even if hardening by covering the pots extended the period of survival, such plants could not endure for more than a fortnight after being gradually exposed to *ex vitro* conditions. With over 95% survival after three months, *in vitro* hardening was shown to be the best method for micropropagated *W. arborea* plantlets.

CONCLUSION

Wrightia tinctoria Roxb. and *Wrightia arborea* (Densst.) Mabb. (Syn. *Wrightia tomentosa* Roem. & Schult.) have medicinal and commercial importance and are overexploited. Many scientists have developed successfully micropropagation protocols using different explants like cotyledons, epicotyls, hypocotyls of *in vitro* seedling and nodal and internodal segments of adult trees to conserve these species. They also developed hardening protocols and tested plant growth and development for some months in the field. Further, the micropropagated plants have to be studied and planted in the native and forest areas and observed for growth and acclimatization. As these plants have many pharmacological properties, some of the bioactive compounds were also studied. Individual compounds have to be isolated and characterised which can be used by the pharmaceutical industries for drug discovery. Also, enhancement of bioactive compound production using tissue culture techniques can be performed, which can be used in healthcare.

REFERENCES

[1] Dao DBS, Brown R, Wern M. Natural history society. FOC 1811; 16: 174.

[2] Anuj KS, Upadhyaya SK, Sanjay C, Shrikant S. A comparativephytochemical analysis of *Wrightia* species of family Apocynaceae by spot tests. IntJ Phytopharmacy 2017; 7(2): 14-7.

[3] Rajyalakshmi GR, Jyoti H. *In vitro* antihelminthic activity of *Wrightia tinctoria.* Int J Pharm Tech Res 2013; 5(2): 308-10.

[4] Purohit SD, Kukda G. Micropropagation of an adult tree *Wrightia tinctoria.* Indian J Biotechnol 2004; 3: 216-20.

[5] Purohit SD, Kukda G. *In vitro* propagation of *Wrightia tinctoria.* Biol Plant 1994; 36(4): 519-26. [http://dx.doi.org/10.1007/BF02921172]

[6] Kairamkonda M, Godishala V, Nanna R. Early flowering somaclonal variant of an endangered forest tree species *Wrightia tinctoria* R. Br: A potential ethno medicinal plant. J Cell and Tissue Res 2011; 11(3): 2949-53.

[7] Aftab Jahan Begum KA, Roja S, Rathore TS, Gopinath SM. Biotechnological intervention for the *in vitro* regeneration of an endangered forest species- *Wrightia tinctoria.* researchgate 2015. [http://dx.doi.org/10.13140/RG.2.2.31900.90248]

[8] Arulanandam LJP, Kumar SG, Mahadevi S. Micropropagation of *Wrightia tinctoria R.*Br., : A traditional medicinal plant. Acta Biomed Sci 2017; 4(2): 63-6.

[9] Mridula MR, Nair AS. Rapid micropropagation of *Wrightia tinctoria* (Roxb.) R Br: A medicinal tree. Int J Botany Studies 2018; 3(1): 126-31.

[10] Priya KY, Jalaja N, Asha S. *In vitro* regeneration and conservation of *Wrightia tinctoria* Roxb., a medicinally important woody plant. Re. J Biotechnol 2022; 17(5): 101-8.

[11] Murashige T, Skoog F. A revised medium for rapid growth and bioassays with tobacco tissue cultures. Physiol Plant 1962; 15(3): 473-97. [http://dx.doi.org/10.1111/j.1399-3054.1962.tb08052.x]

[12] Gamborg OL, Miller RA, Ojima K. Nutrient requirements of suspension cultures of soybean root cells. Exp Cell Res 1968; 50(1): 151-8. [http://dx.doi.org/10.1016/0014-4827(68)90403-5] [PMID: 5650857]

[13] Lloyd GB, Mc Cown BH. Commercially feasible micropropagation of mountain laurel (*Kalmia latifolia*) by use of shoot tip culture. Proc Int Plant Prop Soc 1980; 30: 421-7.

[14] Schenk RU, Hildebrandt AC. Medium and techniques for induction and growth of monocotyledonous and dicotyledonous plant cell cultures. Can J Bot 1972; 50(1): 199-204. [http://dx.doi.org/10.1139/b72-026]

[15] White PR. The cultivation of animal and plant cells. New York: Ronald Press 1963.

[16] Bonga JM. Clonal propagation of mature trees: Problems and possible solutions. Tissue culture in forestry. Martinus Nijhoff Publishers 1982; pp. 249-71. [http://dx.doi.org/10.1007/978-94-017-3538-4]

[17] Sharma SA. Census of rare and endemic flora of South- East Rajasthan. In: Jain SK, Rao RR, Eds. Threatened Plants of India. Howrah: Botanical Survey of India 1983; pp. 63-70.

[18] Bhattacharya S. Anticancer Botanicals. Daya Publishing House 2006; p. 8.

[19] Jamir NS, Limasemba T. Traditional knowledge of Lotha-Naga tribes in Wokha district, Nagaland. Indian J Tradit Knowl 2010; 9: 45-8.

[20] Rastogi RP, Dhawan BN. Anticancer and antiviral activities in indian medicinal plants: A review. Drug Dev Res 1990; 19(1): 1-12. [http://dx.doi.org/10.1002/ddr.430190102]

[21] Mishra M, Sujana KA, Dhole PA. Ethnomedicinal plants used for the treatment of cuts and wounds by tribes of Koraput in Odisha, India. Indian J Plant Sci 2016; 25: 14-9.

[22] Hutt MJ, Houghton PJ. A survey from the literature of plants used to treat scorpion stings. J Ethnopharmacol 1998; 60(2): 97-110. [http://dx.doi.org/10.1016/S0378-8741(97)00138-4] [PMID: 9581999]

[23] Rajendran SM, Agarwal SC, Sundaresan V. Lesser known ethnomedicinal plants of the Ayyakarkoil forest province of Southwestern Ghats, Tamil Nadu, India. Part I. J Herbs Spices Med Plants 2004; 10(4): 103-12. [http://dx.doi.org/10.1300/J044v10n04_10]

[24] Devi N, Gupta AK, Prajapati SK. Indian tribe's and villager's health and habits: Popularity of Apocynaceae plants as medicine. Int J Green Pharmacy 2017; 11(2): S256-79.

[25] Purohit SD, Kukda G, Sharma P, Tak K. *In vitro* propagation of an adult tree *Wrightia tomentosa* through enhanced axillary branching. Plant Sci 1994; 103(1): 67-72. [http://dx.doi.org/10.1016/0168-9452(94)03954-2]

[26] Purohit SD, Kukda G, Tak K. Micropropagation of *Wrightia tomentosa* (Roxb.) Roem et Schult. J Sustain For 1996; 3(4): 25-35. [http://dx.doi.org/10.1300/J091v03n04_03]

[27] Vyas S, Purohit SD. *In vitro* growth and shoot multiplication of *Wrightia tomentosa* Roem et Schult in a controlled carbon dioxide environment. Plant Cell Tissue Organ Cult 2003; 75(3): 283-6. [http://dx.doi.org/10.1023/A:1025836410533]

[28] Purohit SD, Joshi P, Tak K, Nagori R. Development of high efficiency micropropagation protocol of an adult tree—*Wrightia tomentosa* Plant Biotechnology and Molecular Markers (Ed, Srivastava PS, Alka Narula, Sheela Srivastava),. New Delhi, India: Anamaya Publishers. 2004; pp. 217-27.

[29] Nagalakshmi M, Vishwanath S, Viswanath S. Adventitious shoot regeneration from hypocotyls of *Wrightia arborea* (Dennst.) Mabb.: an endangered toy wood species. JCell and Tissue Res 2014; 14(2): 4339-44.

[30] Joshi P, Suthar D, Purohit SD. Effect of polyamines on *in vitro* growth, shoot multiplication and rooting in *Wrightia tomentosa* Roem et Shult. Int J Recent Sci Res 2014; 5(7): 1270-3.

[31] Solarova J, Souckova D, Ullmann D, Pospisilova J. *In vitro* culture: Environmental conditions and plantlet growth as affected by vessel and stopper types. Hortic Sci (Prague) 1996; 23: 51-8.

[32] Joshi P, Joshi N, Purohit SD. Stomatal characteristics during micropropagation of *Wrightia tomentosa.* Biol Plant 2006; 50(2): 275-8. [http://dx.doi.org/10.1007/s10535-006-0019-z]

[33] Joshi P, Rohini T, Purohit SD. Micropropagation of *Wrightia tomentosa*: Effect of gelling agents, carbon source and vessel type. Int J Biotechnol 2009; 8: 115-20.

[34] Tak K. Tissue culture studies on Some forest tree species of Aravallis in South Rajasthan. 1993.

CHAPTER 15

Decalepis hamiltonii Wight & Arn.: An Overview of its Bioactive Constituents and Conservation Strategies

Pradeep Bhat[1], **Santoshkumar Jayagoudar**[2,*], **Sachet Hegde**[3], **Savaliram G. Ghane**[4] and **Harsha V. Hegde**[1]

[1] *ICMR-National Institute of Traditional Medicine, Nehru Nagar, Belagavi, Karnataka-590010, India*

[2] *Department of Botany, G. S. S. College & Rani Channamma University P. G. Centre, Belagavi, Karnataka-590006, India*

[3] *Department of Botany, Bangurnagar Degree College, Ambewadi, Dandeli, Karnataka-581325, India*

[4] *Department of Botany, Shivaji University, Vidyanagar, Kolhapur, Maharashtra-416004, India*

Abstract: *Decalepis hamiltonii* Wight & Arn. (Family: Apocynaceae) is a climber native to Southern Peninsular India, commonly called Swallow Root. The plant is used in Ayurveda, Siddha and other traditional systems of medicines as a blood purifier, appetizer, rejuvenator, wound healing agent, *etc.* Apart from this, various other medicinal uses and pharmacological properties created a great demand for this plant that has resulted in destructive harvesting practices in the wild. The plant is generally reproduced through seeds; however, in most of the cases, germination is an intricate process due to its poor seed viability and delayed seed production. Hence, its population has gradually declined due to over-harvesting of medicinally important tuber. International Union of Conservation of Nature (IUCN) declared all the species of *Decalepis* as 'Critically Endangered Globally'. In the present chapter, complete information on traditional uses, phytoconstituents and micropropagation of ethnomedicinally important and critically endangered species *D. hamiltonii* is discussed.

Keywords: Apocynaceae, Critically Endangered, IUCN Red List, Swallow Root, Traditional Medicine.

* **Corresponding author Santoshkumar Jayagoudar:** Department of Botany, G. S. S. College & Rani Channamma University P. G. Centre, Belagavi, Karnataka-590006, India; E-mail: santoshjayagoudar@gmail.com

T. Pullaiah (Ed.)

INTRODUCTION

Decalepis hamiltonii Wight & Arn. (Family: Apocynaceae) is a climber commonly called Swallow root in English, Maakali beru in Kannada, Nannarikommulu in Telugu and Magalikizhangu in Tamil. The plant is native to Southern Peninsular India and distributed as patches over the rocky slopes and crevices of deciduous forests of Tamil Nadu, Andhra Pradesh, Telangana and Karnataka states [1]. However, its population has gradually declined due to over-harvesting of its medicinally important tuber. The International Union of Conservation of Nature (IUCN) declared all the species of *Decalepis* as Critically Endangered Globally [2].

D. hamiltonii is used in Ayurveda, Siddha and other traditional systems of medicines as a blood purifier, appetizer, rejuvenator, and wound healing agent [3]. The whole plant is used in the treatment of and bronchial asthma, intrinsic haemorrhage, erysipelas and fever [4]. Tuberous roots contain a highly aromatic flavor, which finds its use as an herbal health drink called 'Nannari' prepared by the 'Yanadi' tribe, the Nallamalai forest of Andhra Pradesh [5]. It also increases appetite and provides relief from digestive problems [4, 6]. Tuberous roots are also consumed as pickles, decoction and juices by the tribes of the Western Ghats of India due to their health-promoting properties [3]. Apart from this, various other medicinal uses, pharmacological activities, antimicrobial and food preservative properties [7] created a great demand for this plant, and that has resulted in destructive harvesting practices in the wild. The plant is generally reproduced through seeds; however, in most of the cases, germination is an intricate process due to its poor seed viability, hard seed coat and delayed seed production [8].

PHYTOCONSTITUENTS

Phytochemical studies of *D. hamiltonii* have shown the presence of phenols, flavonoids, tannins, saponins, triterpenes, aldehydes, ketones, sterols, fatty acids, resinol, cardiac glycosides and volatile flavour compounds [4]. Essential oil isolated from hydrodistillation of aromatic tuberous roots yielded 18 compounds, of which 2-hydroxy-4-methoxybenzaldehyde was the major compound with 37.45% composition, followed by 2-hydroxy-benzaldehyde (31.01%), 4-O-methylresorcylaldehyde (9.12%), benzyl alcohol (3.1%), β-caryophyllene (1.19%), and α-atlantone (2.06%) [9]. Similarly, in another study Nagarajan *et al.* [10] reported 2-hydroxy-4-methoxybenzaldehyde as the major compound with 96% composition in essential oil isolated from aromatic roots, followed by vanillin (0.45%), methyl 2-phenylethyl alcohol (0.081%), salicylate (0.044%), salicylaldehyde (0.018%), benzyl alcohol (0.016%), ethyl salicylate (0.038%) and *p*-anisaldehyde (0.01%). Reddy and Murthy [11] have also reported 2-hydroxy-4-methoxybenzaldehyde as a major compound with 96% composition in essential oil isolated from the root.

Gas Chromatography and Mass Spectroscopy (GC-MS) analysis of the methanol extract of root showed major compounds such as octadecanoic acid, n-hexadecanoic acid, oleic acid, linoleic acid methyl ester, benzaldehyde, 2-hydroxy-4-methoxy and dodecanoic acid [12]. Selvaraj *et al.* [13] reported 10 major compounds from the methanolic root extracts and the major phytoconstituents were β-D-mannofuranoside, methyl (27.425%), followed by 2-undecene 5-methyl (20.362%); 2-furancarboxaldehyde,5-(hydroxymethyl) (15.711%); benzaldehyde, D-glycero-L-gluco-heptose (8.611%), octadecanoic acid (7.575%); 2-hydroxy-4-methoxy (6.650%) and 4-ethyl-2- hydroxycyclopent-2-en-1-one (5.377%). Mohan *et al.* [14] reported several compounds from the methanolic extract of root, such as furfural, methyl-2-furoate, 2-hydroxy-4-methoxy benzaldehyde, vanillin, tetradecane, diethyl phtalate, hexadecane, carbromal, lupeol, norolean-12-en respectively along with other minor constituents. Giridhar *et al.* [15] raised *in vitro* roots and further extracted with dichloromethane, and dissolved in ethanol solvent. GC-MS analysis revealed the presence of a flavor compound named 2-hydroxy-4-methoxy benzaldehyde with a quantity of 40 ± 2.1 μg/g dry weight extract.

VEGETATIVE AND *IN VITRO* MICROPROPAGATION STUDIES

Seed Germination Studies

The hairy seeds of *D. hamiltonii* collected from Sathyamangalam forest area of Tamil Nadu state, India subjected to germination studies [16]. A germination test was conducted in four replications of 100 seeds with three different pre-treatments (24 hours soaking in normal water, hot water and 1000 ppm gibberllic acid) and allowed to germinate using sand and filter paper. Results revealed that seeds soaked in hot water (60°C) for 24 h on moist filter paper as a substrate found to be the most significant method for germination with the highest germination percentage of 83 to 98%. Pretreatment study coupled with microscopic observations revealed that 14% of seeds were hard coated, and they could tolerate the desiccation level up to 10% moisture. The total mortality rate of seeds was at 10°C, and rapid depletion of seed metabolites was reported after 4 months of storage. The short viability of seeds and hard seed coats, coupled with the innate germination process, hinder the plant from the natural regeneration process [16].

In vitro Microrhizome Production

Initially, the callus induction was established from the leaf discs of *D. hamiltonii* inoculated on Murashige and Skoog (MS) medium supplemented with 2 μM 6-Benzylaminopurine (BAP) and 6 μM Naphthalene acetic acid (NAA). Further, the callus mass was transferred to MS medium supplemented with 4 μM Indolebutyric acid (IBA) and 8 μM NAA for differentiation of the callus into microrhizome. Initially, 20 microrhizomes in a cluster were produced within 90 days and further supplementation of yeast extract (0.05%), polyvinylpyrrolidone (0.05%), 10% coconut milk and plant growth regulators (PGRs) enhanced the microrhizome formation with average numbers of 16 [17].

Micropropagation

An efficient and rapid micropropagation method through shoot multiplication was developed by Giridhar *et al.* [18]. MS medium with the influence of 4.4-17.7 μM 6-benzyladenine, 2.3-11.4 μM zeatin, 2.5-7.5 μM 6-(γ,γ-Dimethyl- allylamino) purine [2-iP], 2.3-4.7 μM Kinetin (Kn), 2.8-6.8 μM Thidiazuron (TDZ), and in combination with 0.3-0.9 μM Indole-3-acetic acid (IAA) on *in vitro* multiple shoot production was studied. The maximum number of shoots (6.5 ± 0.4) were induced from shoot tips cultured on agar-based MS medium with 4.9 μM 2-iP. But, both zeatin (9.1 μM) and Kn (4.7 μM) in combination with IAA (0.6 μM) were able to produce maximum multiple shoots (5.0 ± 0.4 and 5.1 ± 0.4). Shoot elongation and adventitious shoot formation were obtained on a medium containing 2.5 μM 2-iP and 0.3 μM gibberellic acid. Elongated shoots were transferred to MS medium supplemented with 9.8 μM IBA. Interaction of phloroglucinol and IBA with salicylic acid stimulated *in vitro* rooting. The rooted shoots were successfully transferred to field conditions for hardening [18].

In vitro shoot regeneration from various explants *viz.* leaf, shoot tip, and nodal segments of *D. hamiltonii* were inoculated on MS media augmented with auxin and cytokinin combinations, such as NAA, IAA, 2-iP, BAP and triacontanol (TRIA) [19]. Direct shoot regeneration was obtained in 3.0 mg/l 2-iP alone and in combination with 0.1 mg/l IAA and/or 1.0 mg/l BAP with an average shootlet length of 6.5 ± 0.17 to 8.0 ± 0.92 cm and shoot regeneration percentage was between 68 and 75%. The callus-induced regeneration was obtained from both nodal and leaf explants with the highest response (85%) in combination with 2.0 mg/l 2i-P, 1.0 mg/l IAA and 2.0 mg/l Kn with multiple shoots showing mean shoot number of 1.83 and average shootlet length of 6.3±0.19 cm [19].

The effect of phenylacetic acid (PAA) on shoot multiplication was studied by Giridhar *et al.* [18]. MS medium containing N^6-benzyladenine (N^6-BA, 2.22–31.08 μM) and PAA (7.34–36.71μM) or NAA (0.27–10.74 μM) was used to initiate shoot formation from nodal explants. The maximum number of shoots per culture was produced on a medium containing 31.08 µM BA and 14.68 µM PAA. While the longest shoot length and the maximum nodes were obtained in the medium containing 22.2 µM BA and 14.68 µM PAA with secondary shoot formation. Rooting was initiated on nutrient medium containing 9.8 µM IBA, and all the plantlets were hardened in soil with a survival rate of 80–90% under field conditions [20].

The influence of polyamine inhibitors and putrescine (PUT) on *in-vitro* rooting of micro shoots was studied [8]. Initially, nodal explants were inoculated on Murashige and Skoog (MS) medium containing 2.7 µM NAA and 8.9 µM BAP. Incorporation PUT alone at 50 µM concentration was able to induce maximum number of roots (8.62±1.93) with maximum length of 9.10 ± 1.65 cm. In the same treatment, the fresh weight of the root was also noted high (5.248±1.71 g) as compared to all other treatments. The addition of cyclohexylamine (CHA), the PUT inhibitor in the medium, significantly reduced the rooting response from microshoots. The best response for root number (2.6±1.1), root length (2.92±0.73 cm), and root weight (3.03±0.75 g) was noted when CHA was added to the medium. Results have clearly demonstrated the crucial role of putrescine in the rooting of *D. hamiltonii* explants. Further, plantlets were transferred to the greenhouse, and 90% survival rate of the hardened plants was reported [8].

A novel protocol for inducing somatic embryogenesis from leaf cultures was developed by Giridhar *et al.* [21]. Callus was obtained from leaf sections inoculated in MS medium supplemented with NAA + BA or 2,4-D + BA. The nodular embryogenic callus was developed from the explants on media containing 2,4-D and BA, whereas a compact callus was developed on media containing NAA and BA. The transfer of explants with primary callus onto MS media containing zeatin (13.68 µM) and BA (10.65 µM) resulted in the induction of the highest number of somatic embryos directly from nodular tissue. Embryogenic calli with somatic embryos were sub-cultured on MS basal medium supplemented with 4.56 µM zeatin and 10.65 mM BA. Mature embryos were developed into complete plantlets through indirect somatic embryogenesis and/or organogenesis in 12–16 weeks [21].

Nodal segments of the plant were used as explants and inoculated in MS media containing various concentrations of BA, 2-ip and NAA. MS medium supplemented with BA (0.886 mg/l) and 2-ip (0.24 mg/l) showed significant shoot length (2.9 ± 0.18 cm), more number of shoots (2.7 ± 0.94) and number of leaves per explant (5.8 ± 1.61). Genetic fidelity of *in-vitro* propagated *D. hamiltonii* using DNA based marker was studied. Random Amplified Polymorphic DNA (RAPD) analysis of *in-vitro* raised plants produced clear, reproducible bands similar to that of the mother plant. Number of monomorphic bands were highest (12) in the case of primer OPD-20 and lowest (3) in the case of primer OPC-16 [22].

The effect of $AgNO_3$ as a supplement for *in vitro* rooting of axillary bud explants was carried out by Bais *et al.* [23]. MS medium fortified with myo-inositol (100 mg/l) + NAA (0.05-2.5 mg/l) + BAP (0.05-2.5 mg/l) was used individually or in combination for proliferation of shoot. The basal rooting medium was supplemented with IAA, sucrose, and agar with different concentrations of Silver nitrate ($AgNO_3$, 10 to 50 μM). It was found that 40 μM $AgNO_3$ stimulated rooting efficiency (89.6 ± 6.72%) after 10 to 12 days of inoculation. Shoots of 4 cm length after 4 week cultures were used as explants for rooting. Of various treatments, MS medium with BAP (2.0 mg/l) and NAA (0.5 mg/l) resulted in maximum number of shoots (12.8 ± 0.96), shoot length (5.8 ± 0.43) and number of leaves (16.4 ± 1.23) on 30th day of experiment. Further, the rooted plantlets were hardened for 20 days and transplanted to the field.

A standardized protocol for *in vitro* propagation through apical bud sprouting and organogenic nodules from shoot explants was developed by Sharma *et al.* [24]. Among different combinations of plant growth regulators and additives, MS medium supplemented with 5 μM BA + 0.5 μM IAA + 30 μM adenine sulphate resulted in a maximum average number of shoots per explant (8.20) with a mean shoot length of 6.54 cm. Microshoots were efficiently rooted on ½ strength of MS medium + NAA (2.5 μM). A survival rate of 95.10% of the plantlets was observed after successful acclimatization in Soilrite.

In vitro regeneration of shoots using the seedling nodal explants was reported by Anitha and Pullaiah [25] and Anitha [26]. Different combinations of cytokinins *viz*. BAP, Kn and TDZ were used individually and in combination with NAA (0.1mg/l). An increased number of shoots (13.20 ±1.74) were developed on MS medium fortified with BAP (3 mg/l) + NAA (0.5 mg/l) + Kn (0.5 mg/l) + coconut milk (CM 10%) + AA (Ascorbic acid 0.15 mg/l) with an average shoot length of 3.24±0.39 cm. MS medium (¼ strength) devoid of hormones with various concentrations of auxins such as NAA, IAA, IBA and in combinations of these were used for rooting. Among all the combinations, NAA (1mg/l) was selected for rooting. The rooted plants were acclimatized to field conditions (60%). Callus studies were performed using different seedling explants like leaf, internode, cotyledon and root. Internodal segments produced a compact pale yellowish green callus (90%), followed by cotyledons producing a friable white to green callus (80%) on MS medium supplied with 2,4-D (2 mg/l). A green nodular callus was produced from cotyledonary segments containing 2,4- D (0.5 mg/l) and BAP (0.1 mg/l).

CONCLUSION

This review highlights the status, distribution, phytoconstituents, micropropagation and regeneration aspects of the critically endangered, edible ethnomedicinal plant *D. hamiltonii*, endemic to peninsular India. The report clearly indicates that the effective protocols for direct shoot regeneration from various explants, callus-mediated organogenesis and somatic embryogenesis of the plant have been explored. However, germplasm preservation, elicitation, adventitious root culture and genetic manipulation of *in vitro* cultures are totally unexplored areas, which need special attention to conserve this economically and medicinally important plant species for its sustainable utilization.

REFERENCES

[1] POWO Plants of the World Online.Facilitated by the Royal Botanic Gardens,Kew. Published on the Internet. 2022. Available from : http://www.plantsoftheworldonline.org

[2] Ved D, Saha D, Ravikumar K, Haridasan K. *Decalepis hamiltonii*. The IUCN Red List of Threatened Species 2015. [http://dx.doi.org/10.2305/IUCN.UK.2015-2.RLTS.T50126587A50131330.en]

[3] Sunitha TG, Dhadde SB, Durg S, *et al.* Immunomodulatory effect of *Decalepis hamiltonii* Wight and Arn. roots extract on rodents. Indian J Pharm Educ 2016; 50(2): S146-52.

[4] Pudutha A, Venkatesh K, Chakrapani P, Chandra Sekhar Singh B. Traditional uses, phytochemistry and pharmacology of an endangered plant - *Decalepis hamiltonii* Wight and Arn. Int J Pharm Sci Rev Res 2014; 47(24): 268-78.

[5] Vijayakumar R, Pullaiah T. An ethno-medico-botanical study of Prakasam district, Andhra Pradesh, India. Fitoterapia 1998; 69: 483-9.

[6] Samydurai P, Saradha M, Ramakrishnan R, Santhosh Kumar S, Thangapandian V. Micropropagation prospective of cotyledonary explants of *Decalepis hamiltonii* Wight & Arn.- An endangered edible species. Indian J Biotechnol 2016; 15: 256-60.

[7] Thangadurai D, Murthy KSR, Prasad PJN, Pullaiah T. Antimicrobial screening of *Decalepis hamiltonii* Wight & Arn. (Asclepiadaceae) root extracts against food related microorganisms. J Food Saf 2004; 24(4): 239-45. [http://dx.doi.org/10.1111/j.1745-4565.2004.00537.x]

[8] Matam P, Parvatam G. Putrescine and polyamine inhibitors in culture medium alter *in vitro* rooting response of *Decalepis hamiltonii* Wight & Arn. Plant Cell Tissue Organ Cult 2017; 128(2): 273-82. [http://dx.doi.org/10.1007/s11240-016-1108-0]

[9] Thangadurai D, Anitha S, Pullaiah T, Reddy PN, Ramachandraiah OS. Essential oil constituents and *in vitro* antimicrobial activity of *Decalepis hamiltonii* roots against foodborne pathogens. J Agric Food Chem 2002; 50(11): 3147-9. [http://dx.doi.org/10.1021/jf011541q] [PMID: 12009977]

[10] Nagarajan S, Jagan Mohan Rao L, Gurudutt KN. Chemical composition of the volatiles of *Decalepis hamiltonii* (Wight & Arn). Flavour Fragrance J 2001; 16(1): 27-9. [http://dx.doi.org/10.1002/1099-1026(200101/02)16:1<27::AID-FFJ937>3.0.CO;2-F]

[11] Reddy MC, Murthy KSR. A review on *Decalepis hamiltonii* Wight & Arn. J Med Plants Res 2013; 7(41): 3014-29. [http://dx.doi.org/10.5897/JMPR2013.5099]

[12] Prakash P, Gayathiri E, Manivasagaperumal R, Krutmuang P. Biological activity of root extract of *Decalepis hamiltonii* (Wight & Arn) against three mosquito vectors and their non-toxicity against the mosquito predators. Agronomy 2021; 11(7): 1267. [http://dx.doi.org/10.3390/agronomy11071267]

[13] Selvaraj S, Rajkumar P, Thirunavukkarasu K, Santhiya A, Kumaresan S. Spectral and phytochemical analysis on the extraction of *Decalepis hamiltonii*. ICRASS 2016; 102-7.

[14] Mohan B, Nayak JB, Sunil Kumar R, Shiva Kumari LP, Mohan CH, Rajani B. Phytochemical screening, GC-MS analysis of *Decalepis hamiltonii* Wight & Arn. An endangered medicinal plant. J Pharmacogn Phytochem 2016; 5(5): 10-6.

[15] Giridhar P, Rajasekaran T, Ravishankar GA. Production of a root-specific flavour compound, 2-hydroxy-4-methoxy benzaldehyde by normal root cultures of *Decalepis hamiltonii* Wight and Arn (Asclepiadaceae). J Sci Food Agric 2005; 85(1): 61-4. [http://dx.doi.org/10.1002/jsfa.1939]

[16] Anandalakshmi R, Prakash MS. Seed germination and storage characteristics of *Decalepis hamiltonii*: implications for regeneration. For Trees Livelihoods 2010; 19(4): 399-407. [http://dx.doi.org/10.1080/14728028.2010.9752681]

[17] Thangavel K, Ravichandran P, Ebbie MG, Manimekalai V *In vitro* microrhizome production in *Decalepis hamiltonii*. Afr J Biotechnol 2014; 13(11): 1308-13. [http://dx.doi.org/10.5897/AJB2013.12969]

[18] Giridhar P, Gururaj HB, Ravishankar GA. *In vitro* shoot multiplication through shoot tip cultures of *Decalepis hamiltonii* Wight & Arn., a threatened plant endemic to Southern India. *In Vitro* Cell Dev Biol Plant 2005; 41(1): 77-80. [http://dx.doi.org/10.1079/IVP2004600]

[19] Ranganatha M, As A, Sharma A, N Rao N. *In vitro* shoot regeneration of swallow root (*Decalepis hamiltonii*) : A steno-endemic red listed medicinal plant. Asian J Pharm Clin Res 2020; 13(4): 188-91. [http://dx.doi.org/10.22159/ajpcr.2020.v13i4.36714]

[20] Giridhar P, Ramu DV, Reddy BO, Rajasekaran T, Ravishankar GA. Influence of phenylacetic acid on clonal propagation of *Decalepis hamiltonii* Wight & Arn. An endangered shrub. *In Vitro* Cell Dev Biol Plant 2003; 39(5): 463-7. [http://dx.doi.org/10.1079/IVP2003448]

[21] Giridhar P, Kumar V, Ravishankar GA. Somatic embryogenesis, organogenesis, and regeneration from leaf callus culture of *Decalepis hamiltonii* Wight & Arn., an endangered shrub. *In Vitro* Cell Dev Biol Plant 2004; 40(6): 567-71. [http://dx.doi.org/10.1079/IVP2004567]

[22] Kumar P, Raviraja Shetty G, Souravi K, Rajasekharan PE. Evaluation of genetic fidelity of *in vitro* propagated *Decalepis hamiltonii* Wight & Arn. using DNA Based Marker. IOSR J Pharm Biol Sci 2015; 10(3): 86-9.

[23] Bais HP, George J, Ravishankar GA. *In vitro* propagation of *Decalepis hamiltonii* Wight & Arn., an endangered shrub, through axillary bud cultures. Curr Sci 2000; 79(4): 408-10.

[24] Sharma S, Shahzad A, Ahmad A, Anjum L. *In vitro* propagation and the acclimatization effect on the synthesis of 2-hydroxy-4-methoxy benzaldehyde in *Decalepis hamiltonii* Wight and Arn. Acta Physiol Plant 2014; 36(9): 2331-44. [http://dx.doi.org/10.1007/s11738-014-1606-9]

[25] Anitha S, Pullaiah T. *In vitro* propagation of *Decalepis hamiltonii*. J Trop Med Plants 2002; 3: 227-32.

[26] Anitha S. *in vitro* propagation studies of *Decalepis hamiltonii* Wight & Arn. and *Sterculia foetida* Linn. Ph.D. thesis, Sri Krishnadevaraya University, Anantapur 2001.

CHAPTER 16

Micropropagation of Pharmaceutically Important Plant *Tinospora cordifolia* (Willd.) Hook. f. & Thomson: An Overview

Sachet Hegde[1,*], **Pradeep Bhat**[2], **Santoshkumar Jayagoudar**[3], **Savaliram G. Ghane**[4] and **Harsha V. Hegde**[2]

[1] *Department of Botany, Bangurnagar Degree College, Ambewadi, Dandeli, Karnataka-581325, India*

[2] *ICMR-National Institute of Traditional Medicine, Nehru Nagar, Belagavi, Karnataka-590010, India*

[3] *Department of Botany, G. S. S. College & Rani Channamma University P. G. Centre, Belagavi, Karnataka-590006, India*

[4] *Department of Botany, Shivaji University, Vidyanagar, Kolhapur, Maharashtra-416004, India*

Abstract: *Tinospora cordifolia* (Willd.) Hook. f. & Thomson belongs to the family Menispermaceae. The origin of the species is from Indian subcontinent to Indo-China. The plant has antipyretic, antiperiodic, anti-inflammatory, antirheumatic, spasmolytic, hypoglycaemic and hepatoprotective properties. Due to over-exploitation of this medicinally important species, attention has been given to its conservation through *in-vitro* micropropagation techniques. The present chapter emphasizes the ample research work on *in-vitro* regeneration and enhancement of secondary metabolites from *T. cordifolia.* Additional information on the importance of this genus in various systems of medicine, active components, biological activities and its sustainable utilization for the welfare of mankind is also discussed in the present chapter.

Keywords: Conservation, Callus Culture, *In vitro* Regeneration, Micropropagation, Menispermaceae, Traditional Medicine.

INTRODUCTION

Plants are not only the manufacturers of food but also produce various types of alkaloids, terpenoids, saponins, phenols and many other secondary metabolites, which are often used as medicines either in raw or processed form. Interestingly,

* **Corresponding author Sachet Hegde:** Department of Botany, Bangurnagar Degree College, Ambewadi, Dandeli, Karnataka–581325, India; E-mail: sachet.scorpion@gmail.com

T. Pullaiah (Ed.)

nearly one-quarter of modern medicines comprise of plant sources, such as herbal extracts [1]. Due to the over-exploitation, the medicinal plants are under enormous threat of declining population size, genetic diversity and loss of habitat [2]. Demand for active constituents of plant origin in several pharmaceuticals, fragrances, flavors and color industries exceeded several billion dollars per year [3]. In this scenario, the *in vitro* plant tissue culture method for propagation and also as a source of producing important bioactive secondary metabolites paves the way for conservation and mass multiplication of important medicinal plants. In recent years, plant tissue culture has become of major industrial importance not only in the area of propagation but also in the production of quality materials and large-scale natural metabolites. In this non-conventional method of propagation, thousands of plantlets can be produced from a small piece of tissue (explants) in a relatively short duration under controlled conditions, irrespective of season and weather. Hence, it is also called a low cost-high volume system for the production of biomass and several bioactives present in it. Similarly, as the commercial values for plant secondary metabolites are evolving, eventually, tissue culture technique plays a major role in providing a continuous supply of healthy plant materials to several pharmaceutical industries. Therefore, rapid regeneration and production of high quality, uniform planting material through reliable micropropagation methods are important for the economically important medicinal plants are the needs of the time [4].

Tinospora cordifolia (Willd.) Hook.f. & Thomson belongs to the family Menispermaceae. The origin of the species is from the Indian subcontinent to Indo-China and distributed over East Himalaya, Myanmar, Bangladesh, India, Maldives, Sri Lanka and Vietnam [5]. It is generally a deciduous woody climber or climbing shrub and can also grow from its detached stem [6].

In India, *T. cordifolia* is described in the ancient Ayurvedic texts like *Charaka Samhita, Sushruta Samhita* and also in *The Ayurvedic Pharmacopoeia of India. T. cordifolia* is used to treat diabetes, gastrointestinal disorders, dyspepsia, flatulence, gastritis, jaundice, diarrhea and hemorrhoids [7]. The plant has antipyretic, antiperiodic, anti-inflammatory, antirheumatic, spasmolytic, hypoglycaemic and hepatoprotective properties. The whole plant extract increases the urine output, and the stem juice is used to treat fever. The decoction of the whole plant is used in rheumatic pain, bilious fever and as a febrifuge. The starch extracted from the stem has anti-diarrhoeal, antidysenteric and antacid properties [7 - 9]. Plant stem acts as a diuretic, stimulates bile secretion, enriches the blood, helps in relieving constipation, vomiting, burning sensation, cures jaundice, and a mixture of stem and root juice is used as an antidote for snake bite and scorpion sting [9]. It has known to possess immunomodulatory, anti-diabetic, anti-toxic, anti-arthritic, anti-osteoporotic, anti-HIV, anti-cancer, anti-microbial, antioxidant, anti-tumor, anti-hypoglycemic, anti-dipressant, cardioprotective, anti-ulcer, anti-diarrheal, analgesic, aphrodisiac, neuroprotective, anti-inflammatory, gastroprotective, radio protective, hepatoprotective, antipsychotric, anti-asthmatic anti-malarial, anti-allergic, larvicidal and anti-severe acute respiratory syndrome coronavirus-2 (SARS-CoV-2) properties [10 - 18]. Active constituents such as berbarine, boldine, choline, clerodane derivatives, columbamine, columbin, cordifolide A, cordifolide B, cordifolide C, cordifolioside A, cordifolioside B, cordifolioside C, cordifoliside D, cordifoliside E, cordioside, cordioside, ecdysterone, furanoid diterpene glucoside, giloinsterol, furanolactone and several other compounds have been reported from *T. cordifolia* [10, 12 - 15, 19 - 23].

IMPORTANCE OF PROPAGATION METHODS

It is reported that *T. cordifolia* is one of the key ingredients used in about 69 Ayurvedic formulations, and it has a huge demand in local as well as international markets due to its wide therapeutic values. Since the demand for raw materials of *T. cordifolia* has increased from 2000 to 5000 MT with an annual growth of 9.1%, the National Medicinal Plant Board (NMPB), India has prioritized this species for its mass multiplication program throughout the country [24].

Even though it can grow by its detached stem, there are a few reports on vegetative propagation [6]. Vegetative propagation through stem cuttings with four lateral buds had significantly better survival rates than the stem cuttings with one to three buds. However, single bud stem cutting proved to be economical for one hector plantation compared to available conventional methods, even though the survival depends on its morphological traits and age of the stem cutting. In the traditional propagation method (stem cuttings with single lateral nodes and planted in a horizontal position as in sugarcane), it consumes a large space in the nursery [25].

IN-VITRO STUDIES

T. cordifolia species may be propagated through stem cuttings; however, there is little information available about the propagation through seeds. The stem part may have the rejuvenating capacity and can even be grown by vegetative multiplication. However, currently, there is a need to grow the plant along with enhancing its active constituents for pharmaceutical industries. Because of the increasing demand due to its medicinal values, it is important to conserve these species and, at the same time, the demand for its medicinal purpose to be met. Partially, it is being fulfilled by agro-industries by growing them in the field through stem cuttings. However, the enhancement of secondary metabolites through callus culture or hairy root culture techniques seems to be the more suitable option for scaling up the required biologically active components, as it produces a uniform batch of plantlets and ensures continuity in the production irrespective of season and agro-climatic conditions.

In vitro Regeneration

Sivakumar *et al.* [26] established a protocol for rapid *in vitro* clonal propagation using nodal explants. Murashige and Skoog medium (MS medium) fortified with 4.36 μM Kinetin (Kn) produced two shoots per culture with 2.32 cm of length and 70% survival rate in comparison to Benzyl adenine (BA) and 6-(γ,γ-Dimethylallylamino)purine [2-iP]. The study also reported the browning of tissues and discoloration of media due to the exudation of phenolics after a month of culture. They were augmented with 20% silver nitrate, with the same concentration of Kn in the MS medium to obtain 100% response within 16 days of culture. For shoot proliferation and elongation, nodal shoots were cultured on MS medium containing 8.82 micro molar (μM) BA, and it produced 4.6 to 4.8 shoots per explant with 3 - 4.82 cm length. These elongated shoots were transferred to ½ strength MS + 3% sucrose + 6.43 μM Indolebutyric acid (IBA), produced 5.2 roots per plant after 27 days of culturing and acclimatized with survival rate of 80%.

The production of berberine was confirmed from *in vitro* regenerated plants through Quadrupole Time of Flight Liquid Chromatography Mass Spectrometry (QToF-LC-MS) [27]. A high frequency of multiple shoot formation was reported from the nodal segments cultured on MS medium augmented with 1 mg/l 6-Benzylaminopurine (BAP) and 0.5 (mg/l) 2-iP. In the study, authors reported 8 shoots per explant with the highest shoot length of 9.3 cm. Further, it was subjected to rhizogenesis using ½ strength of MS medium augmented with 0.5 mg/l IBA and resulted in 89% response with root length of 8.3 cm. Berberine content in *in-vitro* plants was found maximum (19.8 μg/g) as compared to the stem (9.3 μg/g) and leaves (8.4 μg/g) of field grown plants.

The establishment of *in vitro* propagation techniques and the presence of berberine content were reported by Sudan *et al.* [28]. For the establishment of contamination-free cultures, explants were treated with 1% Bavistin (w/v), followed by the treatment with 0.1% mercuric chloride ($HgCl_2$) solution for 5 minutes. The initiation of axillary buds from nodal explants was obtained by culturing the explants on MS medium supplemented with 2 mg/l BAP alone and also in combination with 0.5 mg/l Naphthalene acetic acid (NAA) + 40 mg/100 ml gentamycin. Findings revealed the formation of 5 shoots per explant. Authors also reported berberine from the methanol extract using thin layer chromatography (TLC).

Madhu [29] grew nodal explants on MS medium containing 1.0 mg/l BAP and obtained 1-2 shoots per explant. Use of MS medium and incorporation of 1.0 mg/l BAP and 0.5 mg/l Kn initiated 3-4 multiple shoots per explant. Meanwhile, 3 shoots per nodal explants were initiated in MS medium supplemented with 1mg/l BAP + 0.5 mg/l NAA, and they were hardened and acclimatized with a survival rate of 70%.

Bhalerao *et al.* [30] developed a protocol for *in vitro* callus formation and direct organogenesis through leaf and stem explants and compared the chemo-profiling of the plant growing on different supporting trees. Leaf explants inoculated on MS medium fortified with 2,4-Dichlorophenoxyacetic acid (2,4-D) alone or in combinations with Kn developed into a callus but failed in organogenesis. While the nodal explants transferred on MS medium fortified with 8 μM Kn alone or 12 μM Kn and 2 μM BAP proliferated into a single shoot per explant. These shoots were then transferred to the rooting medium (MS + 8 μM NAA). *T. cordifolia* was grown with the support of different tree species such as *Azardirachta indica, Butea monosperma, Acacia leucophloea* and *Prosopis juliflora* from identical geo-climatic conditions, and comparative chemical profiling using Thin layer chromatography (TLC), High performance thin layer chromatography (HPTLC), Infrared spectroscopy (IR) and Ultraviolet (UV) spectroscopic methods was carried out. Both leaf and stem extracts of *T. cordifolia* grew with the support of *A. indica* showed better results in chemical profiling than that of the plants grown with the support of other trees.

Using young and mature leaves and cotyledonary explants, Mridula *et al.* [31] developed an efficient protocol for direct and indirect organogenesis. A significantly higher responsive rate of organogenic callus (97-100%) was obtained on MS medium fortified with 2 mg/l IAA. IAA and NAA offered calli with whitish-yellow compact nodules. At the same time, the callus produced on MS + 2,4-D was yellowish with friable nature. Incorporation of IBA in the medium produced callus with roots. MS medium was further modified with B5 (Gamborg Medium) salts to induce shoot buds from the obtained callus. When it was supplemented with 2 mg/l BA + 1mg/l Kn + 100mg/l ascorbic acid + 0.5 mg/l IAA, cultures were noted with 24, 19 and 16 shoots per cotyledonary, young leaves and mature leaf explants, respectively. Modified MS supplemented with 2 mg/l BA + 1 mg/l Kn + 100 mg/l ascorbic acid + 0.75 mg/l NAA resulted in 10, 14 and 20 shoots per mature leaf, young leaf and cotyledonary explants, respectively. However, these calli failed to respond with individual auxin or cytokinin treatments. In the case of the modified MS medium, these explants showed significant direct organogenesis with the appreciable response rate and 14.5 and 11 shoots were produced from young leaves and cotyledonary explants, respectively. Shoot buds initiated from direct organogenesis were transferred to

MS medium containing 1 mg/l gibberellin A3 (GA_3) with average length of 8.2, 7.6 and 7.4 cm from cotyledon, young leaf and mature leaf explants, respectively. Meanwhile, MS medium, along with 0.5 mg/l GA_3, showed an average shoot length of 8.3 cm in the case of cotyledon young leaf explants. Elongated shoots were further cultured on ½ strength MS medium augmented with 0.5 mg/l IBA, which reported induction of roots. Well rooted shoots were further acclimatized using red soil: sand: vermiculite at 1:1:3 ratio [31].

Anika and Aditi [32] studied the effect of PGRs on *in vitro* regeneration using the nodal and internodal explants and also optimized the sterilization time (4 min) using 0.1% $HgCl_2$. Nodal explants cultured on MS medium fortified with 1 mg/l BAP, regenerated significantly. Whereas the highest frequency of regeneration was observed from internodal explants cultured on MS medium containing 1 mg/l IAA. Meanwhile, the highest regeneration frequency (70%) was noted from the nodal segments cultured on MS medium augmented with 1 mg/l IAA and 1.5 mg/l BAP. Likewise, the response of internodal explants was up to 78% when the MS medium was supplemented with 1 mg/l BAP + 2 mg/l IAA. These studies also reported that various concentrations of individual auxins and their synergic effect with cytokinins altered the regeneration frequency [32].

An efficient direct organogenesis protocol was developed by Mittal *et al.* [33] using mature nodal explants. MS medium supplemented with 4.44 μM BA in combination with 2.45 μM N6-2-iso-pentenyl adenine produced eight shoots per explant with 9.3 cm of the average length. To absorb the phenolic exudation, activated charcoal, polyvinyl pyrrolidone (PVP) and ascorbic acid were used. Of these, 1% PVP, along with plant growth regulators (PGRs), successfully controlled the secretion of phenolics. The rooting response was 89% and the highest root length of 8.3 cm was reported when these shoots were cultured on ½ strength MS medium fortified with 2.45 μM IBA. Well-rooted plantlets were acclimatized and hardened using garden soil and cocopeat (2:1).

An *in-vitro* clonal propagation through nodal and leaf explants was also studied [34] and reported that leaf explants failed to produce callus on any concentrations and combinations of PGRs used. Nodal explants notably produced multiple shoots (7.6 shoots per explant) on MS medium fortified with 2 mg/l BA with a regeneration frequency of 89%. The study also reported that the addition of Indole-3-acetic acid (IAA), NAA or 2,4-D along with either BA or Kn did not improve the response, while the medium supplemented with Kn alone improved the shoot elongation with longer internodes. Around ten roots per shoot were produced when the shoots were cultured on MS medium amended with 0.5 mg/l NAA.

Handique and Choudhury [35] found that combinations of 2 mg/l BAP, 4 mg/l Kn and 0.2 mg/l Thidiazuron (TDZ) in MS medium offered maximum average of ten shoots per explant. They were transferred to ½ strength MS media fortified with 2 mg/l IBA for rooting.

In *in-vitro* regeneration and tuberization of microshoots of *T. cordifolia*, the effect of phloroglucinol (PG) was studied by Jani *et al.* [36]. Nodal explants were cultured on MS medium supplemented with 6.98 μM Kn and gave a maximum response (52.2%) with an average of 3 shoots per explant, 3.9 cm shoot length and an average of 4.2 leaves per shoot. Authors also reported that the plant has the tendency of producing a single shoot per explant. The synergic effect of PG with Kn showed an increase in shoot bud induction (52.2 to 84.8%), and multiple shoot production was also increased to 60.3 from 12.2% with an average of 7.5 shoots per explant. The lower concentration of PG supplemented media induced multiple shoots and the higher concentration of PG induced direct roots. Maximum rooting response (81.1%) was observed in ½ strength MS medium augmented with 7.4 μM IBA + 793.7 μM PG. The study also reported the production of tuberous roots from the PG augmented medium [36].

Raghu *et al.* [37] developed a protocol for *in vitro* clonal propagation using mature nodes. Shoots were initiated from MS as well as Woody Plant Medium (WPM) fortified with 2.32 μM Kn. An average multiplication rate of six shoots per explant was obtained on WPM augmented with 8.87 μM BA. Shoots were then transferred to WPM containing 2.22 μM BA and 8.87 μM Kn for further elongation. MS medium (½ strength) fortified with 2.85 μM IAA was used for further rooting the shoots. Of the MS and WPM media used, the other media reported superior results with 68% of bud break, while for axillary proliferation and elongation of shoots, BA and Kn were superior, respectively.

Nodal and internodal segments of the plant were used to develop a protocol for *in-vitro* propagation using MS, Gamborg and Nitsch media [38]. Among all the media, the explants showed good responses in MS medium with good callogenesis, multiple shoot formation and rooting. The study reported that single shoot proliferation on MS medium containing 5 mg/l BA combined with 1 mg/l Kn. Multiple shoots were obtained on the medium containing BA + Kn + adenine; however, the response of explants with the medium containing 0.5 mg/l IAA + 5 mg/l Kn was found more effective. The shoots were rooted on medium augmented with 0.2 mg/l or 0.4 mg/l NAA, but in the case of internodal explants, medium containing IAA was found the most ideal. For callogenesis, B5 medium fortified with 3 ppm 2,4-D proved to be the superior.

Mridula *et al.* [39] developed an *in vitro* propagation protocol using young and mature shoot tip explants from two different aged plants (15 d young seedlings and 3-year-old plants). Shoot bud initiated on MS medium supplemented with 1 mg/l Kn showed 90.1% response, but the highest number of shoots were produced on the media supplemented with 2 mg/l BA + 1mg/l Kn + 0.5mg/l IAA when young shoot tips were used. The study also reported the effect of additives on shoot production. Addition of 0.1 mg/l GA_3 along with 2 mg/l BA + 1 mg/l Kn + 0.5 mg/l IAA enhanced the shoot formation significantly with 91.3 and 92.1% response from young and mature shoot tips, respectively. Eleven shoots per explant with 6.6 cm of length were reported to be the highest shoot proliferation from the same combination. To check the phenolic exudation from the explants they have used various antioxidants, of which 100 mg/l ascorbic acid was the most successful. Authors also studied the effect of different media for bud break and multiple shoot formation, in which MS media was the most supportive as compared to modified MS media and WPM. For rooting, half-strength MS medium augmented with 0.5 mg/l IBA produced maximum rooting responses (9.1 and 77%) from young and mature shoot tip derived shoots. Plantlets were successfully acclimatized and hardened on red soil and vermiculite (1:3), with survival rate of about 80%.

Anita and Sharma [40] studied micropropagation through stem explants on an MS medium. Before culturing, collected stems were washed and kept in normal water to ensure the leaching of phenolic compounds. Shoot regeneration frequency was significant in medium supplemented with 4 mg/l Kn than in 3 and 2 mg/l Kn with 60, 40 and 10%, respectively. Whereas high frequency (4-5 shoots/explant) of microshoots was proliferated in the medium containing 4 and 5 mg/l BAP. For multiple shoot regeneration, the synergic effect of 3 mg/l BAP and 6 mg/l Kn played a major role. Roots were induced on medium augmented with 6 mg/l NAA with an average frequency of 90% and 10-11 roots per shoot. The survival rate of 75-80% was reported in the study when these rooted shoots acclimatized to the field conditions.

Rapid *in-vitro* clonal propagation through shoot tip explants on MS and WPM was studied by Sharma *et al.* [41]. The maximum frequency of shoots was observed in WPM supplemented with 2 mg/l BAP with an average of 4.8 shoots per explant. The synergic effect of 2 mg/l BAP and 1 mg/l Kn using WPM resulted in an average of four shoots per explant. *In vitro* developed shoots were then rooted on ½ strength MS medium fortified with 1 mg/l IBA with 40% frequency of induction, and 1-2 roots per shoots with an average length of 1.8 cm were produced. These rooted plantlets were acclimatized on sterile soil and sand mixture in equal proportions and showed a 70% survival rate.

Chatterjee and Ghosh [42] developed an efficient protocol for rapid *in vitro* multiplication through shoot tip culture on MS medium. About sixteen shoots per explant were obtained on medium augmented with 5 mg/l BAP. Rooting frequency was 82% on MS medium supplemented with 1mg/l IBA and a survival frequency of 90% was reported when transferred to greenhouse conditions. They have also reported that the growth of the plant can be slowed down for conservation purposes for up to 8 months on MS medium augmented with sorbitol and mannitol (2% each) at 10 °C temperature with the response rate of 95%.

Callus Culture Studies

Callus and cell suspension cultures were established using the stem explants [43]. The optimum callus was obtained in MS medium with 0.2 mg/l Kn and 1.0 mg/l 2,4-D. Later, the callus obtained was transferred to a liquid MS medium containing 0.2 mg/l Kn + 20 mg/l 2,4-D, followed by agitation. Roots of the field grown plant and *in vitro* grown biomass (callus and cell suspension) were extracted with methanol and subjected to TLC and HPLC analysis. Two compounds, *viz*., berberine and jatrorrhizine, were found to be slightly higher in field grown roots than callus and cell suspension cultures [43].

Khalilsaraie *et al*. [44] induced the embryogenic callus from mature leaves on both liquid and semi-solid MS basal, as well as MS medium augmented with 2 mg/l 2,4-D and 1 mg/l BAP. They have also reported the significant germination rate of embryos on the medium supplemented with high concentrations of sucrose and low salt strength. Gas chromatography mass spectroscopy (GC-MS) analysis reported the presence of 7-tetradecene, hydroxylamine, decocyl acrylate, isopropyl myristate, squalene and phenol-2,4 bis (1,1-dimethylethyl) in both the extracts.

Maanhvizhi and Revathi [45] reported the maximum callus from the stem explants grown on MS medium fortified with 2 mg/l of 2,4-D + 0.3 mg/l of Kn. Similar results were obtained from the leaf explant on MS medium supplemented with 1.5 mg/l 2,4-D + 0.3 mg/l BAP after 3-4 weeks of culturing. *In-vitro* anti-inflammatory activity was carried out using callus, leaf and stem extracts derived from wild plants. Of which, nearly 78.15% of human red blood cells were protected in hypertonic solution (100 µg/ml) from the leaf callus extract.

The comparative microbicidal efficacy of *in-vitro* and *in-vivo* raised cells was analysed by Mittal *et al*. [46]. MS medium containing 1 mg/l IBA and 0.1% PVP was used to inoculate the leaf and internodal parts, and the callus was obtained. Callus fractions significantly inhibited the growth of all the pathogens tested. The proliferation efficiency of the internodal callus was reported higher as compared to the leaf callus within 3-4 weeks.

Priti [47] estimated the total alkaloid from the stem of wild plants, callus and *in-vitro* regenerated plants on MS media incorporated with combinations of PGRs. They reported higher total alkaloid contents in callus tissues as well as in *in vitro* regenerated plants as compared to that of wild plants.

Aditi *et al.* [48] induced the callus from nodal, internodal and leaf explants on an MS medium containing various concentrations and combinations of BAP and NAA (0.5 to 3 mg/l). However, various concentrations of Kn (0.5 to 3 mg/l) alone produced callus that, too, only from leaf explants.

CONCLUSION

T. cordifolia needed attention for conservation and sustainable utilization because of its huge demand both in traditional medicine systems and pharmaceutical industries. In the context of the Indian subcontinent, *T. cordifolia* is having huge demand, and extensive research is being carried out on this species. It is getting more and more attention for its conservation strategies and to enhance the production of active constituents according to the demand of herbal industries. These goals can be achieved through the combination of tissue culture and other advanced biological techniques.

REFERENCES

[1] Pezzuto J. Taxol production in plant cell culture comes of age. Nat Biotechnol 1996; 14(9): 1083. [http://dx.doi.org/10.1038/nbt0996-1083] [PMID: 9631053]

[2] Lemma DT, Banjaw DT, Megersa HG. Micropropagation of medicinal plants. Int J Plant Breed Crop Sci 2020; 7(2): 796-802.

[3] Debnath M, Malik C, Bisen P. Micropropagation: A tool for the production of high quality plant-based medicines. Curr Pharm Biotechnol 2006; 7(1): 33-49. [http://dx.doi.org/10.2174/138920106775789638] [PMID: 16472132]

[4] Oseni OM, Pande V, Nailwal TK. A review on plant tissue culture, a technique for propagation and conservation of endangered plant species. Int J Curr Microbiol Appl Sci 2018; 7(7): 3778-86. [http://dx.doi.org/10.20546/ijcmas.2018.707.438]

[5] POWO. Plants of the World Online. 2022. Available from: http://www.plantsoftheworldonline.org

[6] De Wet H. An ethnobotanical and chemotaxonomic study of South African Menispermaceae. S Afr J Bot 2008; 74(1): 2-9. [http://dx.doi.org/10.1016/j.sajb.2007.07.001]

[7] Chi S, She G, Han D, Wang W, Liu Z, Liu B. Genus *Tinospora*: Ethnopharmacology, phytochemistry, and pharmacology. Evid Based Complement Alternat Med 2016. [http://dx.doi.org/10.1155/2016/9232593]

[8] Udayan PS, George S, Tushar KV, Balachandran I. *Tinospora sinensis* (Lour.) Merr. from Sickupara, Kolli Hills forest, Namakkal District, Tamil Nadu. Zoos' Print J 2004; 19(9): 1622-3. [http://dx.doi.org/10.11609/JoTT.ZPJ.1180.1622-3]

[9] Meenu MT, Radhakrishnan KV. Menispermaceae family of plants and its action against infectious diseases: A review. Map J Sci 2020; 19(7): 33-71.

[10] Ghosh S, Saha S. *Tinospora cordifolia*: One plant, many roles. Anc Sci Life 2012; 31(4): 151-9. [http://dx.doi.org/10.4103/0257-7941.107344] [PMID: 23661861]

[11] Mishra A, Kumar S, Pandey AK. Scientific validation of the medicinal efficacy of *Tinospora cordifolia.* Scient World J 2013; 1-8. [http://dx.doi.org/10.1155/2013/292934] [PMID: 24453828]

[12] Reddy NM, Reddy NR. *Tinospora cordifolia* chemical constituents and medicinal properties: A review. Sch Acad J Pharm 2015; 4(8): 364-9.

[13] Kumar VD, Geetanjali B, Avinash KO, Kumar JR, Chandrashekrappa GK, Basalingappa KM. *Tinospora cordifolia*: The antimicrobial property of the leaves of Amruthaballi. J Bacteriol Mycol 2017; 5(5): 363-71.

[14] Tiwari P, Nayak P, Prusty SK, Sahu PK. Phytochemistry and pharmacology of *Tinospora cordifolia*: A review. System Rev Pharm 2018; 9(1): 70-8. [http://dx.doi.org/10.5530/srp.2018.1.14]

[15] Sharma P, Dwivedee BP, Bisht D, Dash AK, Kumar D. The chemical constituents and diverse pharmacological importance of *Tinospora cordifolia.* Heliyon 2019; 5(9): e02437. [http://dx.doi.org/10.1016/j.heliyon.2019.e02437] [PMID: 31701036]

[16] Paul A, Raj VS, Vibhuti A, Pandey R. Larvicidal efficacy of *Andrographis paniculata* and *Tinospora cordifolia* against *Aedes aegypti*: A dengue vector. Pharmacognosy Res 2020; 12(4): 352-60. [http://dx.doi.org/10.4103/pr.pr_35_20]

[17] Upadhyay G, Tewari LM, Tewari G, *et al.* Evaluation of antioxidant potential of stem and leaf extracts of Himalayan *Tinospora cordifolia* Hook.f. & Thomson. Open Bioactive Compd J 2021; 9(1): 2-8. [http://dx.doi.org/10.2174/1874847302109010002]

[18] Leena S, Asha B, Azam J, *et al.* Phytotherapy for treatment of cytokine storm in COVID-19. Front Biosci 2021; 26(5): 51-75. [http://dx.doi.org/10.52586/4924]

[19] Vibha J, Ashwini K. *Tinospora cordifolia* (Willd.) Miers ex Hook.f. & Thoms. (Menispermaceae): A plant of immense medicinal value used in traditional medical system: A review. World J Pharm Res 2020; 9(12): 722-8.

[20] Bajpai V, Singh A, Chandra P, Negi MPS, Kumar N, Kumar B. Analysis of phytochemical variations in dioecious *Tinospora cordifolia* stems using HPLC/QTOF MS/MS and UPLC/QqQ $_{LIT}$ -MS/MS. Phytochem Anal 2016; 27(2): 92-9. [http://dx.doi.org/10.1002/pca.2601] [PMID: 26627195]

[21] Bala M, Verma PK, Awasthi S, Kumar N, Lal B, Singh B. Chemical prospection of important ayurvedic plant *Tinospora cordifolia* by UPLC-DAD-ESI-QTOF-MS/MS and NMR. Nat Prod Commun 2015; 10(1): 1934578X1501000. [http://dx.doi.org/10.1177/1934578X1501000113] [PMID: 25920217]

[22] Pan L, Terrazas C, Lezama-Davila CM, *et al.* Cordifolide A, a sulfur-containing clerodane diterpene glycoside from *Tinospora cordifolia.* Org Lett 2012; 14(8): 2118-21. [http://dx.doi.org/10.1021/ol300657h] [PMID: 22497272]

[23] Ahmed SM, Manhas LR, Verma V, Khajuria RK. Quantitative determination of four constituents of *Tinospora* sps. by a reversed-phase HPLC-UV-DAD method. Broad-based studies revealing variation in content of four secondary metabolites in the plant from different eco-geographical regions of India. J Chromatogr Sci 2006; 44(8): 504-9. [http://dx.doi.org/10.1093/chromsci/44.8.504] [PMID: 16959127]

[24] Rakshe A, Mokat D. On vegetative propagation through stem cuttings in medicinally lucrative *Tinospora* species. J Pharmacogn Phytochem 2018; 7(2): 2313-8.

[25] Saran PL, Patel R, Meena RP, Kalariya KA, Choudhary R. Mini cutting technique: an easy and cost-

effective way of *Tinospora cordifolia* multiplication. Indian J Agric Sci 2019; 89(2): 38-41. [http://dx.doi.org/10.56093/ijas.v89i2.87003]

[26] Sivakumar V, Dhana Rajan MS, Sadiq AM. Hypoglycaemic activity of tissue cultured and field grown plants of *Tinospora cordifolia* in experimental rats. Int J Pharma Bio Sci 2014; 1(1): 08-18.

[27] Mittal J, Sharma MM. Enhanced production of Berberine in *in vitro* regenerated cell of *Tinospora cordifolia* and its analysis through LC-MS QToF 3 Biotech 2017; 7(25): 1-12.

[28] Sudan SS, Bhavika B, Gusain T, Ankita L, Pant M. *In vitro* propagation of *Tinospora cordifolia* and estimation of berberine content by chromatographic analysis. Eco Env & Cons 2020; 26: 58-62. [http://dx.doi.org/10.1007/s13205-016-0592-6]

[29] Madhu K. Evaluation of methanolic extracts of *in vitro* grown *Tinospora cordifolia* (Willd) for antibacterial activities. Asian J Pharm Clin Res 2012; 5(3): 172-5.

[30] Bhalerao BM, Vishwakarma KS, Maheshwari VL. *Tinospora cordifolia* (Willd.) Miers ex Hook. f. & Thoms. plant tissue culture and comparative chemo-profiling study as a function of different supporting trees. Indian J Nat Prod Resour 2013; 4(4): 380-6.

[31] Mridula K, Parthibhan S, Senthil Kumar T, Rao MV. *In vitro* organogenesis from *Tinospora cordifolia* (Willd.) Miers : A highly valuable medicinal plant. S Afr J Bot 2017; 113: 84-90. [http://dx.doi.org/10.1016/j.sajb.2017.08.003]

[32] Malik A, Arya A. Studies on effect of hormone on *in vitro* regeneration of *Tinospora cordifolia.* Int J Curr Microbiol Appl Sci 2019; 8(10): 1386-93. [http://dx.doi.org/10.20546/ijcmas.2019.810.162]

[33] Mittal J, Mishra Y, Singh A, Batra A, Sharma MM. An efficient micropropagation of *Tinospora cordifolia* (Willd.) Miers ex Hook. f. and Thoms : A NMPB prioritized medicinal plant. Indian J Biotechnol 2017; 16: 133-7.

[34] Nitnaware KM, Patinge P, Nikam TD. *In vitro* propagation of *Tinospora cordifolia* through nodal culture. Flora Fauna 2014; 20(1): 248-51.

[35] Handique PJ, Choudhury SS. Micropropagation of *Tinospora cordifolia*: A prioritized medicinal plant species of commercial importance of NE India. Icfai Univ J Genet Evol 2009; 2: 1-8.

[36] Suman Kumar JNJ, Jha SK, Nagar DS, Chauhan RS, Hegde H. Phloroglucinol plays role in shoot bud induction and *in vitro* tuberization in *Tinospora cordifolia* : A medicinal plant with multi therapeutic application. Adv Tech Biol Med 2015; 3(2): 125. [http://dx.doi.org/10.4172/2379-1764.1000125]

[37] Raghu AV, Geetha SP, Martin G, Indra B, Ravindran PN. *In vitro* clonal propagation through mature nodes of *Tinospora cordifolia* (Willd.) Hook. F. & Thoms. : An important Ayurvedic medicinal plant *in vitro* . Cell Dev Biol- Plants. 2006; 42: pp. 584-8. [http://dx.doi.org/10.1079/IVP2006824]

[38] Khanapurkar RS, Paul NS, Desai DM, Raut MR, Gangawane AK. *In vitro* propagation of *Tinospora cordifolia* (Wild.) Miers ex Hook. f. Thoms. J Bot Res 2012; 3(1): 17-20.

[39] Mridula KR, Parthiban S, Kumar TS, Rao AS, Rao MV. *In vitro* micropropagation of *Tinospora cordifolia* (Willd.) Miers from shoot tip explants. Agric Nat Resour 2019; 53: 449-56. [http://dx.doi.org/10.34044/j.anres.2019.53.5.02]

[40] Anita S, Sharma HP. Micropropagation and phytochemical screening of *Tinospora cordifolia* (Willd.) Miers ex Hook. f. & Thoms.: A medicinal plant. Int J Adv Pharm Biol Chem 2015; 4(1): 114-21.

[41] Sharma H, Vashistha BD, Singh N, Kumar R. *Tinospora cordifolia* (Willd.) Miers ex Hook. F & Thoms. (Menispermaceae): Rapid *in vitro* propagation through shoot tip explants. Int J Recent Sci Res 2015; 6(2): 2714-8.

[42] Chatterjee T, Ghosh B. Efficient stable *in vitro* micropropagation and conservation of *Tinospora cordifolia* (Willd.) Miers: An anti-diabetic indigenous medicinal plant. Int J Bio-Resour Stress Manag

2016; 7(4): 814-22.
[http://dx.doi.org/10.23910/IJBSM/2016.7.4.1537a]

[43] Chintalwar GJ, Gupta S, Roja G, Bapat VA. Proto- berberine alkaloids from callus and cell suspension cultures of *Tinospora cordifolia.* Pharm Biol 2003; 41(2): 81-6.
[http://dx.doi.org/10.1076/phbi.41.2.81.14243]

[44] Khalilsaraie MF, Saima N, Meti NT, Bhadekar RK, Nerkar DP. Cytological study and anti microbial activity of embryogenic callus induced from leaf cultures of *Tinospora cordifolia* (Willd.) Miers. J Med Plants Res 2011; 5(14): 3002-6.

[45] Maanhvizhi E, Revathi K, *In vitro* anti-inflammatory activity and tissue culture studies on *Tinospora cordifolia.* Int J Curr Res Med Sci 2016; 3(10): 76-81.
[http://dx.doi.org/10.22192/ijcrms.2016.02.10.008]

[46] Mittal J, Singh A, Batra A, Sharma MM. Comparative microbicidal efficacy of fractionated extracts from *in vitro* and *in vivo* raised cells of *Tinospora cordifolia* against MDR pathogens. Braz Arch Biol Technol 2016; 59.
[http://dx.doi.org/10.1590/1678-4324-2016150508]

[47] Priti RS. Estimation of total alkaloids in wild and *in vitro* regenerated *Tinospora cordifolia.* Int J Pharm Sci Res 2019; 10(2): 2777-84.

[48] Aditi S, Sah SK, Anuji P, Sabari R, Maharajan N. *In vitro* study of *Tinospora cordifolia* (Willd.) Miers (Menispermaceae). Botanica Orientalis-J Plant Sci 2009; 6: 103-5.

CHAPTER 17

A Systematic Review of Phytoconstituents and Tissue Culture Studies of the genus *Hoya* R. Br.

Santoshkumar Jayagoudar[1], **Pradeep Bhat**[2,*], **Sachet Hegde**[3], **Savaliram G. Ghane**[4] and **Harsha V. Hegde**[2]

[1] *Department of Botany, G. S. S. College & Rani Channamma University P. G. Centre, Belagavi, Karnataka-590006, India*

[2] *ICMR-National Institute of Traditional Medicine, Nehru Nagar, Belagavi, Karnataka-590010, India*

[3] *Department of Botany, Bangurnagar Degree College, Ambewadi, Dandeli, Karnataka-581325, India*

[4] *Department of Botany, Shivaji University, Vidyanagar, Kolhapur, Maharashtra-416004, India*

Abstract: The genus *Hoya* (Family: Apocynaceae) has more than 500 species, comprising mainly of epiphytes and geographically distributed in South America, Southeast Asia, Indo-Malesia and Australian regions. Most of the species are cultivated for their ornamental, aromatic and showy flowers. Philippines is one of the countries with the highest diversity of *Hoya* species. As seed setting is very rare in most of the species, it necessitated the development of conservation strategies through *ex situ* conservation methods using vegetative or micropropagation techniques. Present chapter provides detailed information on the traditional uses, phytoconstituents, conservation status and micropropagation studies of the ornamental and medicinally important genus *Hoya*.

Keywords: *Hoya*, Endangered, Conservation, Bioactives, Ornamental Plants, Horticulture.

INTRODUCTION

The genus *Hoya* R. Br. with more than 500 species, is one of the largest genera in the family Apocynaceae [1]. The species are mainly epiphytic lianas, often twining or climbing by adventitious roots and known as wax plants due to the appearance of their leaves and flowers. Of all, more than 300 species of *Hoya* are

* **Corresponding author Pradeep Bhat:** ICMR-National Institute of Traditional Medicine, Nehru Nagar, Belagavi, Karnataka-590010, India; E-mail: bhat.pradeep08@gmail.com

T. Pullaiah (Ed.)

distributed in tropical and sub-tropical Asia to West Pacific [1, 2]. *H. imperialis* and *H. coronaria* are prevalently cultivated for their beautiful, aromatic and ornamental flowers in Europe, America and Australia. However, the propagating methods of these plants have not been reported yet. Wild populations of *Hoya* in Brunei Darussalam have become highly threatened because of habitat loss and overexploitation [3].

The Philippines Island is considered one of the eight hottest biodiversity hotspots in the world and has the highest proportion of endemic and threatened vascular plants [4, 5]. Moreover, it is one of the countries with the highest diversity of *Hoya* species, and reported from all the altitudes of the Archipelago at different types of habitats such as limestone cliffs, boulders and swamp forests. More than hundred species of *Hoya* have been recorded throughout the country. In Philippines, Mindoro is the 7th largest island, and interestingly, 18 *Hoya* species have been reported from this island alone [6]. Among these, 15 species are endemic to Philippines. It is also a notable fact that among these 18 species, three are included in the endangered category, such as *Hoya alagensis* Kloppenb., *Hoya halconensis* Kloppenb. and *Hoya paziae* Kloppenb.

Indonesia is another country which has a rich diversity of *Hoya* species. In Ketori forest areas of Indonesia, 14 *Hoya* species were identified, and most of them have ethnomedicinal importance as reported by several indigenous tribes. Further, some of the *Hoya* species were found to be used as vegetables and medicines as well. It was found that the people communities of Ketori forest frequently use *H. coronaria* and *H. meredithii* as raw vegetables and they believe that these species have the ability to decrease high blood pressure. Another species *H. waymaniae* was used to treat stomach and headache. It is an interesting fact that the people of Ketori village practice their own local traditional rules for sustainable utilization of forest products and management of local biodiversity [7].

A compound isolated from *H. multiflora* used to treat several diseases including rheumatistm, abdominal pain, asthma and intestinal inflammations [3]. Leaf extract of *Hoya parasitica* Wall. used to treat fever, body pain, rheumatism, kidney problems, urinary tract disorders and jaundice [8, 9]. Leaf paste of *Hoya globulosa* Hook. f. used in the treatment of bone fractures [10]. Crushed leaves of *Hoya coronaria* Blume. are used to cure cut wounds, and leaf paste of *Hoya potsii* Trail found effective in the treatment of injury, gynecological disorders, rheumatoid arthritis and digestive disorders [11]. *Hoya kerrii* has long been used as a folk medicines for curing inflammatory-related disorders [12]. Cold macerated extract of *Hoya vanuatensis* green young leaves used as an oxytocic agent [6, 13, 14].

PHYTOCONSTITUENTS IN *HOYA*

The attractive flower shapes and pleasant fragrance of some *Hoya* species make them suitable for a new source of aroma. Basir *et al.* [1] studied the biosynthesis of secondary metabolites and volatile fragrances from three *Hoya* species such as *H. cagayanensis* C.M.Burton, *H. lacunosa* Blume and *H. coriacea* Blume, using solid phase micro extraction. Further, gas chromatography mass spectroscopy (GC-MS) analysis, phytochemicals and transcriptomic methods were explored to elucidate the mechanism of perfume synthesis, particularly terpenoids. GC-MS analysis revealed the presence of 23, 14 and 36 compounds in *H. cagayanensis*, *H. lacunosa*, and *H. coriacea*, respectively. Volatile components showed different fragrance profiles from all three *Hoya* species. Monoterpene compounds such as β-ocimene (25.75%) and methyl salicylate (24.67%) were dominated in *H. cagayanensis*, whereas the alcohol component 1-octen-3-ol (26.1%) and the ester compound named (Z)-butyric acid, 3-hexenyl ester (29.36%) were the major compounds in *H. lacunosa* and *H. coriacea*, respectively.

Ebajo Jr *et al.* [15] isolated total seven compounds from dichloromethane extracts of an endemic Philippine ornamental plant, *H. buotii* Kloppenb. Compounds taraxerone, taraxerol, and a mixture of β-sitosterol and stigmasterol were isolated from roots and stems. The mixture of α-amyrin cinnamate and β-amyrin cinnamate was isolated from flowers. At the same time, squalene was also reported from leaves. Ragasa *et al.* [16] isolated number of bioactive compounds from the dichloromethane extracts of *H. cumingiana* Decne. The extract afforded α-amyrin, β-amyrin, bauerenol, lupeol, β-sitosterol, stigmasterol and taraxerol from different parts of the plant. Similarly, dichloromethane extract of stem and leaves of *H. cagayanensis* yielded dihydrocanaric acid, lupeol, lupenone, 2-hydroxyethyl benzoate, β-sitosterol and stigmasterol [17].

VEGETATIVE AND *IN VITRO* MICROPROPAGATION STUDIES

In vitro micropropagation of *H. wightii* ssp. *palniensis*, a highly vulnerable and endemic species of the Western Ghats, Tamil Nadu, India was reported by Revathi Lakshmi *et al.* [18]. Shoot tip explants were cultured on Murashige and Skoog medium (MS medium) containing different concentrations and combinations of cytokinins [Thidiazuron (TDZ), 6-(γ,γ-Dimethylally- lamino)p urine (2-iP), Kinetin (Kn) and Benzyl adenine (BA)] and auxins [Naphthalene acetic acid (NAA), Indolebutyric acid (IBA) and Indole-3-acetic acid (IAA)]. High frequency of shoot proliferation and multiple shoot induction was observed in the media containing Kn (4.65 μM) + IBA (1.47 μM), supplied with 100 mg/L ascorbic acid. Rhizogenesis was observed on MS medium supplemented with IBA

(0.98 μM) and plantlets produced through micropropagation were hardened with a survival rate of 56%.

The standardization of rhizogenesis of *H. wightii* ssp. *palniensis* was carried out by Revathi Lakshmi *et al.* [19]. *In vitro* raised microshoots were used as explants and inoculated on MS medium containing various concentrations of auxins (IAA, NAA and IBA at 0.01, 0.05, 0.1, 0.2 and 0.3 mg/L). It was observed that higher percentage rooting (90%) was noted with MS medium supplemented with 0.2 mg/L IBA and 3% sucrose. The rooted shoots were acclimatized for two months and the overall rate of survival was 80%.

A standardized protocol for indirect plant regeneration from leaf and internode explants of *H. wightii* ssp. *palniensis* was developed by Revathi Lakshmi *et al.* [20]. A higher percentage of callus induction was obtained when MS medium supplemented with 2,4-Dichlorophenoxyacetic acid [2,4-D, 2.0 mg/L] and NAA (1.0 mg/L) was used. A significant rate of shoot bud induction was observed in MS medium containing BA (1.0 mg/L) + IBA (0.5 mg/L). The use of undefined supplements, such as coconut water (15%), influenced the growth of shoot initials into elongated shoots. Matured shoots (> 3 cm length) were rooted on MS basal medium containing 0.2 mg/L IBA. The rooted plants were further transferred to the soil for hardening, and the highest (80%) survival was noted.

Microprogation of *H. kerrii* was carried out by Siddique *et al.* [21]. In the study, roots, petiole, inter-nodal segments and leaves were used as explants. MS medium containing varied concentrations of 2,4-D (1, 2, 3, 4 and 5 mg/L) was used and induction of callus. Medium pH was adjusted to 5.8 ± 0.02 after adding organic components, 3% sugar and salts. Agar (6.9 g) was added as a solidifying agent, and then the medium was autoclaved at 121°C at 15 psi for 20 min. Culture conditions were maintained as per standard procedures. A significant quantity of callus was produced in MS medium containing 5.0 mg/L 2,4-D.

Stem cutting and micropropagation methods were followed to propagate two highly threatened *Hoya* species (*H. imperialis* and *H. coronaria*) distributed in Brunei Darussalam having ornamental value [3]. Micropropagation and stem cutting methods were followed as potential approaches to make the growth assessment studies. IBA and NAA at 0–2000 mg/L concentrations were used to treat the stem cuttings. Researchers further propagated the treated stem cuttings using peat moss and perlite. In *H. imperialis*, mean number of new leaves and relative growth rate (RGR) based on stem diameter were significantly increased (6.3 ± 1.0 and 0.004 ± 0.0007 cm/day) when treated with 500 mg/L NAA and 2000 mg/L IBA. Meanwhile, in *H. coronaria*, a significantly high number of roots were observed when cuttings were treated with 1000 mg/L NAA (16.6 ± 1.4) and

2000 mg/L IBA (17.5 ± 2.7). Field observations were made for up to one year, and a 100% survival rate of the treated cuttings was observed during first 20 weeks in both species. Murashige and Skoog (MS) medium supplemented with IBA or Kn at 0-10.0 mg/L concentration was used for micropropagation of the species. Leaf was used as explant in both the species and inoculated on respective media. It was found that callus induction was not promising with *H. imperialis* and comparatively less responsive towards both the PGRs. Whereas in the case of *H. coronaria*, the most potential callus induction was observed through the higher surface area using the medium enriched with 3.0 mg/L IBA and 3.0 mg/L Kn.

In vitro propagation of diallelic plants *viz. H. carnosa* var. *compacta* and *H. carnosa* var. *rubra* was also carried out [22]. MS medium containing different combinations of 0-57.0 μM IAA + 0-46.0 μM Kn, 0-54.0 μM NAA + 0-46.0 μM Kn and 0-40.0 μM 2,4-D + 0-150.0 μM Kn were used. The callus formation from the leaf explants in both the genotypes noted when 2,4-D and Kn were added to the medium. In the secondary culture, asexual embryos were developed only from the callus initiated on the primary culture medium containing 2,4-D in combination with Kn. It was found that the supplementation of 2,4-D and Kn concentrations (2.3-5.0 μM and 2.5-20.0 μM, respectively) determined the frequency of embryogenesis during primary culture.

CONCLUSION

The popularity of *Hoya* species for their ornamental and medicinal purposes has considerably increased the anthropogenic pressure on this species in the native forests. Since the forest tree trunks are the substrates for the climbing *Hoya* species, deforestation due to unscientific land use and agricultural developments had serious consequences that led to decline in *Hoya* populations. As seed setting is very rare in most of the *Hoya* species, it necessitates the development of conservation strategies *via ex situ* and *in vitro* techniques. The protocols provided herein could help in minimizing the pressure on wild populations of endangered *Hoya* species and contribute to the conservation of valuable biodiversity. Further detailed studies on the mass multiplication of ornamental and medicinally important taxa from the genus are needed. In addition, special efforts must be taken on the production of potent metabolites using conventional and non-conventional techniques.

REFERENCES

[1] Basir S, Akbar MA, Talip N, Baharum SN, Bunawan H. An integrative volatile terpenoid profiling and transcriptomics analysis in *Hoya cagayanensis, Hoya lacunosa* and *Hoya coriacea* (Apocynaceae, Marsdenieae). Horticulturae 2022; 8(3): 224. [http://dx.doi.org/10.3390/horticulturae8030224]

[2] POWO. Plants of the World Online Facilitated by the Royal Botanic Gardens 2022. Available from :

http://www.plantsoftheworldonline.org

[3] Mohd Don SM, Abdul Hamid NM, Taha H, Sukri RS, Metali F. Vegetative propagation of *Hoya imperialis* and *Hoya coronaria* by stem cutting and micropropagation. Trop Life Sci Res 2021; 32(3): 1-23. [http://dx.doi.org/10.21315/tlsr2021.32.3.1] [PMID: 35656369]

[4] Myers N, Mittermeier RA, Mittermeier CG, da Fonseca GAB, Kent J. Biodiversity hotspots for conservation priorities. Nature 2000; 403(6772): 853-8. [http://dx.doi.org/10.1038/35002501] [PMID: 10706275]

[5] Sodhi NS, Posa MRC, Lee TM, Bickford D, Koh LP, Brook BW. The state and conservation of Southeast Asian biodiversity. Biodivers Conserv 2010; 19(2): 317-28. [http://dx.doi.org/10.1007/s10531-009-9607-5]

[6] Villanueva ELC. Buot IEJr. Hoyas of Mindoro island, Philippines: conservation concerns. J Nature Studies 2016; 15(1): 87-97.

[7] Rahayu S. Sustainable Utilization of *Hoya* species and other bioresources in Ketori, Sanggau, West Kalimantan, Indonesia. IOP Conf Ser Earth Environ Sci 2019; 298(1): 012039. [http://dx.doi.org/10.1088/1755-1315/298/1/012039]

[8] Atiqur Rahman M, Uddin SB, Wilcock CC. Medicinal plants used by Chakma tribe in hill tracts districts of Bangladesh. Indian J Tradit Knowl 2007; 6(3): 508-17.

[9] Sarkar KK, Mitra T, Rahman MA, *et al. In vivo* bioactivities of *Hoya parasitica* (Wall.) and *in silico* study against cyclooxygenase enzymes. BioMed Res Int 2022; 2022: 1-20. [http://dx.doi.org/10.1155/2022/1331758] [PMID: 35528171]

[10] Das AK, Dutta BK, Sharma GD. Medicinal plants used by tribes of Cachar district, Assam. Indian J Tradit Knowl 2008; 7(3): 446-54.

[11] Samuel AJSJ, Kalusalingam A, Chellappan DK, *et al.* Ethnomedical survey of plants used by the Orang Asli in Kampung Bawong, Perak, West Malaysia. J Ethnobiol Ethnomed 2010; 6(1): 5. [http://dx.doi.org/10.1186/1746-4269-6-5] [PMID: 20137098]

[12] Sittisart P, Dunkhunthod B, Chuea-nongthon C. Antioxidant and anti-inflammatory activities of ethanolic extract from *Hoya kerrii* Craib. Warasan Khana Witthayasat Maha Witthayalai Chiang Mai 2020; 47(5): 912-25.

[13] Bradacs G, Heilmann J, Weckerle CS. Medicinal plant use in Vanuatu: A comparative ethnobotanical study of three islands. J Ethnopharmacol 2011; 137(1): 434-48. [http://dx.doi.org/10.1016/j.jep.2011.05.050] [PMID: 21679762]

[14] Zheng X, Xing F. Ethnobotanical study on medicinal plants around Mt.Yinggeling, Hainan Island, China. J Ethnopharmacol 2009; 124(2): 197-210. [http://dx.doi.org/10.1016/j.jep.2009.04.042] [PMID: 19409476]

[15] Ebajo VD Jr, Brkljaca R, Urban S, Ragasa CY. Chemical constituents of *Hoya buotii* Kloppenb. J Appl Pharm Sci 2015; 5(11): 69-72.

[16] Ragasa CY, Panajon NM, Aurigue FB, Brkljaca R, Urban S. Chemical constituents of *Hoya cumingiana* Decne. Int J Pharmacogn Phytochem Res 2016; 38(12): 2033-8.

[17] Ragasa CY, Borlagdan MS, Aurigue FB, Brkljaca R, Urban S. Chemical constituents of *Hoya cagayanensis* C. M. Burton. J Appl Pharm Sci 2017; 7(05): 61-5.

[18] Revathi Lakshmi S, Franklin Benjamin JH, Senthil Kumar T, Murthy GVS, Rao MV. *In vitro* propagation of *Hoya wightii* ssp. *palniensis* K.T. Mathew, a highly vulnerable and endemic species of Western Ghats of Tamil Nadu, India. Afr J Biotechnol 2010; 9(5): 620-7. [http://dx.doi.org/10.5897/AJB09.846]

[19] Revathi Lakshmi S, Franklin Benjamin JH, Senthil Kumar T, Suryanarayana Murthy GV, Rao MV. Efficient rhizogenesis of *in vitro* raised microshoots of *Hoya wightii* Hook. f. ssp. *palniensis* K.T.

Mathew : A vulnerable species endemic to Western Ghats. J Biosci Res 2010; 1(3): 137-45.

[20] Lakshmi SR, Benjamin JHF, Kumar TS, Murthy GVS, Rao MV. Organogenesis from *in vitro*-derived leaf and internode explants of *Hoya wightii* ssp. *palniensis* : A vulnerable species of Western Ghats. Braz Arch Biol Technol 2013; 56(3): 421-30. [http://dx.doi.org/10.1590/S1516-89132013000300010]

[21] Siddique R. Micropropagation of *Hoya kerrii* (Valentine Hoya) through callus induction for long term conservation and dissemination. Int J Sci Res 2013; 2(8): 162-4.

[22] Maraffa SB, Sharp WR, Tayama HK, Fretz TA. Apparent asexual embryogenesis in cultured leaf sections of Hoya. Z P flanzenphysiol Bd 1981; 102: 45-55.

CHAPTER 18

In vitro Regeneration and Conservation of the Medicinal and Aromatic genus *Kaempferia*: An Overview

Avijit Chakraborty[1] and **Biswajit Ghosh**[1,*]

[1] *Plant Biotechnology Laboratory, Post Graduate Department of Botany, Ramakrishna Mission Vivekananda Centenary College, Rahara, Kolkata-700118, India*

Abstract: Genus *Kaempferia* comprises about 124 species distributed in Southeast Asia and is well known for it's diverse medicinal, nutritional and industrial values. The plants of the genus are rhizomatous, perennial, and oil-yielding plants; some are also used as spices. The essential oil obtained from the plants has a considerable market value worldwide. The rhizomes of these plants were used in traditional medicine due to the presence of diverse bioactive compounds and used to treat urinary tract infections, fever, cough, hypertension, metabolic disorder, asthma, rheumatism, epilepsy, skin diseases, *etc*. Seed dormancy, seasonal outgrowth and seed made through cross-pollination were found to be non-viable, which are the prime limitations of *ex situ* conservation regarding this genus. To overcome this type of problem, *in vitro* tissue culture is the way to get the plants available over the year without any limitations. This chapter is based mainly on exploring those bioactive compounds containing species of the genus *Kaempferia*, and obtaining an alternative resource of phyto-compounds for use in pharmaceuticals and conserving them through an artificial way to get them throughout the year without exploiting the area and genotypic alteration.

Keywords: Conservation, *Kaempferia*, Micropropagation, Phyto-compounds, Zingiberaceae.

INTRODUCTION

Zingiberaceae is one of the largest and most exploitable plant families, comprising 53 genera and more than 1300 species worldwide and distributed chiefly in tropical and subtropical areas. The plants under this family are perennial, rhizomatous, oil-yielding medicinal herbs used worldwide as spices, food, medicine, and others. Due to the enrichment of essential oil with diverse compounds, this family's plants also have significant aromatic, medicinal, nutriti-

* **Corresponding author Biswajit Ghosh:** Plant Biotechnology Laboratory, Post Graduate Department of Botany, Ramakrishna Mission Vivekananda Centenary College, Rahara, Kolkata-700118, India;
E-mail: ghosh_b2000@yahoo.co.in

T. Pullaiah (Ed.)

onal, and ornamental properties. The genus *Kaempferia* consists of more than 100 species (The World Flora Online, Plants of the World Online, The Plant List) and is distributed throughout Southeast Asia. Species included under this respective genus are *Kaempferia rotunda, K. galanga, K. parviflora, K. angustifolia,K. marginata, K. pandurata, etc.*, explored with a diverse range of bio-activity. The plant species are well known for treating allergies, arthritis, cancer, cardiovascular disorders, inflammation, sexual dysfunction, skin integrity, *etc* [1]. *K. parviflora* is used as a food supplement for treating metabolic disorders in Japan [2]. Rhizomes of *K. galanga* are used as a spice in cooking worldwide and sold as an industrial crop in the market [3]. *K. elegans* is well known for its medicinal uses, also known as an ornamental plant in Vietnam [4]. Other species belonging to the genus are also known for their utilization as food, medicine, spice, and many more. Hence, all the species explored, not or adequately, should have some activities that can be used as an alternative resource in different branches of medicine development.

The plants of the family Zingiberaceae are seasonal and found at a specific period in the environment. This is also a limitation of this plant species, that we cannot obtain them throughout the year. Conservation is the only alternative path that should be followed to prevent the plants from vanishing and to obtain them throughout the year from the environment. The massive depletion of medicinal plants for a continuous supply of phytocompounds is becoming a prime threat to their extinction from their natural habitat. Large-scale *in vitro* production of plants is a useful alternative to maintain Phyto-diversity from depletion in the natural environment. *In vitro* conservation of the species of *Kaempferia* is the safest way to protect the elite germplasm from extinction due to disease, natural disaster, or extensive non-scientific use. As the plants belonging to this genus are seasonal, and outgrowth is visible in a respective season in the year, *in vitro* tissue culture must help to obtain the plant over the year for a continuous supply of phytocompounds.

Medicinal Importance of the Genus *Kaempferia*

The species belonging to the genus *Kaempferia* are well-known for their medicinal uses. Several *Kaempferia* species, like *Kaempferia galanga, Kaempferia parviflora, Kaempferia angustifolia, Kaempferia rotunda*, and *Kaempferia elegans,* are used in folk medicine, mainly in southeast Asia from the ancient era. However, few are used to develop modern medicine under research to combat emerging diseases.

Kaempferia galanga is one of the most well-explored species belonging to the genus *Kaempferia* with diverse medicinal importance. The rhizome of this plant is well-known for essential aromatic oil, comprised of many bioactive compounds with high therapeutic index. *K. galanga* is also well known for its aromatic nature; hence, it is often used as a spice. The nutritive value of this rhizomatous plant is also high and taken as food in different countries. Essential and non-essential minerals and ions like calcium, potassium, manganese, and chromium are also present in the rhizome, which is used enormously. The plant is used for flatulence, laxative, stomachache, tonic, intoxication, antiangiogenic, sedative, diuretic and vasorelaxation activity [5 - 11]. The antimicrobial, antioxidant and cytotoxic activity of this plant has been tested previously against many pathogenic as well as non-pathogenic organisms by different researchers [12 - 14]. The plant is proven to be an excellent antimicrobial agent and causes the elimination of different multi-drug resistant microorganisms. Anti-tumour and anticancer activity of several compounds and the crude extract of *K. galanga* have also been reported earlier [15 - 17].

Another important species is *Kaempferia parviflora* has significant medicinal importance. The rhizome of this plant is well-known for antibacterial, anti-obesity, anti-diabetic, cardiovascular protective, immunoregulatory, neuroprotective, and skin-whitening activity [1, 18 - 20]. The rhizome extract also reduces visceral fat in overweight Japanese adults [21]. Anticancer activity of this plant on different cell lines has also been reported [22]. *K. parviflora* was also reported to enhance sexual performance in traditional use [23, 24]. Furthermore, the compounds polymethoxyflavones have an anti-ageing ability and are used in cosmetics and nutraceutical products [25]. The plant extract is also used for wine fermentation as a base composition [26].

Kaempferia rotunda is another plant species with widespread applications, including treating indigestion, fever and wound healing properties [27]. The plant has been reported for its anticancer activity against different cancer cell lines. *K. rotunda* can prevent the proliferation of colon cancer cells SW480 and SW48 by causing apoptosis in the intrinsic mitochondrial pathway [28]. Strong anticancer activity against breast cancer and pancreatic cancer of different solvent extracts of this plant has been reported by several researchers in the past [29, 30]. Antioxidant, anti-proliferative, anti-inflammatory and antibacterial activity of rhizome extract has been reported for this plant [31 - 33]. Silver nanoparticle synthesis from the plant rhizome and their activities towards tumour growth suppression in mice was reported [32]. This is an essential aromatic oil-yielding plant; the oil is a food preservative, flavouring, and antibacterial agent in different aspects [34]. The plant is also well known for its nematicidal and immunomodulatory activity [35, 36].

Kaempferia pandurata is traditionally used as a food ingredient worldwide. This plant species is also medicinally crucial due to its application for treating dental caries, colic disorder, fungal infection, dry cough, rheumatism and muscular pains [37]. The plant has anticancer activity against breast, skin, colon, and prostate cancer [38 - 42]. Cytotoxic, antibacterial activity and anti-inflammatory activity of different solvent extracts of this plant have been reported [43 - 53].

Several other plant species are under the genus *Kaempferia*, such as *Kaempferia albomaculata, Kaempferia elegans, Kaempferia scaposa, Kaempferia angustifolia, Kaempferia marginata,* which are also reported for their biological importance in the past. Most of the plants included in this genus are well known for their use as food and additives. The medicinal importance of the species belonging to the genus *Kaempferia* is represented in Table **1**. *Kaempferia marginata* is used for its anti-inflammatory and wound-healing properties and for treating fever [54 - 56]. Other important species, *Kaempferia pulchra, Kaempferia koratensis,* and *Kaempferia roscoeana,* are cultivated as ornamental and rhizomes are often used as food and spice and to produce tonic [57]. A few species under this family were explored; most are still underutilized and not appropriately explored. The genus contains more than 100 species, whereas only about 16 were known for their medicinal and industrial significance. Proper exploration of those unexplored plants with proper utilization can cause a revolution in the present and future medical research and industries.

MICROPROPAGATION OF *KAEMPFERIA*

The conventional process of implantation of the species belonging to the genus *Kaempferia* is challenging. The plant species are mostly seasonal, and outgrowth is not seen throughout the year; the seed of these plants is maximally present in non-viable conditions. The lack of sexual reproduction in the species of this genus also limits the process of conventional breeding.

Table 1. Medicinal importance of the species belonging to the genus *Kaempferia*.

S. No.	Plant Species	Importance of the Species	References
1.	*Kaempferia albomaculata*	Rhizomes and young leaves are used as spices and food. The species is cultivated as an ornamental plant in different countries.	[10]
2.	*Kaempferia angustifolia* Roscoe	The whole plant is cultivated as an ornamental species. Rhizomes and young leaves are used to treat flatulence, laxative, stomach ache, tonic, and intoxication.	[10]
3.	*Kaempferia elegans* (Wall.) Baker	Used as a potted plant with ornamental value	[69]

(Table 1) cont.....

S. No.	Plant Species	Importance of the Species	References
4.	*Kaempferia elegans* (Wall.) Baker	The whole plant is cultivated as an ornamental plant. Rhizomes of this plant have cytotoxic and antimicrobial activity.	[148]
5.	*Kaempferia galanga* L..	Rhizomes and young leaves are used as food, and spices are cultivated as ornamental plants. This species is used in flatulence, laxative, stomach ache, and tonic. Rhizomes are often used for their insecticidal, anticancer, sedative, diuretic and vasorelaxation activity.	[5-11, 70]
6.	*Kaempferia grandifolia* Saensouk & Jenjitt.	Leaves or rhizomes are blended with dehusked rice and used as a herpes treatment. Young inflorescences are blanched and used as vegetables.	[71]
7.	*Kaempferia koratensis* Picheans.	Rhizomes and young leaves are used as food and spice as well as used for flatulence, laxative, stomach ache, tonic, and intoxication.	[10]
8.	*Kaempferia marginata* Carey ex Roscoe	Anti-inflammatory and wound healing activity	[54-56]
		They are used in the treatment of inflammation. The whole plant decoction is used for the treatment of fever. Ethanol extracts, which exert potent anti-inflammatory properties, are selected for wound healing assay.	[54]
9.	*Kaempferia pandurata* Roxb.	Used in the treatment of periodontal inflammation.	[37]
		Cytotoxic, antibacterial, and anti-inflammatory activities.	[43-53]
		Food-borne pathogens killing properties.	[53]
		Anti-colon cancer, anti-prostate cancer, and anti-skin cancer.	[38-41]
		Matrix metalloproteinase-1 inhibitory activity.	[72]
		Depigmentation of melanocytes.	[73]
		Antiestrogen receptor (-) breast cancer cell line 3,4-methylenedioxyamphetamine-mb-231.	[42]
		Traditionally used as a food ingredient and in folk medicine to treat dental caries, colic disorder, fungal infection, dry cough, rheumatism and muscular pains. The rhizome extracts have also been tested for various pharmacological effects, including antibacterial, anti-inflammatory, antitumor, antidiarrhea, antidysentry, antiflatulence and antiepidermophytid activities.	[37, 145-147]

(Table 1) cont.....

S. No.	Plant Species	Importance of the Species	References
10.	*Kaempferia parviflora* Wall. ex Baker	Suppression of adipocyte hypertrophy, and skeletal muscle hypertrophy and promote the differentiation of brown adipocyte cells.	[74-76]
		Enzyme modulation of mouse hepatic cytochrome P450 enzymes.	[77]
		Anti-cancer effects on ovarian cancer, HeLa cell, melanoma cancer, benign prostate hyperplasia, anti-pancreatic cancer activities.	[22, 78-84]
		Rhizomes of this plant can prevent fatigue and improve the physical fitness of soccer players and adults.	[85-88]
		Anticholinesterase, anti-osteoporotic, anti-glycation, anti-gastric ulcer, anti-arthritis, anti-inflammatory, anti-proliferative, anti-allergic, anti-ageing, and anti-acne activities.	[89-103]
		Antibacterial, anti-obesity, anti-diabetic, cardiovascular protective, immunoregulatory, and neuroprotective activities.	[1, 18-20]
		Antimutagenic and α-glucosidase inhibitory effects, and acetylcholinesterase activity.	[104, 105]
		Improvement of damage in human dermal fibroblasts.	[106]
		Delays the development of cataracts.	[107]
		Inhibits SGLT2- and GLUT2-mediated glucose transport in the human renal proximal tubule.	[108]
		Reduces blood glucose levels in humans.	[109]
		Inhibits the development and progression of atherosclerosis.	[110]
		Mitigates age-related disease and hypogonadism by enhancing testosterone production.	[111]
		Inhibition of *Streptococcus mutans* biofilm formation.	[112]
		Diminishes hyperglycemia and visceral fat accumulation.	[113]
		Prevention of Alzheimer's disease.	[114]
		It is used as food preservative agent, perfuming and flavouring agents. Essential oil is an allelopathic agent and protects from insects and other parasites. Also used as an agent to vitalize, promote health, and as a stimulant. Traditionally used as health-promoting herb, including use for the treatment of peptic and duodenal ulcers, colic disorder, asthma, allergy, gout, impotence, diarrhoea, diabetes, antioxidant, peptic ulcer, and vasorelaxant.	[115, 147]
11.	*Kaempferia pulchra* Ridl.	Inhibitory effects against TREx-HeLa-Vpr cells.	[3]
		The whole plant is cultivated as ornamental, and rhizomes are often used as a tonic.	[10]
12.	*Kaempferia roscoeana* Wall.	Used as spice and food in Thai cuisine. Preliminary screening of dichloromethane-methanol extracts of the whole plants showed cytotoxic activity against the MOLT-3 cancer cell line.	[57]

(Table 1) cont.....

S. No.	Plant Species	Importance of the Species	References
13.	*Kaempferia rotunda* L.	The whole plant is cultivated as ornamental, and rhizomes are used for flatulence, laxative, stomach ache, tonic, and intoxication.	[10]
		Anticancer and cytotoxic activity against breast, colon and pancreatic cancer. Antibacterial and anti-proliferative activities.	[28-30, 116,117]
		Anthelmintic activity.	[118]
		Nematicidal activity.	[36]
		Wound healing activity and anti-oxidant activity.	[33]
		Anti-inflammatory and analgesic activities.	[119]
		Inhibits tumour growth.	[32]
		Immunomodulatory activity.	[35]
		Antihyperglycemic and Antinociceptive activities.	[27]
		Oil is used as a food preservative, flavouring agent, and antibacterial agent.	[34]
		Anti-inflammatory effect.	[31]
14.	*Kaempferia siamensis* Sirirugsa	Young leaves are used as food. Rhizomes are used for flatulence, laxative, stomach ache, and tonic.	[10]
15.	*Kaempferia scaposa* (Nimmo) Benth.	The plant rhizome is used in several indigenous medicinal formulations and antioxidant activity.	[120]
16.	*Kaempferia sikkimensis* King ex Baker	The paste made from leaves and rhizomes is applied to bone fractures, and a bandage is made with the help of bamboo strips.	[121, 122]

Furthermore, the multiplication of *Kaempferia* rhizomes is very slow, and the environmental parameters also negatively affect the rhizome to multiply faster. Hence, plant tissue culture must be an alternative and effective choice for large-scale disease-free plant production in a short area and time throughout the year (Fig. **1**). Field-grown *in vivo* plant-derived rhizomes and tissue cultured *in vitro* plants of *Kaempferia angustifolia* and *Kaempferia galanga* are represented in Fig. (**2**). The plants are rhizomatous, and the rhizomes rely on the soil with diverse symbiotic bacterial and fungal associations [58]. Therefore, plant tissue culture primarily aims to produce non-infected sterile plants. So, the successful establishment of a plant in, *in vitro* conditions is needed to remove all the microorganisms through surface disinfection before introduction into the medium. There are about 124 species known to belong to the *Kaempferia* genus, from which only five plants, including *Kaempferia galanga, K. marginata, K.*

parviflora, K. rotunda, and *K. angustifolia* have been reported for *in vitro* conservation till date (Table **2**).

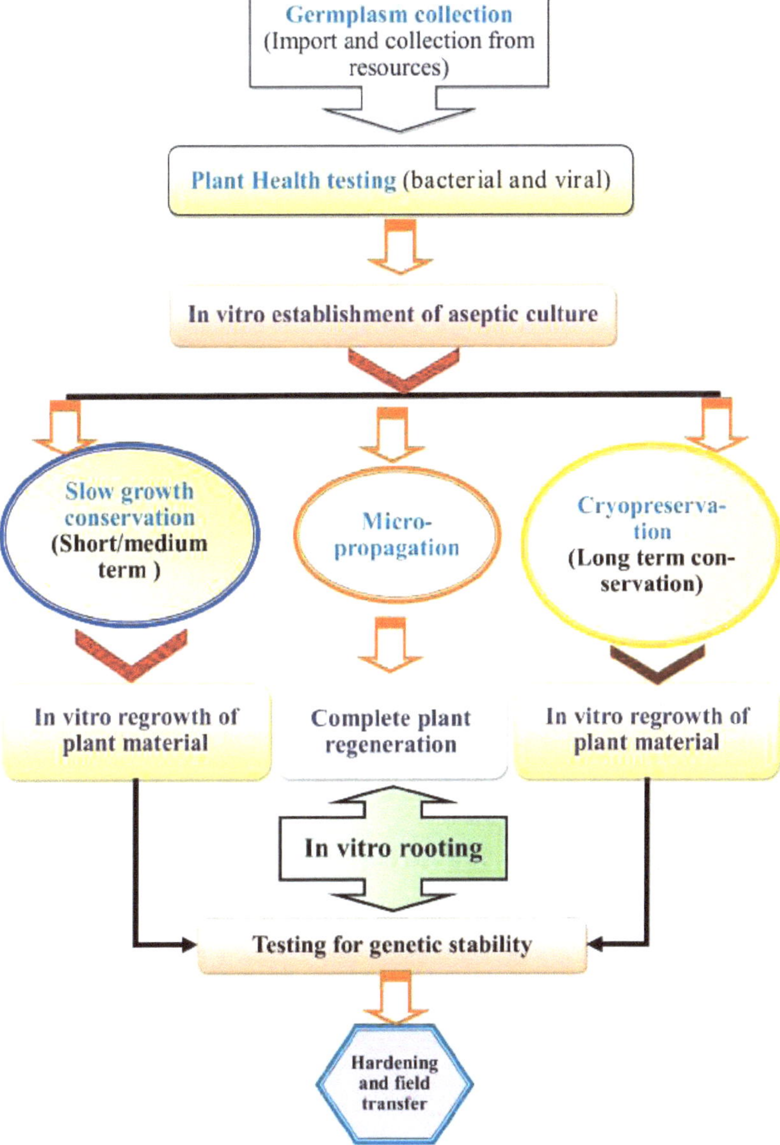

Fig. (1). Schematic representation of the procedure of germplasm selection and it's conservation through different techniques.

Fig. (2). The plant species of *Kaempferia angustifolia* and *Kaempferia galanga* in different conditions. (A-B) Field-grown plants of *K. angustifolia* and *K. galanga*. (C-D) Rhizomes of *K. angustifolia* and *K. galanga*. (E-F) *In vitro* conservation through plant tissue culture of *K. angustifolia* and *K. galanga*.

Table 2. Micropropagation of different species belonging to the genus *Kaempferia*.

Plant Species	Medium Composition	Optimum Result	Collection	References
Kaempferia angustifolia Roxb.	MS medium was supplemented with 2.0 mg/l BAP, 2.0 mg/l Kn and 1.0 mg/l NAA	6.60 shoots per explant	Baruipur, West Bengal, India	[150]
Kaempferia galanga L.	MS medium supplemented with 2.2 μM BAP and 13.9 μM Kn	7.80 shoots and 17.60 roots per explant	Calicut, Kerala, India	[123]
	MS medium supplemented with 0.5 mg/l Kn and 1.5% sucrose solidified with 0.7% agar	90% survival	Kerala, India	[67]
	MS medium supplemented with 12 μM BAP, 3 μM NAA, and 3% sucrose	Maximum shoots per explant	India	[124]
	MS medium supplemented with 1.5 mg/l 2,4-D and 1 mg/l BAP	Maximum callus induction	Khulna, Bangladesh	[65]
	MS medium supplemented with 1.0 mg/l BAP and 0.1 mg/l NAA	20.50 ± 1.80 shoots per explant	Khulna, Bangladesh	[59]
	MS medium supplemented with 1.7 μM silver nitrate, 8.87 μM BAP and 2.46 μM IBA	8.3 shoots per explant	Calicut, Kerala, India	[125]
	MS medium supplemented with 2.0 mg/l BAP and 0.3 mg/l IAA	Maximum shoots per explant	Sri Lanka	[126]
	MS medium supplemented with 2.0 mg/l BAP and 0.2 mg/l NAA	19.40 shoots per explant	Tamilnadu, India	[60]
	MS medium supplemented with 0.54 μM NAA and 8.87 μM BAP	Maximum shoots per explant	Western Ghat forest, India	[127]
	MS medium supplemented with 1 mg/l BAP and 0.5 mg/l IAA	11.50 shoots per explant	Orissa, India	[128]
	MS medium supplemented with 0.1 mg/l NAA and 1.0 mg/l BAP	6.7 ± 0.51 shoots per explant	Karnataka, India	[129]
	MS medium supplemented with 1 mg/l BAP and 0.5 mg/l IAA	Maximum shoots per explant	Orissa, India	[130]
	MS medium supplemented with 1.0 mg/l 2, 2, 4-D and 30 g/l sucrose	Maximum Callus induction	Malaysia	[66]

(Table 2) cont.....

Plant Species	Medium Composition	Optimum Result	Collection	References
Kaempferia galanga L.	MS medium supplemented with 5 mg/ l BAP and 30 gm/l sucrose	7.4 ± 1.0 shoots per explant	Malaysia	[131]
	MS medium supplemented with 2.0 mg/l KIN and 1.0 mg/l NAA	7.0 ± 0.25 shoots and 6.7±0.291 roots per explant	Imphal, Manipur	[132]
	MS medium supplemented with 2.0 mg/ml BAP and 1.0 mg/l Kn	10.85 shoot per explants	India	[133]
	MS Media supplemented with 3 mg/l BAP and 4 mg/l Kn	6.52 shoots per explant	Siliguri, India	[64]
	MS Media supplemented with 3 mg/l BAP, 3 mg/l Kn and 1 mg/l NAA	15.6 ± 0.2 shoots per explant	Orrisa, India	[63]
	MS medium supplemented with 1 mg/l BAP and 0.5 mg/l IAA	11.33 shoots per explant	Orissa, India	[134]
	MS medium supplemented with 4.0 mg/l BAP along with 1.0 mg/l each of NAA and Kn	10.60 multiple shoots per explants	Kerala, India	[62]
	MS medium supplemented with 1 mg/l BAP and 0.1 mg/l IAA for shoot multiplication 0.2 mg/l of IBA for root formation	19.36 shoots per explant and 905 roots per explant	Thrissur, Kerala, India	[61]
	MS medium supplemented with 2 mg/l of BAP and 0.2 mg/l of NAA	Maximum shoots per explant	Tamil Nadu, India	[135]
	MS medium supplemented with 2.0 mg/l BAP	9 shoots per bud	Thailand	[136]
	MS medium supplemented with 5.7 μM IAA alone and a combination of 0.5 μM IAA plus 4.65 μM	3.67 shoots per explant	Manipur, India	[137]
	MS medium supplemented with 2.0 mg/l BAP and 0.5 mg/l IAA	12.0 ± 0.02 shoots per explant	Colombo, Sri Lanka	[138]
Kaempferia marginata Carey ex Roscoe	MS medium added with 2 mg/l BAP and 1 mg/l TDZ	4.90 shoots per explant	Mahasarakham Province, Thailand	[68]

(Table 2) cont.....

Plant Species	Medium Composition	Optimum Result	Collection	References
Kaempferia parviflora Wall. ex Baker	MS medium containing 8 μM BAP and 0.5 μM of TDZ for shoot and MS with 2 μM of IBA for root	12.20 shoots per explant	--	[149]
	MS media supplemented with 35.52 μM BAP	22.4 ± 1.84 shoots per explant	Putra, Malaysia	[139]
	MS media supplemented with 1.5 mg/l BAP	1.5 shoots per explant	Selangor, Malaysia	[140]
	MS media supplemented with 0.2 mg/l 2,4-D 0.2 mg/l NAA	callus induction of 20%	Pahang, Malaysia	[141]
	MS media supplemented with 1 mg/l BAP and 1 mg/l NAA, with 60 g/l sucrose	fresh weight of microrhizomes 265 mg/plantlet	Selangor, Malaysia	[142]
	MS medium supplemented with BAP 7 mg/l	4.2 shoots per explant	Thailand	[143]
	MS medium supplemented with 35.52 μM BAP	47.3 ± 3.4 shoots per initial explant	Loei Province, Thailand	[144]
Kaempferia rotunda L.	MS medium supplemented with 0.5mg/l NAA and 1.0 mg/l BAP	5.00 shoots per explant	Kerala, India	[80]
	MS medium supplemented with 2.5 mg/l 2,4-D and 0.5 mg/l BAP	Embryogenic callus induction	Kerala, India	[145]
	MS medium supplemented with 2.69 μM NAA and 2.22 μM BAP	6.10 shoots per explant	Manipur, India	[137]
	MS medium supplemented with 2.5mg/l 2,4-D and 0.5mg/l BAP	Maximum callus regeneration	Karnataka, India	[146]

Kaempferia galanga is the species reported most among all the species under the genus *Kaempferia* till now. Many researchers reported earlier on this plant's shoot, root and callus induction. In most cases, Murashige and Skoog (MS) medium was used to culture this plant by adding different plant growth hormones. Majorly, two cytokinins were used in the maximum studies, 6-benzylaminopurine (BAP) and kinetin (Kn), with and without in combination with indole-3-acetic acid (IAA) and naphthalene acetic acid (NAA) to influence the multiplication in *in vitro* conditions. A higher number of shoots is reported in the presence of MS medium supplemented with 1.0 mg/l BAP and 0.1 mg/l NAA [59]. 19.40 shoots per explant in the presence of MS medium supplemented with 2.0 mg/l BAP and 0.2 mg/l NAA was reported by Kalpana and Anbazhagan [60]. Following the results of previous studies, 19.36 shoots per explant were reported by Bindu [61] in the MS medium with 1 mg/l BAP and 0.1 mg/l IAA. Hence, the maximum number of shoot multiplication is seen when the explant is introduced in the MS medium with a combination of BAP and NAA or BAP with IAA. One cytokinin and one auxin proved to be the best combination for the multiplication of plants in the *in vitro* condition of this plant. Previous reports on shoot multiplication of *K. galanga* are also represented in Table **2**. A higher concentration of BAP (4mg/l) along with 1.0 mg/l each of NAA and Kn proved to be a lower influencer for the multiplication of this plant reported by Preetha *et al.* [62] with a shoot multiplication of 10.60 multiple shoots per explants. In comparison, when the plant explant was introduced in MS medium supplemented with 3 mg/l BAP, 3 mg/l Kn and 1 mg/l NAA, the multiplication rate increased to 15.6 ± 0.2 shoots per explant [63]. Another report suggests that when the plantlets are introduced into MS Media supplemented with 3 mg/l BAP and 4 mg/l Kn, the multiplication is hampered at a higher rate and has resulted in 6.52 shoots per explant [64]. In conclusion, it can be assumed increase in the concentration of all the combining phytohormones up to a level can increase the shoot proliferation. However, the best growth rate is observed when the hormones are combined at a lower concentration. As per our knowledge, there are only two reports on callus induction from this plant. Maximum callus induction occurs in MS medium with 2,4-dichlorophenoxyacetic acid (2, 4-D) [65, 66].

Micropropagation of the plant *Kaempferia parviflora* was done fewer times and reported for maximum shoot induction of 47.3 ± 3.4 shoots and 22.4 ± 1.84 shoots per explant in the MS medium supplemented with 35.52 µM BAP [139, 144]. Maximum callus induction (20%) from this plant explant is reported in MS media supplemented with 0.2 mg/l 2, 4-D and 0.2 mg/l NAA by Nazreena *et al.* [141]. *In vitro* microrhizomes formation is seen when the explant was introduced in MS media in combination with 1 mg/l BAP and 1 mg/l NAA with 60 g/l sucrose.

Kaempferia rotunda is another important medicinal species belonging to the genus *Kaempferia*, also reported four times for tissue culture. Two reports on shoot multiplication and two on callus induction are reported. 5.00 shoots per explant and 6.10 shoots per explant were reported in MS medium supplemented with 0.5 mg/l NAA and 1.0 mg/l BAP, and in MS medium supplemented with 2.69 µM NAA and 2.22 µM BAP, respectively [67]. Embryonic callus regeneration was reported by Mustafaanand *et al.* [145] and Chandana *et al.* [146] using the same medium composition of MS medium supplemented with 2.5mg/l 2, 4-D and 0.5mg/l BAP. Park *et al.* [147] developed rapid micropropagation system for *Kaempferia parviflora* from rhizome buds on MS medium containing BAP and TDZ.

Kaempferia marginata and *Kaempferia angustifolia* were reported once each for micropropagation. Micropropagation of *Kaempferia marginata* reported 4.90 shoots per explant in the MS medium added with 2 mg/l BAP and 1 mg/l thidiazuron (TDZ) [68]. According to Haque and Ghosh [148], 6.60 shoots per explant is when *Kaempferia angustifolia* was introduced into MS medium supplemented with 2.0 mg/l BAP, 2.0 mg/l Kn and 1.0 mg/l NAA.

CONCLUSION

Many important plant species are present in the genus *Kaempferia* that remained unexplored; due to that, conservation is also limited in research. In future medicine development, exploring those species and their conservation is necessary. However, the members of this genus are poor in the production of seeds and flowers, as well as seasonal growth is observed. Conventional breeding of the *Kaempferia* genus is not applicable in the field, and to solve this problem, *in vitro* regeneration is the most effective alternative. Hence, the need for *in vitro* propagation is crucial in the future to conserve elite genotypes as well as chemotypes.

REFERENCES

[1] Hashiguchi A, San Thawtar M, Duangsodsri T, Kusano M, Watanabe KN. Biofunctional properties and plant physiology of *Kaempferia* spp.: Status and trends. J Funct Foods 2022; 92(105029): 105029. [http://dx.doi.org/10.1016/j.jff.2022.105029]

[2] Nakao K, Murata K, Deguchi T, *et al.* Xanthine oxidase inhibitory activities and crystal structures of methoxyflavones from *Kaempferia parviflora* rhizome. Biol Pharm Bull 2011; 34(7): 1143-6. [http://dx.doi.org/10.1248/bpb.34.1143] [PMID: 21720029]

[3] Win NN, Ito T, Matsui T, *et al.* Isopimarane diterpenoids from *Kaempferia pulchra* rhizomes collected in Myanmar and their Vpr inhibitory activity. Bioorg Med Chem Lett 2016; 26(7): 1789-93. [http://dx.doi.org/10.1016/j.bmcl.2016.02.036] [PMID: 26916438]

[4] Pham NK, Nguyen HT, Nguyen QB. A review on the ethnomedicinal uses, phytochemistry and pharmacology of plant species belonging to *Kaempferia* L. genus (Zingiberaceae). Sci Asia 2021; 48(1): 1-24.

[5] Huang L, Yagura T, Chen S. Sedative activity of hexane extract of *Keampferia galanga* L. and its active compounds. J Ethnopharmacol 2008; 120(1): 123-5. [http://dx.doi.org/10.1016/j.jep.2008.07.045] [PMID: 18761077]

[6] Liu XC, Liang Y, Shi WP, Liu QZ, Zhou L, Liu ZL. Repellent and insecticidal effects of the essential oil of *Kaempferia galanga* rhizomes to *Liposcelis bostrychophila* (Psocoptera: Liposcelidae). J Econ Entomol 2014; 107(4): 1706-12. [http://dx.doi.org/10.1603/EC13491] [PMID: 25195466]

[7] Mohammad SP, Harindran J, Kannaki KS, Revathy R. Diuretic activity of *Kaempferia galanga* Linn rhizome extract in albino rat. World J Pharm Pharm Sci 2016; 5(6): 1161-9.

[8] Ichwan SJA, Husin A, Suriyah WH, Lestari W, Omar MN, Kasmuri AR. Anti-neoplastic potential of ethyl-p-methoxycinnamate of *Kaempferia galanga* on oral cancer cell lines. Mater Today Proc 2019; 16(4): 2115-21. [http://dx.doi.org/10.1016/j.matpr.2019.06.100]

[9] Srivastava N, Mishra S, Iqbal H, Chanda D, Shanker K. Standardization of *Kaempferia galanga* L. rhizome and vasorelaxation effect of its key metabolite ethyl p-methoxycinnamate. J Ethnopharmacol 2021; 271(271): 113911. [http://dx.doi.org/10.1016/j.jep.2021.113911] [PMID: 33571614]

[10] Ragsasilp A, Saensouk P, Saensouk S. Ginger family from Bueng Kan Province, Thailand: Diversity, conservation status, and traditional uses. Biodivers J 2022; 23(5): 2739-52.

[11] Lallo S, Hardianti B, Sartini S, Ismail I, Laela D, Hayakawa Y. Ethyl P-Methoxycinnamate: An active anti-metastasis agent and chemosensitizer targeting NFκB from *Kaempferia galanga* for melanoma cells. Life (Basel) 2022; 12(3): 337. [http://dx.doi.org/10.3390/life12030337] [PMID: 35330088]

[12] Parvez MA, Khan MM, Islam MZ, Hasan SM. Antimicrobial activities of the petroleum ether, methanol and acetone extracts of *Kaempferia galanga* rhizome. J Life Earth Sci 2005; 1(1): 25-9.

[13] Dash PR, Nasrin M, Ali MS. *In vivo* cytotoxic and *In vitro* antibacterial activities of *Kaempferia galanga.* J Pharmacogn Phytochem 2014; 3(1): 172-7.

[14] Shamsol Z, Iskandar MI, Ariffin ZZ, Safian MF. Chemical constituents, antioxidant and antimicrobial activities of *Kaempferia galanga* rhizome essential oils. Int J Soc Sci Res 2021; 3(4): 405-19.

[15] Ichwan SJ, Sazeli S, Lestari W. Analysis of the anti-cancer effect of ethyl-p-methoxycinnamate extracted cekur (*Kaempferia galanga*) on cancer cell lines with wild-type and null p53. IIUM J Orofac Sci 2020; 1(1): 28-33.

[16] Zhang L, Liang X, Ou Z, *et al.* Screening of chemical composition, anti-arthritis, antitumor and antioxidant capacities of essential oils from four Zingiberaceae herbs. Ind Crops Prod 2020; 149(112342): 112342. [http://dx.doi.org/10.1016/j.indcrop.2020.112342]

[17] Elshamy AI, Mohamed TA, Swapana N, *et al.* Two new diterpenoids from kencur (*Kaempferia galanga*): Structure elucidation and chemosystematic significance. Phytochem Lett 2021; 44(44): 185-9. [http://dx.doi.org/10.1016/j.phytol.2021.06.023]

[18] Jeong D, Kim DH, Chon JW, *et al.* Antibacterial effect of crude extracts of *Kaempferia parviflora* (Krachaidam) against *Cronobacter* spp. and enterohemorrhagic *Escherichia coli* (EHEC) in various dairy foods: A preliminary study. J Milk Sci Biotechnol 2016; 34(2): 63-8. [http://dx.doi.org/10.22424/jmsb.2016.34.2.63]

[19] Plaingam W, Sangsuthum S, Angkhasirisap W, Tencomnao T. *Kaempferia parviflora* rhizome extract and *Myristica fragrans* volatile oil increase the levels of monoamine neurotransmitters and impact the proteomic profiles in the rat hippocampus: Mechanistic insights into their neuroprotective effects. J Tradit Complement Med 2017; 7(4): 538-52. [http://dx.doi.org/10.1016/j.jtcme.2017.01.002] [PMID: 29034205]

[20] Miyazaki M, Izumo N, Yoshikawa K, *et al.* The anti-obesity effect of *Kaempferia parviflora* (KP) is attributed to leptin in adipose tissue. J Nutrit Health Food Sci 2019; 2019(7): 1-9.

[21] Yoshino S, Tagawa T, Awa R, Ogasawara J, Kuwahara H, Fukuhara I. Polymethoxyflavone purified from *Kaempferia parviflora* reduces visceral fat in Japanese overweight individuals: A randomised, double-blind, placebo-controlled study. Food Funct 2021; 12(4): 1603-13. [http://dx.doi.org/10.1039/D0FO01217C] [PMID: 33475663]

[22] Paramee S, Sookkhee S, Sakonwasun C, *et al.* Anti-cancer effects of *Kaempferia parviflora* on ovarian cancer SKOV3 cells. BMC Complement Altern Med 2018; 18(1): 178. [http://dx.doi.org/10.1186/s12906-018-2241-6] [PMID: 29891015]

[23] Temkitthawon P, Hinds TR, Beavo JA, *et al. Kaempferia parviflora*, a plant used in traditional medicine to enhance sexual performance contains large amounts of low affinity PDE5 inhibitors. J Ethnopharmacol 2011; 137(3): 1437-41. [http://dx.doi.org/10.1016/j.jep.2011.08.025] [PMID: 21884777]

[24] Stein RA, Schmid K, Bolivar J, Swick AG, Joyal SV, Hirsh SP. *Kaempferia parviflora* ethanol extract improves self-assessed sexual health in men: A pilot study. J Integr Med 2018; 16(4): 249-54. [http://dx.doi.org/10.1016/j.joim.2018.05.005] [PMID: 29880257]

[25] Klinngam W, Rungkamoltip P, Thongin S, *et al.* Polymethoxyflavones from *Kaempferia parviflora* ameliorate skin aging in primary human dermal fibroblasts and *ex vivo* human skin. Biomed Pharmacother 2022; 145: 112461. [http://dx.doi.org/10.1016/j.biopha.2021.112461] [PMID: 34839253]

[26] Vichitphan S, Vichitphan K, Sirikhansaeng P. Flavonoid content and antioxidant activity of krachai-dum *(Kaempferia parviflora)*wine Curr J Appl Sci 2007; 7((2-1)): 97-105.

[27] Sultana Z, Imam KM, Azam FM, *et al.* Evaluation of antihyperglycemic and antinociceptive activities of methanolic extract of *Kaempferia rotunda* L. (Zingiberaceae) rhizomes. Adv Nat Appl Sci 2012; 6(8): 1302-6.

[28] Islam F, Gopalan V, Lam AKY, Kabir SR. *Kaempferia rotunda* tuberous rhizome lectin induces apoptosis and growth inhibition of colon cancer cells *in vitro*. Int J Biol Macromol 2019; 141: 775-82. [http://dx.doi.org/10.1016/j.ijbiomac.2019.09.051] [PMID: 31505204]

[29] Lallo S, Lee S, Dibwe DF, Tezuka Y, Morita H. A new polyoxygenated cyclohexane and other constituents from *Kaempferia rotunda* and their cytotoxic activity. Nat Prod Res 2014; 28(20): 1754-9. [http://dx.doi.org/10.1080/14786419.2014.945175] [PMID: 25111413]

[30] Rashel Kabir S, Amir Hossen M, Abu Zubair M, *et al.* A new lectin from the tuberous rhizome of *Kaempferia rotunda*: Isolation, characterization, antibacterial and antiproliferative activities. Protein Pept Lett 2011; 18(11): 1140-9. [http://dx.doi.org/10.2174/092986611797200896] [PMID: 21707523]

[31] Jantan I, Raweh SM, Sirat HM, *et al.* Inhibitory effect of compounds from Zingiberaceae species on human platelet aggregation. Phytomedicine 2008; 15(4): 306-9. [http://dx.doi.org/10.1016/j.phymed.2007.08.002] [PMID: 17913483]

[32] Kabir SR, Dai Z, Nurujjaman M, *et al.* Biogenic silver/silver chloride nanoparticles inhibit human glioblastoma stem cells growth *in vitro* and Ehrlich ascites carcinoma cell growth *in vivo.* J Cell Mol Med 2020; 24(22): 13223-34. [http://dx.doi.org/10.1111/jcmm.15934] [PMID: 33047886]

[33] Atun S, Sundari A. Development of potential kunci pepet rhizoma (*Kaempferia rotunda*) plant as antioxidant. Proc. 3rd Intern Conference on Research, Implementation and Education of Mathematics and Science, Yogyakarta. 2016; 99-104.

[34] Diastuti H, Chasani M, Suwandri S. Antibacterial activity of benzyl benzoate and crotepoxide from *Kaempferia rotunda* L. Rhizome. Ind J Chem 2019; 20(1): 9-15. [http://dx.doi.org/10.22146/ijc.37526]

[35] Devi AR, Kariyil BJ, Raj N, Akhil GH. Immunomodulatory activity of *Kaempferia rotunda* L. rhizome against cyclophosphamide induced immunosuppression in swiss albino mice. J Vet Pharmacol Ther 2021; 20(1): 57-65.

[36] Krishnakumar P, Varghese L. Nematicidal activity of *Lagenandra toxicaria* Dalz and *Kaempferia rotunda* L. rhizome extracts against root-knot nematode, *Meloidogyne incognita* (Kofoid and White) Chitwood and burrowing nematode, *Radopholus similis* Cobb. Indian Phytopathol 2022; 75(4): 1103-10. [http://dx.doi.org/10.1007/s42360-022-00527-3]

[37] Yanti , Anggakusuma , Gwon SH, Hwang JK. *Kaempferia pandurata* Roxb. inhibits *Porphyromonas gingivalis* supernatant-induced matrix metalloproteinase-9 expression *via* signal transduction in human oral epidermoid cells. J Ethnopharmacol 2009; 123(2): 315-24. [http://dx.doi.org/10.1016/j.jep.2009.02.047] [PMID: 19429378]

[38] Yun JM, Kwon H, Mukhtar H, Hwang JK. Induction of apoptosis by Panduratin A isolated from *Kaempferia pandurata* in human colon cancer HT-29 cells. Planta Med 2005; 71(6): 501-7. [http://dx.doi.org/10.1055/s-2005-864149] [PMID: 15971119]

[39] Yun JM, Kweon MH, Kwon H, Hwang JK, Mukhtar H. Induction of apoptosis and cell cycle arrest by

a chalcone panduratin A isolated from *Kaempferia pandurata* in androgen-independent human prostate cancer cells PC3 and DU145. Carcinogenesis 2006; 27(7): 1454-64. [http://dx.doi.org/10.1093/carcin/bgi348] [PMID: 16497706]

[40] Shim JS, Kwon YY, Hwang JK. The effects of panduratin A isolated from *Kaempferia pandurata* on the expression of matrix metalloproteinase-1 and type-1 procollagen in human skin fibroblasts. Planta Med 2008; 74(3): 239-44. [http://dx.doi.org/10.1055/s-2008-1034297] [PMID: 18253916]

[41] Parwata A, Sukardiman MH, Widhiartini A. Inhibition of fibrosarcoma growth by 5-hydroxy-7-ethoxy-flavanons from *Kaempferia pandurata* Roxb. Biomed Pharmacol J 2016; 9(3): 941-8. [http://dx.doi.org/10.13005/bpj/1033]

[42] Fadilah F. Activity study of *Kaempferia pandurata* Roxb. extract as antiestrogen receptor (-) breast cancer cell line 3, 4-methylenedioxyamphetamine-MB-231 by molecular docking and 3-(4, 5-dimethylthiazol-2-yl)-2, 5-diphenyltetrazolium bromide assay. Int J Green Pharmacy 2018; 12(4): 1-7. [http://dx.doi.org/10.22377/ijgp.v12i04.2273]

[43] Darwanto A, Tanjung M, Darmadi MO. Cytotoxic mechanism of flavonoid from Temu Kunci (*Kaempferia pandurata*) in cell culture of human mammary carcinoma. Clin Hemorheol Microcirc 2000; 23((2,3,4)): 185-90.

[44] Hwang JK, Chung JY, Baek NI, Park JH. Isopanduratin A from *Kaempferia pandurata* as an active antibacterial agent against cariogenic *Streptococcus mutans.* Int J Antimicrob Agents 2004; 23(4): 377-81. [http://dx.doi.org/10.1016/j.ijantimicag.2003.08.011] [PMID: 15081087]

[45] Park KM, Choo JH, Sohn JH, Lee SH, Hwang JK. Antibacterial activity of panduratin A isolated from *Kaempferia pandurata* against *Porphyromonas gingivalis.* Food Sci Biotechnol 2005; 14(2): 286-9.

[46] Song MS, Shim JS, Gwon SH, Lee CW, Kim HS, Hwang JK. Antibacterial activity of panduratin A and isopanduratin A isolated from *Kaempferia pandurata* Roxb. against acne-causing microorganisms. Food Sci Biotechnol 2008; 17(6): 1357-60.

[47] Sarmoko ID, Febriansah R, Romadhon AF, *et al.* Cytotoxic effect of ethanolic extract of Temu Kunci (*Kaempferia pandurata*) and Sirihan (*Piper aduncum* L.) on breast cancer line. Proceeding Molecular Targeted Therapy Symposium. 94-102.

[48] Lee CW, Kim HS, Kauk IY, *et al.* The study of inflammatory cytokine-induced pigmentation and antibacterial activity of panduratin A isolated from *Kaempferia pandurata* Roxb. against Acne. Planta Med 2008; 74(9): PA237. [http://dx.doi.org/10.1055/s-0028-1084235]

[49] Tewtrakul S, Subhadhirasakul S, Karalai C, Ponglimanont C, Cheenpracha S. Anti-inflammatory effects of compounds from *Kaempferia parviflora* and *Boesenbergia pandurata.* Food Chem 2009; 115(2): 534-8. [http://dx.doi.org/10.1016/j.foodchem.2008.12.057]

[50] Miksusanti M. Antibacterial of *Kaempferia pandurata* Roxb essential oil tubers on *Listeria monocytogenes* including mechanism and application. In: International Seminar on Chemistry.

[51] Sukandar EY, Fidrianny IR, Kamil A. In situ antibacterial activity of *Kaempferia pandurata* (Roxb.) rhizomes against *Staphylococcus aureus.* Int J Pharma Sci 2015; 7(2): 239-44.

[52] Sukandar EY, Kurniati NF, Anggadiredja K, Kamil A. *In vitro* antibacterial activity of *Kaempferia pandurata* Roxb. and *Curcuma xanthorrhiza* Roxb. extracts in combination with certain antibiotics against MSSA and MRSA. Int J Pharm Pharm Sci 2016; 8(1): 1-4.

[53] Marliyana SD, Mujahidin D, Syah YM, Rukayadi Y. Time-kill assay of 4-Hydroxypanduratin an isolated from *Kaempferia pandurata* against foodborne pathogens. Molekul 2017; 12(2): 166-73. [http://dx.doi.org/10.20884/1.jm.2017.12.2.363]

[54] Muthachan T, Tewtrakul S. Anti-inflammatory and wound healing effects of gel containing

Kaempferia marginata extract. J Ethnopharmacol 2019; 240(111964): 111964. [http://dx.doi.org/10.1016/j.jep.2019.111964] [PMID: 31112755]

[55] Kaewkroek K, Wattanapiromsakul C, Matsuda H, Nakamura S, Tewtrakul S. Anti-inflammatory activity of compounds from *Kaempferia marginata* rhizomes. Songklanakarin J Sci Technol 2017; 39(1): 91-9.

[56] Thanasakdecha S, Tewtrakul S. Wound healing gel containing compound 2α-acetoxysandara-copimaradien-1α-ol from *Kaempferia marginata* rhizomes. J Herb Med 2021; 28(100437): 100437. [http://dx.doi.org/10.1016/j.hermed.2021.100437]

[57] Boonsombat J, Mahidol C, Chawengrum P, *et al.* Roscotanes and roscoranes: Oxygenated abietane and pimarane diterpenoids from *Kaempferia roscoeana.* Phytochemistry 2017; 143(143): 36-44. [http://dx.doi.org/10.1016/j.phytochem.2017.07.008] [PMID: 28759790]

[58] Chakraborty A, Kundu S, Mukherjee S, Ghosh B. Endophytism in Zingiberaceae: Elucidation of beneficial impact. Endophytes and secondary metabolites. Cham: Springer 2019; pp. 187-212. [http://dx.doi.org/10.1007/978-3-319-90484-9_31]

[59] Rahman MM, Amin MN, Ahamed T, *et al. In vitro* rapid propagation of black thorn (*Kaempferia galanga* L.): A rare medicinal and aromatic plant of Bangladesh. J Biol Sci 2005; 5(3): 300-4. [http://dx.doi.org/10.3923/jbs.2005.300.304]

[60] Kalpana M, Anbazhagan M. *In vitro* production of *Kaempferia galanga* (L.) : An endangered medicinal plant. J Phytol 2009; 1(1): 56-61.

[61] Bindu KH, Rohini MR. Crop improvement and improved varieties in medicinal crops. Trainers'. Training 2018; 33: 1-106.

[62] Preetha TS, Hemanthakumar AS, Krishnan PN. A comprehensive review of *Kaempferia galanga* L. (Zingiberaceae): A high sought medicinal plant in Tropical Asia. J Med Plants Stud 2016; 4(3): 270-6.

[63] Mohanty S, Parida R, Sahoo S, Nayak S. *In vitro* conservation of nine medicinally and economically important species of Zingiberaceae from Eastern India. Proc Natl Acad Sci, India, Sect B Biol Sci 2014; 84(3): 799-803. [http://dx.doi.org/10.1007/s40011-013-0251-1]

[64] Bhattacharya M. *In vitro* regeneration of pathogen-free *Kaempferia galanga* L.: A rare medicinal plant. Res Plant Biol 2013; 3(3): 24-30.

[65] Rahman MM, Amin MN, Ahamed T, Ali MR, Habib A. Efficient plant regeneration through somatic embryogenesis from leaf base derived callus of *Kaempferia galanga* L. Asian J Plant Sci 2004; 3(6): 675-8. [http://dx.doi.org/10.3923/ajps.2004.675.678]

[66] Kuen TG, Khaladalla MM, Bhatt A, Keng CL. Callus induction and cell line establishment from various explants of *Kaempferia galanga.* Int J Curr Res 2011; 3(12): 1-4.

[67] Ravindran P. Micropropagation of *Kaempferia* spp. (*K. galanga* L. and *K. rotunda* L.) J Spices Aromat 1997; 6(2): 129-35.

[68] Saensouk P, Muangsan N, Saensouk S, Sirinajun P. *In vitro* propagation of *Kaempferia marginata* Carey ex Roscoe, a native plant species to Thailand. J Anim Plant Sci 2016; 6(5): 1405-10.

[69] Li DM, Zhao CY, Liu XF. Complete chloroplast genome sequences of *Kaempferia galanga* and *Kaempferia elegans*: Molecular structures and comparative analysis. Molecules 2019; 24(3): 474. [http://dx.doi.org/10.3390/molecules24030474] [PMID: 30699955]

[70] He ZH, Yue GGL, Lau CBS, Ge W, But PPH. Antiangiogenic effects and mechanisms of trans-ethyl p-methoxycinnamate from *Kaempferia galanga* L. J Agric Food Chem 2012; 60(45): 11309-17. [http://dx.doi.org/10.1021/jf304169j] [PMID: 23106130]

[71] Saensouk S, Jenjittikul T. *Kaempferia grandifolia* sp. nov. (Zingiberaceae) a new species from Thailand. Nord J Bot 2001; 21(2): 139-42.

[http://dx.doi.org/10.1111/j.1756-1051.2001.tb01349.x]

[72] Shim JS, Choi EJ, Lee CW, Kim HS, Hwang JK. Matrix metalloproteinase-1 inhibitory activity of *Kaempferia pandurata* Roxb. J Med Food 2009; 12(3): 601-7. [http://dx.doi.org/10.1089/jmf.2007.1041] [PMID: 19627209]

[73] Yoon JH, Shim JS, Cho Y, *et al.* Depigmentation of melanocytes by isopanduratin A and 4-hydroxypanduratin A isolated from *Kaempferia pandurata* ROXB. Biol Pharm Bull 2007; 30(11): 2141-5. [http://dx.doi.org/10.1248/bpb.30.2141] [PMID: 17978489]

[74] Okabe Y, Shimada T, Horikawa T, *et al.* Suppression of adipocyte hypertrophy by polymethoxyflavonoids isolated from *Kaempferia parviflora.* Phytomedicine 2014; 21(6): 800-6. [http://dx.doi.org/10.1016/j.phymed.2014.01.014] [PMID: 24629599]

[75] Kobayashi H, Horiguchi-Babamoto E, Suzuki M, *et al.* Effects of ethyl acetate extract of *Kaempferia parviflora* on brown adipose tissue. J Nat Med 2016; 70(1): 54-61. [http://dx.doi.org/10.1007/s11418-015-0936-2] [PMID: 26386971]

[76] Ono S, Yoshida N, Maekawa D, *et al.* 5-Hydroxy-7-methoxyflavone derivatives from *Kaempferia parviflora* induce skeletal muscle hypertrophy. Food Sci Nutr 2019; 7(1): 312-21. [http://dx.doi.org/10.1002/fsn3.891] [PMID: 30680186]

[77] Mekjaruskul C, Jay M, Sripanidkulchai B. Modulatory effects of *Kaempferia parviflora* extract on mouse hepatic cytochrome P450 enzymes. J Ethnopharmacol 2012; 141(3): 831-9. [http://dx.doi.org/10.1016/j.jep.2012.03.023] [PMID: 22465145]

[78] Patanasethanont D, Nagai J, Matsuura C, *et al.* Modulation of function of multidrug resistance associated-proteins by *Kaempferia parviflora* extracts and their components. Eur J Pharmacol 2007; 566(1-3): 67-74. [http://dx.doi.org/10.1016/j.ejphar.2007.04.001] [PMID: 17481606]

[79] Patanasethanont D, Nagai J, Yumoto R, *et al.* Effects of *Kaempferia parviflora* extracts and their flavone constituents on P-glycoprotein function. J Pharm Sci 2007; 96(1): 223-33. [http://dx.doi.org/10.1002/jps.20769] [PMID: 17031860]

[80] Matsuda H, Murata K, Hayashi H, Matsumura S. Suppression of benign prostate hyperplasia by *Kaempferia parviflora* rhizome. Pharmacognosy Res 2013; 5(4): 309-14. [http://dx.doi.org/10.4103/0974-8490.118827] [PMID: 24174827]

[81] Ninomiya K, Matsumoto T, Chaipech S, *et al.* Simultaneous quantitative analysis of 12 methoxyflavones with melanogenesis inhibitory activity from the rhizomes of *Kaempferia parviflora.* J Nat Med 2016; 70(2): 179-89. [http://dx.doi.org/10.1007/s11418-015-0955-z] [PMID: 26711832]

[82] Potikanond S, Sookkhee S, Na Takuathung M, *et al.* *Kaempferia parviflora* extract exhibits anti-cancer activity against HeLa cervical cancer cells. Front Pharmacol 2017; 8(8): 630. [http://dx.doi.org/10.3389/fphar.2017.00630] [PMID: 28955234]

[83] Thaklaewphan P, Ruttanapattanakul J, Monkaew S, *et al.* *Kaempferia parviflora* extract inhibits TNF-α-induced release of MCP-1 in ovarian cancer cells through the suppression of NF-κB signaling. Biomed Pharmacother 2021; 141(111911): 111911. [http://dx.doi.org/10.1016/j.biopha.2021.111911] [PMID: 34328090]

[84] Sun S, Kim MJ, Dibwe DF, *et al.* Anti-austerity activity of Thai medicinal plants: Chemical constituents and anti-pancreatic cancer activities of *Kaempferia parviflora.* Plants 2021; 10(2): 229. [http://dx.doi.org/10.3390/plants10020229] [PMID: 33503922]

[85] Wattanathorn J, Muchimapura S, Tong-Un T, Saenghong N, Thukhum-Mee W, Sripanidkulchai B. Positive modulation effect of 8-week consumption of *Kaempferia parviflora* on health-related physical fitness and oxidative status in healthy elderly volunteers. Evid Based Complement Alternat Med 2012; 2012: 1-7.

[http://dx.doi.org/10.1155/2012/732816] [PMID: 22899957]

[86] Chatchawan U, Eungpinichpong W, Sripanidkulchai B, Chatchawan U. Effect of *Kaempferia parviflora* extract on physical fitness of soccer players: A randomized double-blind placebo-controlled trial. Med Sci Monit Basic Res 2015; 21: 100-8. [http://dx.doi.org/10.12659/MSMBR.894301] [PMID: 25957542]

[87] Yoshino S, Awa R, Miyake Y, *et al.* Daily intake of *Kaempferia parviflora* extract decreases abdominal fat in overweight and preobese subjects: A randomized, double-blind, placebo-controlled clinical study. Diabetes Metab Syndr Obes 2018; 11: 447-58. [http://dx.doi.org/10.2147/DMSO.S169925] [PMID: 30214264]

[88] Sripanidkulchai B, Promthep K, Tuntiyasawasdikul S, Tabboon P, Areemit R. Supplementation of *Kaempferia parviflora* extract enhances physical fitness and modulates parameters of heart rate variability in adolescent student-athletes: A randomized, double-blind, placebo-controlled clinical study. J Diet Suppl 2022; 19(2): 149-67. [http://dx.doi.org/10.1080/19390211.2020.1852356] [PMID: 33272042]

[89] Rujjanawate C, Kanjanapothi D, Amornlerdpison D, Pojanagaroon S. Anti-gastric ulcer effect of *Kaempferia parviflora.* J Ethnopharmacol 2005; 102(1): 120-2. [http://dx.doi.org/10.1016/j.jep.2005.03.035] [PMID: 16023318]

[90] Tewtrakul S, Subhadhirasakul S, Kummee S. Anti-allergic activity of compounds from *Kaempferia parviflora.* J Ethnopharmacol 2008; 116(1): 191-3. [http://dx.doi.org/10.1016/j.jep.2007.10.042] [PMID: 18077118]

[91] Wongsrikaew N, Kim H, Vichitphan K, Cho SK, Han J. Antiproliferative activity and polymethoxyflavone composition analysis of *Kaempferia parviflora* extracts. J Korean Soc Appl Biol Chem 2012; 55(6): 813-7. [http://dx.doi.org/10.1007/s13765-012-2175-5]

[92] Nakata A, Koike Y, Matsui H, Shimada T, Aburada M, Yang J. Potent SIRT1 enzyme-stimulating and anti-glycation activities of polymethoxyflavonoids from *Kaempferia parviflora.* Nat Prod Commun 2014; 9(9): 1934578X1400900. [http://dx.doi.org/10.1177/1934578X1400900918] [PMID: 25918795]

[93] Sawasdee P, Sabphon C, Sitthiwongwanit D, Kokpol U. Anticholinesterase activity of 7-methoxyflavones isolated from *Kaempferia parviflora.* Phytother Res 2009; 23(12): 1792-4. [http://dx.doi.org/10.1002/ptr.2858] [PMID: 19548291]

[94] Kobayashi S, Kato T, Azuma T, Kikuzaki H, Abe K. Anti-allergenic activity of polymethoxyflavones from *Kaempferia parviflora.* J Funct Foods 2015; 13: 100-7. [http://dx.doi.org/10.1016/j.jff.2014.12.029]

[95] Thao NP, Luyen BTT, Lee SH, Jang HD, Kim YH. Anti-osteoporotic and antioxidant activities by rhizomes of *Kaempferia parviflora* Wall. ex Baker. Nat Prod Sci 2016; 22(1): 13-9. [http://dx.doi.org/10.20307/nps.2016.22.1.13]

[96] Jin S, Lee MY. *Kaempferia parviflora* extract as a potential anti-acne agent with anti-inflammatory, sebostatic and anti-*Propionibacterium acne* activity. Int J Mol Sci 2018; 19(11): 3457. [http://dx.doi.org/10.3390/ijms19113457] [PMID: 30400322]

[97] Kobayashi H, Suzuki R, Sato K, *et al.* Effect of *Kaempferia parviflora* extract on knee osteoarthritis. J Nat Med 2018; 72(1): 136-44. [http://dx.doi.org/10.1007/s11418-017-1121-6] [PMID: 28823024]

[98] Kongdang P, Jaitham R, Thonghoi S, Kuensaen C, Pradit W, Ongchai S. Ethanolic extract of *Kaempferia parviflora* interrupts the mechanisms-associated rheumatoid arthritis in SW982 culture model *via* p38/STAT1 and STAT3 pathways. Phytomedicine 2019; 59(59): 152755. [http://dx.doi.org/10.1016/j.phymed.2018.11.015] [PMID: 31005814]

[99] Nemidkanam V, Kato Y, Kubota T, Chaichanawongsaroj N. Ethyl acetate extract of *Kaempferia*

parviflora inhibits *Helicobacter pylori*-associated mammalian cell inflammation by regulating proinflammatory cytokine expression and leukocyte chemotaxis. BMC Complem Med Therap 2020; 20(1): 124.
[http://dx.doi.org/10.1186/s12906-020-02927-2] [PMID: 32321502]

[100] Phung HM, Lee S, Hong S, Lee S, Jung K, Kang KS. Protective effect of polymethoxyflavones isolated from *Kaempferia parviflora* against TNF-α-induced human dermal fibroblast damage. Antioxidants 2021; 10(10): 1609.
[http://dx.doi.org/10.3390/antiox10101609] [PMID: 34679744]

[101] Takuathung MN, Potikanond S, Sookkhee S, *et al.* Anti-psoriatic and anti-inflammatory effects of *Kaempferia parviflora* in keratinocytes and macrophage cells. Biomed Pharmacother 2021; 143(112229): 112229.
[http://dx.doi.org/10.1016/j.biopha.2021.112229] [PMID: 34649355]

[102] Ongchai S, Chiranthanut N, Tangyuenyong S, Viriyakhasem N, Kongdang P. *Kaempferia parviflora* extract alleviated rat arthritis, exerted chondroprotective properties *in vitro,* and reduced expression of genes associated with inflammatory arthritis. Molecules 2021; 26(6): 1527.
[http://dx.doi.org/10.3390/molecules26061527] [PMID: 33799537]

[103] Prasanth MI, Malar DS, Brimson JM, *et al.* DAF-16 and SKN-1 mediate Anti-aging and Neuroprotective efficacies of "thai ginseng" *Kaempferia parviflora* Rhizome extract in *Caenorhabditis elegans.* Nutr Healthy Aging 2022; 7(1-2): 23-38.
[http://dx.doi.org/10.3233/NHA-210148]

[104] Azuma T, Tanaka Y, Kikuzaki H. Phenolic glycosides from *Kaempferia parviflora.* Phytochemistry 2008; 69(15): 2743-8.
[http://dx.doi.org/10.1016/j.phytochem.2008.09.001] [PMID: 18922550]

[105] Seo SH, Lee YC, Moon HI. Acetyl-cholinesterase inhibitory activity of methoxyflavones isolated from *Kaempferia parviflora.* Nat Prod Commun 2017; 12(1): 1934578X1701200.
[http://dx.doi.org/10.1177/1934578X1701200107] [PMID: 30549816]

[106] Lee S, Jang T, Kim KH, Kang KS. Improvement of damage in human dermal fibroblasts by 3, 5, 7-trimethoxyflavone from black ginger (*Kaempferia parviflora*). Antioxidants 2022; 11(2): 425.
[http://dx.doi.org/10.3390/antiox11020425] [PMID: 35204307]

[107] Miyata Y, Tatsuzaki J, Yang J, Kosano H. Potential therapeutic agents, polymethoxylated flavones isolated from *Kaempferia parviflora* for cataract prevention through inhibition of matrix metalloproteinase-9 in lens epithelial cells. Biol Pharm Bull 2019; 42(10): 1658-64.
[http://dx.doi.org/10.1248/bpb.b19-00244] [PMID: 31582653]

[108] Thipboonchoo N, Soodvilai S. Effects of *Kaempferia parviflora* extract on glucose transporters in human renal proximal tubular cells. J Physiol Biomed Sci 2017; 30(2): 57-61.

[109] Ahmad SM, Muhammad NN, Johan J, Mohd RO, Roszymah H. Evaluating the effect of Volten VR4® *Kaempferia parviflora* extracts on blood glucose levels in human Type-2 diabetes mellitus and healthy individual: A case-control study. J Med Assoc Thai 2021; 104(10): 1610-6.
[http://dx.doi.org/10.35755/jmedassocthai.2021.10.12751]

[110] Horigome S, Yoshida I, Ito S, *et al.* Inhibitory effects of *Kaempferia parviflora* extract on monocyte adhesion and cellular reactive oxygen species production in human umbilical vein endothelial cells. Eur J Nutr 2017; 56(3): 949-64.
[http://dx.doi.org/10.1007/s00394-015-1141-5] [PMID: 26704713]

[111] Horigome S, Maeda M, Ho HJ, Shirakawa H, Komai M. Effect of *Kaempferia parviflora* extract and its polymethoxyflavonoid components on testosterone production in mouse testis-derived tumour cells. J Funct Foods 2016; 26(26): 529-38.
[http://dx.doi.org/10.1016/j.jff.2016.08.008]

[112] Mala S, Thaweboon S, Luksamijarukul P, Thaweboon B, Saranpuetti C, Kaypetch R. Effect of *Kaempferia parviflora* on *Streptococcus mutans* Biofilm Formation and its Cytotoxicity. Key Eng

Mater 2018; 773: 328-32.
[http://dx.doi.org/10.4028/www.scientific.net/KEM.773.328]

[113] Promson N, Puntheeranurak S. *Kaempferia parviflora* extract diminishes hyperglycemia and visceral fat accumulation in mice fed with high fat and high sucrose diet. J Physiol Biomed Sci 2014; 27(1): 13-9.

[114] Youn K, Lee J, Ho CT, Jun M. Discovery of polymethoxyflavones from black ginger (*Kaempferia parviflora*) as potential β-secretase (BACE1) inhibitors. J Funct Foods 2016; 20(20): 567-74.
[http://dx.doi.org/10.1016/j.jff.2015.10.036]

[115] Begum T, Gogoi R, Sarma N, Pandey SK, Lal M. Direct sunlight and partial shading alter the quality, quantity, biochemical activities of *Kaempferia parviflora* Wall., ex Baker rhizome essential oil: A high industrially important species. Ind Crops Prod 2022; 180(180): 114765.
[http://dx.doi.org/10.1016/j.indcrop.2022.114765]

[116] Atun S, Arianingrum R. Anticancer activity of bioactive compounds from *Kaempferia rotunda* rhizome against human breast cancer. Int J Pharmacogn Phytochem Res 2015; 7(2): 262-9.

[117] Ahmed FRS, Amin R, Hasan I, Asaduzzaman AKM, Kabir SR. Antitumor properties of a methyl-β- d -galactopyranoside specific lectin from *Kaempferia rotunda* against Ehrlich ascites carcinoma cells. Int J Biol Macromol 2017; 102(102): 952-9.
[http://dx.doi.org/10.1016/j.ijbiomac.2017.04.109] [PMID: 28461165]

[118] Agrawal S, Bhawsar A, Choudhary P, Singh S, Keskar N, Chaturvedi M. *In-Vitro* anthelmintic activity of *Kaempferia rotunda.* Int J Pharm Life Sci 2011; 2(9): 1062-4.

[119] Timai P. Anti-inflammatory and analgesic activity of ethanolic extract of *Kaempferia rotunda* rhizome in rats. PhD diss. Coimbatore: RVS College of Pharmaceutical Sciences 2017; pp. 1-47.

[120] Jagtap S. Preliminary phytochemical screening and antioxidant activity of rhizome extracts of *Kaempferia scaposa* (Nimmo) Benth. J Acad Ind Res 2015; 3(12): 613-20.

[121] Idrisi MS, Badola HK, Singh R. Indigenous knowledge and medicinal use of plants by local communities in Rangit Valley, South Sikkim, India. NeBio 2010; 1(2): 34-45.

[122] Prawat U, Tuntiwachwuttikul P, Taylor WC, Engelhardt LM, Skelton BW, White AH. Diterpenes from a *Kaempferia* species. Phytochemistry 1993; 32(4): 991-7.
[http://dx.doi.org/10.1016/0031-9422(93)85242-J]

[123] Vincent KA, Mathew KM, Hariharan M. Micropropagation of *Kaempferia galanga* L., a medicinal plant. Plant Cell Tissue Organ Cult 1992; 28(2): 229-30.
[http://dx.doi.org/10.1007/BF00055522]

[124] Shirin F, Kumar S, Mishra Y. *In vitro* plantlet production system for *Kaempferia galanga*, a rare Indian medicinal herb. Plant Cell Tissue Organ Cult 2000; 63(3): 193-7.
[http://dx.doi.org/10.1023/A:1010635920518]

[125] Chithra M, Martin KP, Sunandakumari C, Madhusoodanan PV. Protocol for rapid propagation, and to overcome delayed rhizome formation in field established *in vitro* derived plantlets of *Kaempferia galanga* L. Sci Hortic 2005; 104(1): 113-20.
[http://dx.doi.org/10.1016/j.scienta.2004.08.014]

[126] Wijekoon KM, Senarath WT. *In vitro* propagation of *Kaempferia galanga* (L). Proceedings of International Forestry and Environment Symposium.

[127] Rajasekharan PE, Ambika SR, Ganeshan S. *In vitro* regeneration and conservation of *Kaempferia galanga.* ICFAI J Genet Evol 2009; 2(2): 235-43.

[128] Parida R, Mohanty S, Kuanar A, Nayak S. Rapid multiplication and *In vitro* production of leaf biomass in *Kaempferia galanga* through tissue culture. Electron J Biotechnol 2010; 13(4): 5-6.
[http://dx.doi.org/10.2225/vol13-issue4-fulltext-12]

[129] Hanumantharaju N, Shashidhara S, Rajashekharan PE, Rajendra CE. Evaluation of ethyl-p-methoxy

cinnamate in natural and *In vitro* regenerated plant of *Kaempferia galanga* by HPTLC Method Int J Biotech Bioeng Res 1(2): 131-7.

[130] Mohanty S, Parida R, Singh S, Joshi RK, Subudhi E, Nayak S. Biochemical and molecular profiling of micropropagated and conventionally grown *Kaempferia galanga.* Plant Cell Tissue Organ Cult 2011; 106(1): 39-46. [http://dx.doi.org/10.1007/s11240-010-9891-5]

[131] Bhatt A, Kean OB, Keng CL. Sucrose, benzylaminopurine and photoperiod effects on *in vitro* culture of *Kaempferia galanga.* Linn Plant Biosyst - Int J Deal AspPlant Biol 2012; 146(4): 900-5.

[132] Ibemhal A, Laishram JM, Dhananjoy CH, Naorem B, Toijam R. *In vitro* induction of multiple shoot and root from the rhizome of *Kaempferia galanga* L. NeBio 2012; 3(3): 46-50.

[133] Kochuthressia K, Britto S, Jaseentha MO, Raphael R. *In vitro* antimicrobial evaluation of *Kaempferia galanga* L. rhizome extract. Am J Biotechnol Mol Sci 2012; 2(1): 1-5. [http://dx.doi.org/10.5251/ajbms.2012.2.1.1.5]

[134] Sahoo S, Parida R, Singh S, Padhy RN, Nayak S. Evaluation of yield, quality and antioxidant activity of essential oil of *in vitro* propagated *Kaempferia galanga* Linn. J Acute Dis 2014; 3(2): 124-30. [http://dx.doi.org/10.1016/S2221-6189(14)60028-7]

[135] Anbazhagan M, Balachandran B, Sudharson S, Arumugam K. *In vitro* propagation of *Kaempferia galanga* (L.)-an endangered medicinal plant. Int J Curr Sci Res 2015; 15S: 63-9.

[136] Jitsopakul N, Sangyojarn P, Homchan P, Thammasiri K. Micropropagation for conservation of Zingiberaceae in Surin province, Thailand. International Symposium on Tropical and Subtropical Ornamentals. 75-80.

[137] Chirangini P, Sinha SK, Sharma GJ. *In vitro* propagation and microrhizome induction in *Kaempferia galanga* Linn. and *K. rotunda* Linn. Indian J Biotechnol 2005; 4: 404-8.

[138] Senarath RM, Karunarathna BM, Senarath Wtpsk JG. *In-vitro* propagation of *Kaempferia galanga* (Zingiberaceae) and comparison of larvicidal activity and phytochemical identities of rhizomes of tissue cultured and naturally grown plants. J Appl Biotechnol Bioeng 2017; 2(4): 1-6.

[139] Labrooy C, Abdullah TL, Stanslas J. Influence of N6-benzyladenine and sucrose on *in vitro* direct regeneration and microrhizome induction of *Kaempferia parviflora* Wall. ex Baker, an important ethnomedicinal herb of Asia. Trop Life Sci Res 2020; 31(1): 123-39. [http://dx.doi.org/10.21315/tlsr2020.31.1.8] [PMID: 32963715]

[140] Khairudin NA, Haida Z, Hakiman M. *In vitro* shoot and root induction of *Kaempferia parviflora* (Zingiberaceae) rhizome using 6-benzylaminopurine. J Trop Plant Physiol 2020; 12(2): 23-32.

[141] Zuraida AR, Nazreena OA, Liyana Izzati KF, Aziz A. Establishment and optimization growth of shoot buds-derived callus and suspension cell cultures of *Kaempferia parviflora.* Am J Plant Sci 2014; 5(18): 2693-9. [http://dx.doi.org/10.4236/ajps.2014.518284]

[142] Zuraida A, Izzati K, Nazreena O, Omar N. *In vitro* microrhizome formation in *Kaempferia parviflora.* Annu Res Rev Biol 2015; 5(5): 460-7. [http://dx.doi.org/10.9734/ARRB/2015/13950]

[143] Mongkolchaipak N, Chansuwanit N, Suchantaboot P. Plant tissue culture of *Kaempferia parviflora* Wall. ex Baker. Warasan Krom Witthayasat Kan Phaet 2006; pp. 145-55.

[144] Prathanturarug S, Apichartbutra T, Chuakul W, Saralamp P. Mass propagation of *Kaempferia parviflora* Wall. ex Baker by *in vitro* regeneration. J Hortic Sci Biotechnol 2007; 82(2): 179-83. [http://dx.doi.org/10.1080/14620316.2007.11512217]

[145] Mustafaanand PH. *In-vitro* plant regeneration in *Kaempferia rotunda* Linn. through somatic embryogenesis : A rare medicinal plant. Int J Curr Microbiol 2014; 3(9): 409-14.

[146] Chandana BC, Kumari Nagaveni HC, Lakshmana D, Shashikala SK, Heena MS. Role of plant tissue

culture in micropropagation, secondary metabolites production and conservation of some endangered medicinal crops. J Pharmacogn Phytochem 2018; 7(3): 246-51.

[147] Park HY, Kim KS, Ak G, *et al.* Establishment of a rapid micropropagation system for *Kaempferia parviflora* Wall. ex Baker: Phytochemical analysis of leaf extracts and evaluation of biological activities. Plants 2021; 10(4): 698.
[http://dx.doi.org/10.3390/plants10040698] [PMID: 33916375]

[148] Haque SM, Ghosh B. Micropropagation of *Kaempferia angustifolia* Roscoe-an aromatic, essential oil yielding, underutilized medicinal plant of Zingiberaceae family. J Crop Sci Biotechnol 2018; 21(2): 147-53.
[http://dx.doi.org/10.1007/s12892-017-0051-0]

CHAPTER 19

Micropropagation of *Stemona tuberosa* Lour. –A Review

K. Sri Rama Murthy[1,*] and **D. Raghu Ramulu**[2]

[1] *R & D Center for Conservation Biology and Plant Biotechnology, Shivashakti Biotechnologies Limited, S. R. Nagar, Hyderabad-500038, Telangana, India*

[2] *Department of Botany, Government College (A), Anantapur-515001, Andhra Pradesh, India*

Abstract: The present review summarises the *in vitro* multiple plantlet regeneration of *Stemona tuberosa.* MS medium fortified with 7 mg/L Kn was found to be the optimum for multiple shoot induction from axillary buds. Excision and culture of nodal segments from the *in vitro* shoots on medium containing 7 mg/L Kn and 4 mg/L TDZ showed a maximum number of shoot multiplication. Shoots developed were rooted best on ½ strength MS with 1 mg/L IAA. Plantlets established in pots exhibited 85% survival.

Keywords: Micropropagation, Multiple Shoots, *Stemona tuberosa.*

INTRODUCTION

Stemona tuberose Lour., belonging to the family Stemonaceae, is mainly distributed in India, Bangladesh, Nepal, Myanmar, Cambodia, Vietnam, Thailand, Taiwan and China. It is a twiner with tuberous rhizome, leaves ovate with basal nerves, flowers bisexual, axillary, anther dorsifixed, petal-like connective long, fruit ovoid-oblong capsule, and 5-8-seeded (Figs. **1A** & **B**).

The tuberous roots of *S. tuberosa* have antibacterial, antiparasitic and expectorant properties. They are used in the treatment of coughs, ascariasis and oxyuriasis. The tuberous roots show bacteriostatic activity and are used in phthisis and cough. The drug soothes the respiratory centers without affecting the heart [1].

Herbal extracts from the species of *Stemona* have been used for treating of respiratory diseases and as anthelmintics in Asian countries for thousands of years [2, 3]. *Stemona tuberosa* has been officially listed in the 2005 edition of the Chinese Pharmacopoeia as antitussive traditional Chinese medicinal herb [4]. In

* **Corresponding author K. Sri Rama Murthy:** R & D Center for Conservation Biology and Plant Biotechnology, Shivashakti Biotechnologies Limited, S. R. Nagar, Hyderabad-500038, Telangana, India; Tel: +91-9618633355; E-mails: drmurthy@gmail.com and raghuramuludevarakonda@gmail.com

T. Pullaiah (Ed.)

India, various *Stemona* species are used as insecticides and traditional medicines, *e.g.*, treatment of skin diseases, killing head lice and scabicide [5, 6]. The plant is used to cure different human diseases, *viz.*, whooping cough, chronic bronchitis, dermatitis, eczema, urticaria, amoebic dysentery, psoriasis, trichomonas vaginitis, pinworm disease, and pulmonary tuberculosis [7, 8]. *S. tuberosa* has two alkaloids, neostenine and neotuberoStemonine, which showed antitussive activities [9].

Four new dehydrotocopherols (chromenols) have been isolated from different species by Brem *et al.* [10]. Based on TLC tests and microplate assays with the free radical DPPH, the antioxidant capacities of all chromenol derivatives were comparable with that of α-tocopherol.

The Stemonaceae is so far the only source of the *Stemona* alkaloids. Pilli *et al.* [3] discussed the biological activity and natural sources of *Stemona* alkaloids. The biological activities of some *Stemona* alkaloids have been evaluated in order to find the active principles of *Stemona* species. Tubero Stemonine is the first *Stemona* alkaloid to have its biological activity tested. The anthelminthic activity of this alkaloid was detected against the motility of some helminthic worms, such as *Angiostrongylus cantonensis, Dipylidium caninun*, and *Fasciola hepatica.* TuberoStemonine has been reported as an effective insecticide equivalent with azadirachtin when tested against the larva of *Spodoptera littoralis*. The action of tuberoStemonine on the neuromuscular transmission in crayfish was also investigated, revealing that this alkaloid depressed glutamate-induced responses at similar concentrations to those of established glutamate inhibitors.

Five new Stemoninine-type alkaloids, bisdehydroStemoninine (1), isobisdehydroStemoninine (2), bisdehydroneoStemoninine (3), and bisde- hydro Stemoninines A (4) and B (5), have been isolated by Lin *et al.* [11] from the crude-alkaloid extract of the roots of *S. tuberosa*. Alkaloid 1 displayed significant antitussive activity in the citric acid-induced guinea pig cough model.

Three new croomine-type *Stemona alkaloids,* along with ten known constituents, were isolated by Lin *et al.* [12] from the roots of *S. tuberosa*. The antitussive activity of the major alkaloids was tested using the citric acid-induced guinea pig cough model. Croomine (**8**) exhibited a dose-dependent inhibition of coughing with an ID_{50} value of 0.18 mmol/kg.

Lin *et al.* [13] isolated alkaloids from the roots of *S. tuberosa*, which include stemoenonine (1), 9a-O-methylstemoenonine (2), oxystemoenonine (3), 1,9a-seco-stemoenonine (4), and oxyStemoninine (5), Stemoninoamide (6) and Stemoninine (7). Compounds 6 and 7 exhibited strong antitussive activity after oral and intraperitoneal administrations.

Twelve dihydrostilbenes, stilbostemins N-Y (1-12), and a phenanthraquinone, stemanthraquinone (13), were isolated from roots of *S. tuberosa*, along with five known dihydrostilbenes by Lin *et al.* [14]. Dihydrostilbene 8 exhibited strong activity against *Bacillus pumilus* (MIT 12.5-25 microg/mL). Many tested compounds exhibited moderate antibacterial activities. *In silico* analysis of compounds from *S. tuberosa* inhibited N1 neuraminidase of H5N1 avian virus [15].

IN VITRO PROPAGATION

The main source of plant material is from natural habitat. This plant material is insufficient to meet the growing demands. Moreover, it leads to the disappearance of the plant species in the natural habitat. Over exploitation, habitat destruction, pollination limitations, loss of potential dispersers, scattered distribution and reproduction inability are the main causes for decreasing natural populations [16]. Hence, in a short span of time, it is going to be endangered and later extinct. It is an urgent need to multiply this endangered medicinal plant both *in vivo* and *in vitro*. Field cultivation is unsuccessful because of time consumption. *Ex-situ* conservation efforts have a limited impact on halting the decline in the population. Hence, *in vitro* methods are the only alternative and effective method to produce this plant on a large scale to meet the growing demand globally. Fresh rhizomes can be collected from the wild and used as a propagating material for either *in vivo* or *in vitro* propagation. *In vitro* propagation is a more effective means than *in vivo* methods.

In vitro regeneration is a competent mean of *ex-situ* conservation of plant diversity by using minimum plant material without disturbing the wild habitat. Moreover, *in vitro* propagation can overcome the problems of seasonal variation, reproductive inefficiency, self-incompatibility and susceptibility to diseases.

Stemona spp. can be vegetatively propagated by planting tuberous roots with attached buds; however, this process takes time. Moreover, sexual propagation by seeds is very poor [17]. Therefore, a number of micropropagation protocols have been developed by different researchers in *Stemona* spp., which includes *S. japonica* [18], *S. collinsae* [19], *S. curtisii* [20], *S. tuberosa* [21 - 23], *Stemona* sp [24], and *S. hutanguriana* [25].

Murthy *et al.* [23] collected the live plants of *S. tuberosa* from the college botanical garden for explants source. Nodal explants are considered the best explants for shoot multiplication experiments as they have pre-existing axillary buds. The nodes were excised and washed in the running tap water, followed by a fungicide and bactericide, each 0.3% for 10 min and with Tween 20 (5% v/v for 4 min, Loba Chemie Pvt. Ltd, Mumbai). Then, explants were treated with surface disinfectant $HgCl_2$ (0.1% w/v for 2 min). Montri *et al.* [22] used ethanol and sodium hypochlorite as a surface sterilant. The plant material was dipped in 70% ethanol solution for 1 min. and then immersed in 5% sodium hypochlorite solution for another 5 min. Later, the sterilized nodal segments are thoroughly washed with sterilized double distilled water. Prior to inoculation onto the nutrient medium, the cut ends are removed. Murashige and Skoog (MS) semi solid medium (with pH 5.7±1) is maintained for all experiments. Different growth regulators at different concentrations, either alone or in combination, are tested. Cultures were maintained at 24±2°C under 16 hr photoperiod with 3000 lux light intensity using fluorescent lights and 95% RH. For culture, 250 ml bottles and 25×150 mm tubes are used depending on the type of experiment. Sub culturing is practiced every 15 days. During sub-culture, the healthy shoots are excised and cultured on the same nutrient medium. Nodal segment excised from these shoots were used as an explants for further screening experiments for best hormonal combinations/morphogenetic responses [22].

Multiple Shoot Induction

Media Evaluation

Three media such as MS, B5 and Woody Plants Medium (WPM) (HiMedia, India) used for medium evaluation. For shoot proliferation from *S. tuberosa*, nodal segments were cultured on different basal media at full strength, such as MS medium [26] with high salt concentration, B_5 medium [27] with medium salt concentration and WPM [28] with low salt concentration with addition of BAP 2 mg/L (Sigma-Aldrich, India) and 3% sucrose. Murthy *et al.* [23] reported that induction of axillary buds was observed on three media, but the percentage of response, number of shoots/explant and average shoot length were higher on MS medium followed by B_5 and WPM [23] (Figs. **1-C**).

Nodal explants of *S. tuberosa* produced 5.80 + 0.38 shoots/explant with length 7.07 ± 0.23 cm on MS medium, 5.00 ± 0.33 shoots/explant with 6.14 ± 0.37 cm length on B5 medium and 4.30 ± 0.30 Shoots/explant with 4.88 ± 0.42 cm shoot length on WPM medium. In the case of *S. tuberosa* S1 clone, Montri *et al.* [22] reported that the full-strength MS gave the highest shoot number and shoot length.

Explant Evaluation

The morphogenic response of explants depends on the type of explants, plant growth regulators and physiological conditions. Normally, juvenile tissue shows more response than mature tissue. Various explants like node, internode and shoot tip of *Stemona tuberosa* are cultured on MS, B_5 and WPM media supplemented with 2 mg/L BAP and 3% sucrose [23]. Criteria used for explants selection were based on a percentage of survival, number of shoots formed and length of the shoots.

Fig. (1). A. Tuberous roots of *S. tuberosa*, **B.** Flowering of *S. tuberosa*, **C.** *In vitro* shoot imitation from nodal segments of on different media MS + BAP 2.0 mg/l and , WPM + BAP 2.0 mg/l, **D.** Effect of Kn 7.0 mg/l on multiple shoot induction from nodal segment, **E.** Effect of MS +BAP 2.0 mg/l on multiple shoot formation from nodal segment, **F.** Effect of BAP 2.0 mg/l in combination with TDZ 4.0 mg/l on multiple shoot induction from nodal segments, **G.** Effect of TDZ 4.0 mg/l in combination with Kn 7.0 mg/l on multiple shoot induction from nodal segments of *S. tuberosa* **H.** Effect of TDZ 4.0 mg/l in combination with NAA 0.5 mg/l on multiple shoot induction from nodal segments of *S. tuberosa*.

Among the various explants tested, the positive morphogenic response was observed only from the node and shoot tip explants. Sprouting of axillary buds was noticed after 1 to 2 weeks of inoculation on the MS medium supplemented with 6- Benzylaminopurine (BAP) 2 mg/L and 3% sucrose. Shoot tip and nodal segments produce more number of shoots, but concerning the quality, bud breaking and percentage of survival, nodal explants are superior to shoot tips. The percentage of survival observed from the nodal explants and shoot tips was 53-80% and 40-53%, respectively [23]. The potentiality of shoot bud regeneration of nodal explants was higher than that of shoot tip explants in *S. tuberosa* survival percentage varied from 53-80% in nodal explants and 40-53% in shoot, tip explants [23].

In *S. tuberosa* the maximum mean number of shoots (5.80±0.38 shoots/explant) with mean length of shoots (7.07±0.23 cm) was obtained from the nodal explants on MS medium with BAP 2mg/L. Whereas from shoot tip explants, the maximum number of shoots (5.40±0.47 shoots/explants) with a maximum length of shoots (6.46 ± 0.51 cm) was obtained on MS medium with BAP 2 mg/L (Figs. **1D** & **E**) [23].

Induction of Multiple Shoots

Various concentrations of cytokinins like BAP, Kinetin (Kn) (Sigma-Aldrich, India) and Thidiazuron (TDZ) (Sigma-Aldrich, India) of different concentrations and in different combinations have been studied by Montri *et al.* [22, 23] for the induction of multiple shoots from nodal explants. Of various concentrations of BAP tested for shoot proliferation from mature nodal explants, BAP 2 mg/L produced a mean number of 5.80 ± 0.38 shoots/explant with 7.07 ± 0.23 cm of shoot length. Among the various concentrations of Kn used for shoot multiplication, a better response was observed with 7 mg/L by inducing 6.10 ± 0.37 shoots/explant with 8.27 ± 0.39 cm length. An increase or decrease in the concentration of Kn decreased the shoot number.

MS medium fortified with TDZ for shoot induction from nodal explants showed a mean number of 6.70 ± 0.36 shoots/explant, and with shoot length 6.43 ± 0.31 cm was at 4 mg/L. Of the various hormonal concentrations tried, Kn 7 mg/L produced the maximum shoot number and shoot length. Whereas Montri *et al.* [20, 22] found that the shoot proliferation was significantly greater on BAP and TDZ in the case of *S. curtisii* and *S. tuberosa* S1 clones, respectively.

Different combinations of cytokinins, such as BAP + TDZ and TDZ + Kn, were used to culture nodal explants for shoot regeneration [23] (Figs. **1F-G**). The response of nodal explants with combinations was comparatively better than single cytokinin. Nodal explants cultured on MS basal medium containing BAP 1 mg/L + TDZ 5 mg/L produced 4.00 ± 0.33 shoots/ explant with 6.48 ± 0.23 shoot length. BAP 2 mg/l +TDZ 4 mg/L showed the highest frequency of shoot formation with 5.60 ± 0.37 shoots/explant and shoot length of 7.07 ± 0.23 cm. Whereas BAP 5 mg/L + TDZ 1 mg/L produced 3.20 ± 0.32 shoots/explant with 4.49 ± 0.31 cm shoot length [23].

In TDZ + Kn combination, TDZ 4 mg/L + Kn 7 mg/L gave maximum shoot number 7.10 ± 0.37 shoots/explant with 8.35 ± 0.28 cm shoot length. Multiple shoots developed with a combination of TDZ and Kn grow faster, while those initiated in the BAP and TDZ combination grow slower [23].

Individual cytokinins and different combinations of cytokinins will alone were enough for good multiple shoot regeneration, but the addition of Gibberellic acid (GA_3) (Sigma-Aldrich, India) in combination with cytokinins can improve the shoot length and produce healthy shoots. The gibberellins enhance the shoot length by elongating the internodal regions.

Various concentrations of GA_3 (0.5 to 3 mg/L) were tested along with various cytokinins [23]. The cytokinins tested along with GA_3 here are BAP, Kn and TDZ. In various combinations of GA_3 with BAP tested, the highest multiple shoots 5.90 ± 0.34 and shoot length 7.93 ± 0.28 cm were observed on BAP 2 mg/L and GA_3 1 mg/L. Whereas in the case of Kn and GA_3 of all the combinations tested, Kn 7 mg/L and GA_3 1 mg/L gave good responses with 6.50 ± 0.26 shoots/explant and shoot length 8.58 ± 0.21 cm. Among various combinations of TDZ and GA_3 tested, the best shoot proliferation with good shoot length was observed. Among all the combinations tested, TDZ 4 mg/L and GA_3 1 mg/L showed the highest number of shoots 7.00 ± 0.29 and shoot length 7.60 ± 0.29 cm. When GA_3 was added in the medium, it produced healthy and long shoots [23].

Differences in the genotypes in *in vitro* response of different *Stemona* species have been observed [24, 18, 20, 22]. It has been suggested that the differences in the number of shoots and roots produced among genotypes during *in vitro* culture are mediated by differences in the endogenous balance of auxins and cytokinins.

Various cytokinins (BAP, Kn and TDZ) in combination with different auxins such as Indole-3-Acetic-Acid (IAA) (Sigma-Aldrich, USA), Indole-3-Butyric Acid (IBA) (Sigma-Aldrich, USA) and Naphthalene Acetic Acid (NAA) (Sigma-Aldrich, USA) were used with MS media for optimizing shoot regeneration [23]. Out of various combinations of BAP with auxins tested best response was noticed with BAP 2 mg/L + NAA 0.5 mg/L in *S. tuberosa*. The maximum number of shoots produced per explant on this combination is about 7.00 ± 0.25 shoots/explant, and the shoot length is 6.34 ± 0.38 cm (Fig. **1H**).

Out of various combinations of Kn with auxins tested Kn 7 mg/L + NAA 0.5 mg/L showed maximum numbers of shoots 7.20 ± 0.24 shoots/explant with 6.07 ± 0.42 cm of shoot length. Among various combinations of TDZ and auxins tested, response in terms of multiple shoot regeneration was observed on MS medium supplemented with TDZ 4 mg/L + NAA 0.5 mg/L with an average of 8.40 ± 0.30 shoots/explant. From nodal segments of *S. tuberosa* with an average shoot length 7.56 ± 0.47 cm [23]. However, in the case of *S. japonica* and *S. tuberosa* S1 clones, the combinations of BA (Sigma-Aldrich, India) and IBA showed the best response [18, 22].

The ability of *S. tuberosa* to form tuberous roots *in vitro* was estimated by using various concentrations and combinations of cytokinins. 3% sucrose, temperature and photoperiod were maintained constant throughout the study, which played a major role in the induction of tubers [23].

In vitro Rooting

After *in vitro* regeneration of plantlets, the important aspect in tissue culture is rooting and acclimatization. To select the best hormonal concentration for root induction, the *in vitro* grown shoots with 3 to 4 nodes were cultured on the media by supplementing different auxins (IAA, IBA and NAA) with different concentrations. The best auxin and its concentration were selected based on the number of roots formed and the root length. After 5-6 weeks, when regenerated shoots attained a length of more than 5 cm, they were excised and inoculated on different ½ strength media such as MS, B_5 and WPM by supplementing with IAA 1 mg/L (Fig. **2A**). Of the three media used for root induction, ½ MS was found to be superior over B_5 and WPM. ½ MS medium showed 66% response with 5.20 ± 0.41 of roots/shoot and with 6.19 ± 0.56 cm root length. Whereas in the B_5 medium, the number of roots produced is 4.70 ± 0.39 with root length 5.33 ± 0.20 cm and in the WPM medium, the number of roots is 2.80 ± 0.44 with root length 4.47 ± 0.54 cm [23].

In vitro shoots were rooted maximum on half-strength MS medium when compared to half-strength B_5 and WPM medium. Half-strength MS medium supplemented with three auxins such as IAA, IBA and NAA at different concentrations (0.1 to 3 mg/L) showed varied effect of rooting. Difference was noticed in the nature of roots induced depending on the auxins used in the medium. Out of the 3 auxins tested, IAA was found to be effective in root induction when compared to the other two auxins [23] (Fig. **2B**).

Fig. (2). Root imitation and acclimatization of *Stemona tuberosa*. **A.** Root imitation from *in vitro* derived shoots on MS + IAA 1.0 mg/l., **B.** Well-developed roots of *S. tuberosa* on MS + IAA 1.0 mg/l. **C.** *In vitro* tuberous roots formation from *in vitro* derived shoots on MS + TDZ 4.0 mg/l + Kn 7.0 mg/l. **D.** *In vitro* acclimatized plant ready to transfer in to greenhouse conditions.

NAA at 1 mg/L showed a maximum of 5.20 ± 0.41 roots/shoot with 4.92 ± 0.29 cm length. Whereas IBA at 0.5 mg/L showed a maximum response of 4.70 ± 0.42 roots/shoots with about 5.31 ± 0.30 cm of maximum root length. IAA was found to be more effective than NAA and IBA. Of all the concentrations tested, IAA at 1 mg/L in the half-strength MS medium proved to be the best. 1 mg/L IAA produced 5.60 ± 0.40 roots/shoot with the highest root length 6.19 ± 0.56 cm. By increasing or decreasing the concentration of IAA to 1 mg/L, there is a decrease in the frequency of rooting and root length [23].

Similar results were also found in the case of *S. curtisii* [20]. However, Chotikadachanarong *et al.* [24] found that NAA was effective in the induction of roots in some *Stemona* species. Whereas in the case of *S. tuberosa* S1 clone, the auxins had no positive effect on root induction and the shoots were successfully rooted *in vitro* after two months when they were cultured on solid MS medium without plant growth regulators [22].

In the *S. tuberosa*, the simultaneous formation of shoots and roots was observed in a medium supplemented with cytokinins combinations, *i.e.*, TDZ and Kn. However, the roots formed in these media were not normal, bulged and tuberous in nature. The *in vitro* tuberous roots formed in *S tuberosa* were similar to that of *in vivo* tubers in morphology. The *in vitro* tubers formed from the roots are small when compared to *ex vitro* but similar in their shape and colour. Most of the tubers developed were spindle-shaped and light creamish white-green in their early stages and slowly turned brown in later stages. The maximum number and size of the tuberous roots were observed on medium containing 3% sucrose, TDZ 4 mg/L and Kn 7 mg/L with a mean number of tubers 3.00 ± 0.36 and with a mean length 4.38 ± 0.25 cm. By increasing or decreasing this concentration, the formation of tubers was also decreased [23] (Fig. **2C**).

Acclimatization

The transplantation stage continues to be a major step in the micropropagation of plants. The rate of success in micropropagation depends on the number of plants survive after acclimatization. Well-rooted plantlets from the culture vessels were gently removed, keeping the roots intact. The plantlets were transferred to a container of warm water and gently rinsed to remove the traces of agar-media of the roots, and then the regenerates were planted in plastic cups with 1:1 (soil and vermin compost) sterile mix. The plants were wrapped in plastic paper to ensure high humidity and irrigated every two days with ½ strength MS liquid medium without any vitamins and sucrose. Lowering sucrose percent or omitting it can give faster and more successful acclimatization. To reduce the humidity, small holes were made in the plastic paper. Plants were allowed to adjust to the reduced humidity, and later, the diameter of the holes increased to reduce the humidity levels. Finally, plastic paper was removed to expose the plants to the external environment. To complete this process, 45 days are required. After 45 days, the hardened plantlets were transferred to earthen pots and maintained under shade for one week. Then, plants were exposed to sunlight for a few hours for a week, and then plants were transferred to soil and watered with tap water. The rooted plants were successfully established in soil with a 75% survival rate (Fig. **2D**) [23].

CONCLUSION

Plant tissue culture techniques are crucial for acquiring secondary metabolites of *Stemona* alkaloids for satisfying the expanding needs of the pharmaceutical sector in an eco-friendly manner. *Stemona tuberosa* is difficult to cultivate in all climatic zones, and plant tissue culture techniques are frequently used for the large-scale generation of secondary metabolites. The present chapter gives an overview of micropropagation methods for large-scale cultivation of this medicinally important species. Further research is needed on the *in vitro* production of secondary metabolites from this plant and biotransformation methods for increasing the secondary metabolite content.

REFERENCES

[1] Pullaiah T. Encyclopaedia of World Medicinal Plants. publications, NewDelhi 2019; p. 2651.

[2] Greger H. Structural relationships, distribution and biological activities of *Stemona* alkaloids. Planta Med 2006; 72(2): 99-113.
[http://dx.doi.org/10.1055/s-2005-916258] [PMID: 16491444]

[3] Pilli RA, Rosso GB, De Oliveira MDCF. The *Stemona* alkaloids. Alkaloids Chem Biol 2005; 62(30): 77-173.
[http://dx.doi.org/10.1016/S1099-4831(05)62002-0] [PMID: 16265922]

[4] Pharmacopoeia commission of People's Republic of China. The pharmacopoeia of the people's republic of China, Part 1. Beijing, China: Chemical Industry Publishing House 2005; p. 100.

[5] Valkenburg JLCH, Bunyapraphatsara N. Plant resources of South-East Asia, No 12 (2) Prosea foundation, Indonesia 2002.

[6] Chuakul W, Saralamp P, Paonil W, Temsiririrkkul R, Clayton T. Medicinal plants in Thailand. Bangkok, Thailand: Amarin printing and publishing, 1997; 2.

[7] Geng J, Huang W, Ren T, Ma X. Practical Traditional Chinese Medicine & Pharmacology Medicinal Herbs. Beijing, China: New World Press 1997.

[8] Zhou Y, Jiang RW, Hon PM, *et al.* Analyses of *Stemona* alkaloids in *Stemona tuberosa* by liquid chromatography/tandem mass spectrometry. Rapid Commun Mass Spectrom 2006; 20(6): 1030-8.
[http://dx.doi.org/10.1002/rcm.2409] [PMID: 16489582]

[9] Chung H-S, Hon P-M, Lin G, But PP, Dong H. Antitussive activity of *Stemona* alkaloids from *Stemona tuberosa.* Planta Med 2003; 69(10): 914-20.
[http://dx.doi.org/10.1055/s-2003-45100] [PMID: 14648394]

[10] Brem B, Seger C, Pacher T, *et al.* Antioxidant dehydrotocopherols as a new chemical character of *Stemona* species. Phytochemistry 2004; 65(19): 2719-29.
[http://dx.doi.org/10.1016/j.phytochem.2004.08.023] [PMID: 15464160]

[11] Lin LG, Zhong QX, Cheng TY, *et al.* Stemoninines from the Roots of *Stemona tuberosa.* J Nat Prod 2006; 69(7): 1051-4.
[http://dx.doi.org/10.1021/np0505317] [PMID: 16872143]

[12] Lin LG, Pak-Ho Leung H, Zhu JY, *et al.* Croomine- and tuberoStemonine-type alkaloids from roots of *Stemona tuberosa* and their antitussive activity. Tetrahedron 2008; 64(44): 10155-61.
[http://dx.doi.org/10.1016/j.tet.2008.08.046]

[13] Lin LG, Li KM, Tang CP, *et al.* Antitussive Stemoninine alkaloids from the roots of *Stemona tuberosa.* J Nat Prod 2008; 71(6): 1107-10.

[http://dx.doi.org/10.1021/np070651+] [PMID: 18452334]

[14] Lin LG, Yang XZ, Tang CP, Ke CQ, Zhang JB, Ye Y. Antibacterial stilbenoids from the roots of *Stemona tuberosa.* Phytochemistry 2008; 69(2): 457-63. [http://dx.doi.org/10.1016/j.phytochem.2007.07.012] [PMID: 17826806]

[15] Manohar A. In silico analysis of compounds from *Stemona tuberosa* as an inhibitor for N1 neuraminidase of H5N1 avian virus. Braz Arch Biol Technol 2013; 56(1): 21-5. [http://dx.doi.org/10.1590/S1516-89132013000100003]

[16] Chen G, Sun W, Wang X, Kongkiatpaiboon S, Cai X. Conserving threatened widespread species: a case study using a traditional medicinal plant in Asia. Biodivers Conserv 2019; 28(1): 213-27. [http://dx.doi.org/10.1007/s10531-018-1648-1]

[17] Rungrojsakul M. Morphological characteristic, chromosome number and propagation of *Stemona* spp.m.s 2001.

[18] Yang Z-D, Huang S-X, Lu T-L. Tissue culture and rapid propagation of medicinal plant *Stemona japonica.* Zhongcaoyao 2002; 33: 843-6.

[19] Rungruchkanont K, Pongrat A. Micropropagation of *Stemona collinsae* Craib. UBU J 2008; 10: 1-13.

[20] Montri N, Wawrosch CH, Kopp B. Micropropagation of *Stemona curtisii* Hook. f., a Thai medicinal plant. Acta Hortic 2006; (725): 341-6. [http://dx.doi.org/10.17660/ActaHortic.2006.725.43]

[21] Singlaw C, Kongbangkerd A, Promthep K, Saenpote P. Effect of cytokinins on *in vitro* shoot proliferation of *Stemona tuberosa* Lour. NU Sci J 2008; 5: 221-9.

[22] Montri N, Wawrosch CH, Kopp B. *In vitro* propagation of *Stemona tuberosa* Lour., an antitussive medicinal herb. Acta Hortic 2009; (812): 165-72. [http://dx.doi.org/10.17660/ActaHortic.2009.812.18]

[23] Murthy KSR, Chandrasekhara Reddy M, Kondamudi R, Pullaiah T. Micropropagation of *Stemona tuberosa* Lour. : An endangered and rare medicinal plant in Eastern Ghats, India. Indian J Biotechnol 2013; 12: 420-4.

[24] Chotikadachanarong K, Dheeranupattana S, Jatisatienr A. Micropropagation and alkaloid production in *Stemona* sp. Acta Hortic 2005; 676(676): 67-72. [http://dx.doi.org/10.17660/ActaHortic.2005.676.7]

[25] Prathanturarug S, Pheakkoet R, Jenjittikul T, Chuakul W, Saralamp P. *In vitro* propagation of *Stemona hutanguriana* W.Chuakul, an endangered medicinal plant. Physiol Mol Biol Plants 2012; 18(3): 281-6. [http://dx.doi.org/10.1007/s12298-012-0116-8] [PMID: 23814443]

[26] Murashige T, Skoog F. A revised medium for rapid growth and bioassays with tobacco tissue cultures. Physiol Plant 1962; 15(3): 473-97. [http://dx.doi.org/10.1111/j.1399-3054.1962.tb08052.x]

[27] Gamborg OL, Miller RA, Ojima K. Nutrient requirements of suspension cultures of soybean root cells. Exp Cell Res 1968; 50(1): 151-8. [http://dx.doi.org/10.1016/0014-4827(68)90403-5] [PMID: 5650857]

[28] Lloyd G, McCown B. Commercially feasible micropropagation of Mountain Laurel, *Kalmia latifolia,* by use of shoot-tip culture. Comb Proc Int Plant Prop Soc 1981; 30: 421-7.

CHAPTER 20

In vitro propagation of *Oxalis corniculata* L.

V. Kumaresan[1,2], S. Parthibhan[1,4], R. Lavanaya[1,3] and M.V. Rao[1,*]

[1] *Department of Botany, Bharathidasan University, Tiruchirappalli-620024, Tamil Nadu, India*

[2] *Department of Botany, Arignar Anna Govt. Arts College, Attur-636121, Tamil Nadu, India*

[3] *Department of Botany, Thanthai Periyar Govt. Arts & Science College, Tiruchirappalli-620023, Tamil Nadu, India*

[4] *Department of Botany, Kalaignar Karunanidhi Govt. Arts College for Women (A), Pudukkottai-622001, Tamil Nadu, India*

Abstract: The Oxalidaceae family is known for small herbs, shrubs and small trees with economic and medicinal properties in folklore medicines. The genus *Oxalis* is distributed worldwide and is famous for tuberous and ornamental cultivars. The present study established reproducible *in vitro* protocols for mass multiplication of *Oxalis corniculata* L. *via.* micropropagation and indirect organogenesis using different explants. Murashige and Skoog (MS) medium augmented with various cytokinins, auxins and gibberellic acid and combinations with respect to the different protocols. In micropropagation, shoot tip and node explants cultured on a medium with 6-benzyl adenine (BA) 3.0 mg/L, 6-furfuryladenine (Kn) 1.0 mg/L and naphthalene acetic acid (NAA) 0.5 mg/L produced the highest average of 35.1 and 28.5 shoots after 25 days of culture, respectively. Gibberellic acid (GA_3) treatment was satisfactory in shoot elongation, and rooting of shoots was best on indole-3-butyric acid (IBA) 3.0 mg/L than indole-3-acetic acid (IAA) and NAA. In indirect organogenesis, internode, leaf and petiole explants produced green, compact nodular calli at varying frequencies on medium fortified with auxins. The maximum frequencies of shoot regeneration and shoot numbers were observed on a medium containing BA 1.0 mg/L and IBA 0.5 mg/L. Further, the shoot elongation was achieved with BA and GA_3, and rooting was best achieved on IBA 3.0 mg/L with Kn 0.5 mg/L. All the plantlets were successfully hardened and acclimatized under the greenhouse condition with maximum survival of 95%. The current protocols established *via* meristem and callus mediated cultures would help in bioprospecting of this less explored medicinal plant.

Keywords: Callus, Micropropagation, Medicinal Plant, MS Medium, Organogenesis, *Oxalis corniculata*, Rooting, Shoot Elongation, Shoot Tip.

[*] **Corresponding author M.V. Rao:** Department of Botany, Bharathidasan University, Tiruchirappalli-620024, Tamil Nadu, India; Tel: +91-9442147460; E-mail: mvrao_456@yahoo.co.in

T. Pullaiah (Ed.)

INTRODUCTION

Plant-based remedies often have minimal unintended effects and are relatively cost-effective. Over 40% of medicines now prescribed in the world contain chemicals derived only from plants, and therefore, demand for traditional herbal drugs is increasing rapidly. At this time of increased demand for high-value medicinal plants, an alternative system is necessitated to protect the natural medicinal plant resources.

The family Oxalidaceae consists of 6 genera and about 770 species with herbs, shrubs, or small trees [1, 2]. The genus *Oxalis* consists of about 500 species worldwide and is known for tuberous plants and ornamental cultivars [3]. *Oxalis corniculata* L. is one among the species, distributed throughout India, South Africa and tropical and sub tropical America. It is commonly known as Indian sorrel in English; Puliyarai in Siddha and Tamil; Chaangeri, Amlapatrikaa, Amlikaa, Chukraa, Chukrikaa, Chhatraamlikaa in Ayurveda; Ambutaa bhaaji, Amutaa saag in Unani; and Tinpatiyaa and Ambilonaa in folk medicine [4].

The leaves of the plant are consumed as raw and cooked food, and the whole plant is used to cure dyspepsia, piles, anemia, tympanites, fever, dysentery, scurvy, snake bite, scorpion sting, and skin diseases [5]. The plant contains major phytochemicals such as carbohydrates, tannins, flavonoids, polyphenols, steroids, alkaloids, volatile oil, fatty acid and glycosides, and several other chemical constituents including tartaric acid, citric acid, oxalic acid, cinnamic acid, malic acid, c-glycosyl flavonoids, ascorbic (I), dehydroascorbic (II), pyruvic, glyoxalic acids, *etc*., and possesses diverse pharmacological properties including antioxidant, anti-cancer, anti-inflammatory, antimicrobial, diuretic, febrifuge, cardio-relaxant, *etc* [6].

In recent times, biotechnological advancement offers attractive opportunities for the production of desired plants and medicinal products using *in vitro* culture systems such as callus cultures, cell suspension cultures and organ cultures, and genetic manipulation. Pieces of literature on *in vitro* propagation of Oxalidaceae members are infrequent and limited among the genera. The present study aimed to establish reproducible protocols *via*. meristem culture and indirect organogenesis using various explants.

MATERIALS AND METHODS

Source of Plant Material

Oxalis corniculata plants were collected from the natural habitats at Kolli Hills, Tamil Nadu, India and maintained in the garden of the Department of Botany,

Bharathidasan University, Tiruchirappalli, Tamil Nadu, India. Plants were collected at different periods from July to September, May to June and December to February and maintained in the garden.

Explant Sterilization

The shoot segments were procured from plants maintained in the garden and cut into small pieces of about 5-7 cm. The excised shoot segments were initially washed in running tap water with 1-2 drops of Teepol (liquid detergent) (Reckitt Benckiser (India) Pvt. Ltd., Gurugram, India) for 5 min. Then, the shoot segments were disinfected with 70% ethanol for 30 s, followed by surface sterilization in 0.1% (w/v) aqueous solution of mercuric chloride ($HgCl_2$) for 5 min, and rinsed with sterile distilled water at least 4 times.

From the sterilized shoot segments, nodes and shoot tips were excised to 0.5-1.0 cm long and cultured upright into the shoot induction medium for micropropagation. Internode, leaf and petiole explants were prepared to 0.5 cm segments and cultured on a callus induction medium by the abaxial surfaces of the leaf, petiole and internodes exposed to the media.

Basal Medium

Murashige and Skoog (MS) medium [7] fortified with 3% sucrose and 0.8% agar (Type I) was used as the basal medium. The pH of the medium was adjusted to 5.7±0.2 before being solidified with agar. About 10-15 ml aliquots of the media were dispensed into the culture tubes and plugged with non-absorbent cotton plugs before being autoclaved at 1.06 kg/cm^2 and 121°C for 15 min.

All the chemicals and media used in the study were purchased from HiMedia®, Bengaluru, India.

Culture Conditions

All cultures incubated under the culture room maintained at 25±2°C, provided with 16-h photoperiod and 35 μM m^{-2} S^{-1} light intensity supplied by cool white fluorescent lamps (Philips, Mumbai, India), and 55-60% relative humidity.

Shoot Bud Induction and Multiplication

The sterilized node and shoot tip explants were cultured on the basal medium containing different plant growth regulators (PGRs). Cytokinins, 6-benzyl adenine (BA) and 6-furfuryladenine (Kn) at 0.5-5.0 mg/L were used either individually or in combination for shoot bud induction. Then, the shoot buds were subcultured on shoot multiplication medium containing BA (3.0 mg/L) and Kn (1.0 mg/L) supplemented with different auxins, indole-3-acetic acid (IAA), naphthalene acetic acid (NAA) and indole-3-butyric acid (IBA) at 0.1-2.0 mg/L.

Callus Induction and Shoot Regeneration

The sterilized internode, leaf and petiole explants were cultured on the basal medium supplemented with individual auxins 2,4-dichlorophenoxyacetic acid (2,4-D), IAA, IBA and NAA at 0.1-3.0 mg/L, for callus induction.

The callus obtained from all the explants were subcultured on BA and Kn at 0.1-3.0 mg/L either alone or in combination with the auxins IAA, IBA and NAA, each at 0.5 mg/L, for shoot regeneration.

Shoot Elongation

In vitro derived shoots of about 1 cm from the node and shoot tip explants were further subcultured on MS medium containing BA (3.0 mg/L) with gibberellic acid (GA_3) at 0.1-2.0 mg/L for shoot elongation.

Microshoots of about 0.5 cm obtained from callus on all the cultures were subcultured on MS medium fortified with BA (1.0 mg/L) and GA_3 at 0.1-2.0 mg/L for shoot elongation.

Rooting

All the elongated shoots were excised from the cultures on micropropagation and indirect organogenesis experiments and treated on MS medium fortified with IAA, IBA and NAA at 0.1-3.0 mg/L either individually or in combination with Kn at 0.1-2.0 mg/L, respectively, for rooting.

Hardening and Acclimatization

The well-rooted plantlets from both experiments were removed from the cultures and washed thoroughly with sterile distilled water. Then, the plantlets were introduced in plastic cups containing a mixture of sterilized vermiculite, sand and red soil at a 2:1:1 (w/w/w) ratio, and maintained under 70-80% relative humidity. Plantlets were regularly watered using ¼ strength MS basal salt solution devoid of sucrose and myo-inositol once in 3-day intervals for 2 weeks and regularly moved from the culture room condition to normal room temperature at regular intervals. Finally, after a month interval, all the plantlets were transferred to the greenhouse condition in earthen pots containing garden soil and kept under shade for another 4 weeks before being transferred to the field.

Experimental Design and Data Analysis

Each treatment consisted of a minimum of 25 replicates, and all the experiments were repeated thrice. All the cultures were periodically examined for morphological changes and recorded. The experimental design was random and factorial. The data on the frequency of shoot proliferation or callus induction, shoot number, shoot elongation and rooting were subjected to mean and mean separation analysis by using Duncan's Multiple Range Test (DMRT) [8].

RESULTS

Micropropagation from Node and Shoot Tip of *O. corniculata*

Shoot Bud Induction

Around 95% of aseptic cultures were obtained from plants collected from July to September, whereas severe contaminations were obtained from the other two periods of plant collection. The shoot bud induction was found better on nodal explants compared to the shoot tip explants. The MS medium devoid of any growth regulators (control) failed to induce bud break in both the explants, even after four weeks of the culture period. The explants remained fresh and green for about four weeks and became dried.

On the other hand, axillary bud enlargement following shoot bud induction was observed from both explants on MS medium fortified with different PGRs. Shoot bud sprouts from nodal explant and shoot tip were observed after 2 weeks of incubation (Figs. **1a** & **b**). Of the two cytokinins, BA 3.0 mg/L was found to be optimum for maximum bud break with the highest frequencies of 98.6% and 90.6% and highest shoot number of 15.1 and 9.6 per node and shoot tip explants, respectively (Table **1**) and (Figs. **1c** & **d**). There was a linear correlation between the increase in concentration and shoot induction response up to the optimal level.

Multiple Shoot Formation

The synergistic treatment of BA 3.0 mg/L with Kn (0.05-3.0 mg/L) improved shoot proliferation frequency and shoot number significantly after 25 days of culture. An increase of 100% and 93% bud break with 19.3 and 12.6 shoot buds per node and shot tip was observed on MS medium supplemented with BA (3.0 mg/L) and Kn 1.0 mg/L (Table **1**). The mean length of shoots was also increased in both the explants.

Fig. (1). Micropropagation of *Oxalis corniculata* L. from node and shoot tip explants. (a-d) Shoot bud initiation and multiplication from node and shoot tip explants after 25 days (1.5x), (e & f) Multiple shoot formation from node and shoot tip explants after 25 days (2.0x, 1.5x), (g & h) Shoot elongation from node and shoot tip derived shoots after 15 days (0.75x), (i & j) Root formation from node and shoot tip derived shoots after 20 days (1.0x), (f) Hardened plantlet after 30 days (0.5x).

Table 1. Shoot bud induction and shoot multiplication response of axillary and apical bud explants of *Oxalis corniculata* grown on MS medium supplemented with cytokinins after 25 days.

Cytokinins (mg/L)	Node			Shoot Tip		
	Shoot Bud Induction (%)	Shoot Number	Shoot Length (cm)	Shoot Bud Induction (%)	Shoot Number	Shoot Length (cm)
BA	-	-	-	-	-	-
0.5	80.0^{e}	4.7^{k}	1.9^{fg}	64.0^{hi}	4.1^{kl}	2.4^{i}
1.0	88.0^{e}	7.1^{h}	1.4^{h}	76.0^{f}	6.9^{gh}	2.0^{j}
2.0	96.0^{b}	13.0^{e}	1.3^{hi}	78.6^{de}	8.3^{e}	1.6^{j}
3.0	98.6^{ab}	15.1^{cd}	1.0^{j}	90.6^{b}	9.6^{c}	1.4^{k}
4.0	86.6^{d}	8.2^{gh}	0.9^{jk}	80.0^{d}	5.3^{j}	1.0^{l}
5.0	78.0^{ef}	5.1^{jk}	0.7^{l}	68.0^{h}	3.8^{l}	0.9^{lm}

(Table 1) cont.....

Cytokinins (mg/L)	Node			Shoot Tip		
	Shoot Bud Induction (%)	Shoot Number	Shoot Length (cm)	Shoot Bud Induction (%)	Shoot Number	Shoot Length (cm)
Kn	-	-	-	-	-	-
0.5	61.3^{h}	4.4^{kl}	2.8^{c}	53.3^{j}	2.5^{n}	3.5^{ef}
1.0	73.3^{g}	6.8^{hi}	3.2^{b}	60.0^{hi}	4.2^{k}	4.0^{e}
2.0	58.6^{hi}	5.2^{j}	2.6^{d}	46.6^{k}	3.4^{lm}	3.7^{ef}
3.0	50.6^{i}	3.5^{l}	2.1^{f}	40.0^{l}	2.3^{no}	3.2^{g}
4.0	41.3^{j}	2.7^{lm}	1.8^{fg}	33.3^{m}	2.0^{o}	2.6^{h}
5.0	36.6^{k}	1.9^{lm}	1.3^{hi}	26.6^{n}	1.5^{op}	2.0^{j}
BA + Kn	-	-	-	-	-	-
3.0 + 0.05	84.0^{de}	8.7^{g}	2.3^{e}	80.0^{d}	6.3^{i}	3.6^{ef}
3.0 + 0.1	90.6^{c}	12.4^{f}	2.8^{c}	86.6^{bc}	7.4^{g}	4.2^{d}
3.0 + 0.5	100.0^{a}	16.9^{b}	3.2^{b}	90.6^{b}	9.3^{cd}	4.9^{b}
3.0 + 1.0	100.0^{a}	19.3^{a}	3.8^{a}	93.3^{a}	12.6^{a}	5.2^{a}
3.0 + 2.0	96.0^{b}	15.2^{c}	2.1^{f}	78.6^{de}	10.9^{b}	4.7^{c}
3.0 + 3.0	80.0^{e}	12.8^{ef}	2.0^{fg}	73.3^{fg}	8.2^{ef}	3.8^{e}

Values are a mean of 25 replicates per treatment and repeated thrice. Values with the same superscript are not significantly different at a 5% probability level, according to DMRT.

Further, treatment with the different auxins showed varied responses with the type of auxin used. NAA at 0.5 mg/L showed the best response of 100% shoot multiplication from a node, while in shoot tip explants, the frequency of response remained the same. However, shoot number and shoot length were highly improved with 35.1 and 28.5 number of shoots with 6.9 and 6.3 cm per nodal and shoot tip explants, respectively. Compared to NAA, both IBA and IAA showed poor response and less number of shoots than the BA with Kn treatment. Conversely, a significant interaction was observed between NAA and BA to enhance shoot length (Table **2**) and Figs. (**1e**, **f**). Likewise, IBA and IAA at lower concentrations showed improved shoot length compared to the previous treatment, while the shoot number was decreased with increasing auxin concentrations.

Shoot Elongation

Proliferated multiple shoot buds were carefully separated from the clumps and individual shoots (about 1 cm) and cultured on shoot elongation medium, MS containing BA (3.0 mg/L) and different concentrations of GA_3. Among the various combinations used, GA_3 at 0.5 mg/L supported maximum shoot elongation of 8.9 and 9.7 cm from node and shoot tip explant derived shoots within 15 days of the culture period (Table **3**) (Figs. **1g**, **h**). Above this optimal level, the shoot elongation frequencies and shoot lengths were decreased considerably.

Table 2. Shoot bud induction and shoot multiplication response of axillary and apical bud explants of *O. corniculata* grown on MS medium + BA (3.0 mg/L) + Kn (1.0 mg/L) and auxins after 25 days.

Auxins (mg/L)	Node			Shoot Tip		
	Freq. of Response (%)	No. of Shoots/ Explants	Shoot Length (cm)	Freq. of Response (%)	No. of Shoots/ Explants	Shoot Length (cm)
NAA	-	-	-	-	-	-
0.1	93.3^{b}	28.9^{b}	5.9^{b}	86.6^{b}	20.4^{b}	5.4^{d}
0.5	100.0^{a}	35.1^{a}	6.3^{a}	93.3^{a}	28.5^{a}	6.9^{a}
1.0	90.6^{bc}	26.2^{c}	5.7^{bc}	80.0^{c}	19.1^{bc}	5.6^{bc}
2.0	86.6^{c}	22.8^{d}	4.4^{e}	76.0^{cd}	15.7^{d}	5.0^{d}
IAA	-	-	-	-	-	-
0.1	76.0^{f}	12.1^{f}	4.1^{f}	70.6^{de}	11.8^{ef}	4.9^{de}
0.5	83.3^{cd}	14.8^{e}	4.8^{d}	73.3^{d}	13.5^{e}	5.7^{b}
1.0	70.6^{g}	11.5^{fg}	3.9^{f}	66.6^{e}	9.7^{g}	4.3^{f}
2.0	56.0^{i}	9.4^{g}	3.2^{h}	60.0^{f}	8.4^{h}	4.0^{gh}
IBA	-	-	-	-	-	-
0.1	73.3^{fg}	10.9^{fg}	4.3^{ef}	70.6^{de}	9.6^{gh}	5.2^{d}
0.5	66.6^{gh}	8.6^{gh}	3.6^{g}	56.6^{fg}	7.8^{hi}	4.6^{f}
1.0	62.6^{h}	6.7^{h}	3.0^{hi}	50.6^{h}	5.2^{j}	4.1^{g}
2.0	50.6^{j}	4.6^{hi}	2.7^{j}	42.6^{i}	4.9^{jk}	3.9^{gh}

Values are a mean of 25 replicates per treatment and repeated thrice. Values with the same superscript are not significantly different at a 5% probability level, according to DMRT.

Table 3. Shoot elongation response of multiple shoots raised from axillary and apical bud explants of *O. corniculata* on MS medium supplemented with BA (3.0 mg/L) and GA_3 after 15 days.

GA_3 (mg/L)	Node		Shoot Tip	
	Freq. of Shoot Elongation (%)	Shoot Length (cm)	Freq. of Shoot Elongation (%)	Shoot Length (cm)
0.1	70.6^{b}	7.8^{cd}	86.6^{b}	8.9^{c}
0.5	80.0^{a}	8.9^{a}	93.3^{a}	9.7^{a}
1.0	76.0^{ab}	8.5^{b}	90.6^{ab}	9.5^{ab}
1.5	62.6^{c}	7.9^{c}	80.0^{c}	8.6^{cd}

(Table 3) cont.....

GA$_3$ (mg/L)	Node		Shoot Tip	
	Freq. of Shoot Elongation (%)	Shoot Length (cm)	Freq. of Shoot Elongation (%)	Shoot Length (cm)
2.0	56.0^{d}	7.3^{e}	76.0^{cd}	8.0^{e}

Values are a mean of 25 replicates per treatment and repeated thrice. Values with the same superscript are not significantly different at a 5% probability level, according to DMRT.

Rooting

The elongated shoots transferred to the rooting medium responded rooting at varied frequencies with callus intervention. In control, negligible frequency of rooting with thin roots were recorded (data not shown). On the other hand, all three tested auxins induced with multiple roots and basal callus, except IBA. IBA 3.0 mg/L induced 100% rooting with an average of 13.4 roots per shoot from the axillary bud, whereas 93% of cultures yielded an average of 10.1 roots per shoot from shoot tip derived shoots (Table **4**) and Figs. (**1i, j**). Rooting on IBA started with a single root, emerged after 10 days that continued its linear growth with lateral roots.

Table 4. Rooting response of *in vitro* raised shoots of *O. corniculata* grown on MS medium supplemented with auxins after 20 days.

Auxins (mg/L)	Node				Shoot tip			
	Rooting response (%)	No. of roots	Root length (cm)	Cultures with callus (%)	Rooting response (%)	No. of roots	Root length (cm)	Cultures with callus (%)
IAA	-	-	-	-	-	-	-	-
0.1	68.8^{j}	4.5j	2.9^{gh}	-	62.2^{l}	3.8^{jk}	2.1^{j}	-
0.5	75.5^{h}	6.2^{l}	3.1^{gh}	20.0^{e}	66.6^{jk}	4.1^{j}	2.7^{h}	22.2^{de}
1.0	80.0^{fg}	7.6^{h}	3.2^{fg}	26.6^{cd}	73.3^{hi}	4.9^{i}	2.9^{fg}	28.8^{c}
2.0	86.6^{e}	8.1^{g}	3.8^{d}	33.3^{b}	80.0^{fg}	5.6^{h}	3.4^{de}	31.1^{b}
3.0	90.0^{d}	9.3^{de}	4.1^{c}	37.7^{a}	84.4^{d}	5.9^{g}	3.5^{d}	40.0^{a}
IBA	-	-	-	-	-	-	-	-
0.1	86.6^{e}	8.2^{fg}	3.5^{e}	-	80.0^{de}	7.2^{f}	3.0^{fg}	-
0.5	91.1^{cd}	8.4^{f}	3.7^{de}	-	82.2^{de}	7.7^{gh}	3.5^{d}	-
1.0	93.3^{c}	9.0^{e}	4.1^{c}	-	88.8^{bc}	8.6^{c}	3.9^{c}	-
2.0	97.7^{ab}	11.5^{b}	4.7^{b}	-	90.0^{b}	9.4^{b}	4.5^{b}	-
3.0	100.0^{a}	13.4^{a}	5.2^{a}	-	93.3^{a}	10.1^{a}	4.7^{a}	-
NAA	-	-	-	-	-	-	-	-
0.1	55.5^{m}	6.2^{l}	1.8^{jk}	-	48.8^{l}	5.7^{gh}	2.0^{jk}	-

(Table 4) cont.....

Auxins (mg/L)	Node				Shoot tip			
	Rooting response (%)	No. of roots	Root length (cm)	Cultures with callus (%)	Rooting response (%)	No. of roots	Root length (cm)	Cultures with callus (%)
0.5	62.2^{l}	7.8^{gh}	1.9^{j}	-	53.3^{k}	6.4^{g}	2.4^{hi}	-
1.0	66.6^{g}	8.2^{fg}	2.5^{i}	17.7^{ef}	60.0^{ij}	6.9^{f}	2.5^{h}	20.0^{de}
2.0	73.3_{hi}	9.4^{d}	3.0^{gh}	24.4^{cd}	71.1^{fg}	7.5^{ef}	3.1^{f}	24.4^{d}
3.0	82.2^{f}	10.5^{c}	3.5^{f}	28.8^{c}	77.7_{e}	8.1^{d}	3.7^{d}	30.0^{bc}

Values are a mean of 25 replicates per treatment and repeated thrice. Values with the same superscript are not significantly different at a 5% probability level, according to DMRT.

In contrast, all the shoots cultured on IAA and NAA, became considerably thickened at the lower end and produced basal callus. In addition, occasional leaf drop noted during root formation showed no interference with the rooting since all the shoots developed new leaves.

II. Indirect Organogenesis from Internode, Leaf and Petiole of *O. corniculata*

Callus Induction

The internode, leaf and petiole explants enlarged near the cut ends and initiated callus on all the auxins in 8-10 days of culture. Subsequent proliferation of the callus on the same medium covered the entire surface of the explant within 4 weeks of culture. Callus morphology and growth response varied with different plant growth regulators used.

Callus induced on 2,4-D was white, friable and fast growing in nature, while NAA at lower concentrations (0.1 mg/L) induced yellow, friable and slow growing. In both instances, the callus became brown after 25 days. In the contrary, green and compact callus was induced on IAA, IBA and NAA (above 0.1 mg/L) treatments in a shorter period of 15 days. Of all the explants, internodes produced a better callus induction response than the leaf and petiole. Among the four auxins, IBA at 3.0 mg/L responded highest callus induction frequencies of 100% for internode, 98.6% for leaf and 96.6% for petiole, followed by 2,4-D, IAA and NAA (Table **5**, Figs. (**2a-c**).

Shoot Regeneration and Multiplication

All the green compact calli induced on the different explants were subcultured on different shoot regeneration medium induced multiple shoot buds at varying frequencies.

Table 5. Callus induction from internode, leaf and petiole explants of *O. corniculata* grown on MS medium supplemented with auxins after 4 weeks.

Auxins (mg/L)	Callus induction (%)		
	Internode	Leaf	Petiole
2,4-D	-	-	-
0.1	90.0^{f}	85.3^{f}	77.3^{g}
0.5	96.6^{bc}	89.6^{de}	84.0^{de}
1.0	97.3^{b}	94.6^{c}	90.0^{c}
2.0	93.3^{de}	88.0^{e}	82.6^{e}
3.0	85.0^{h}	78.6^{gh}	74.6^{g}
IAA	-	-	-
0.1	76.0^{k}	66.6^{j}	64.0^{jk}
0.5	81.3^{j}	70.0^{i}	69.3^{i}
1.0	84.0^{hi}	80.0^{g}	80.0^{f}
2.0	78.6^{j}	74.6^{i}	72.0^{gh}
3.0	70.0^{l}	68.0^{ij}	66.6^{j}
IBA	-	-	-
0.1	84.0^{hi}	78.6^{j}	69.3^{i}
0.5	88.0^{g}	82.6^{f}	73.3^{gh}
1.0	94.6^{d}	90.0^{d}	85.3^{d}
2.0	98.6^{ab}	97.3^{ab}	94.6^{ab}
3.0	100.0^{a}	98.6^{a}	96.6^{a}
NAA	-	-	-
0.1	65.3^{n}	54.6^{n}	50.6^{m}
0.5	69.3^{lm}	60.0^{lm}	56.0^{l}
1.0	70.0^{l}	65.3^{jk}	60.0^{l}
2.0	64.0^{no}	61.3^{l}	50.0^{mn}
3.0	60.0^{p}	57.3^{n}	46.6^{o}

Values are a mean of 25 replicates per treatment and repeated thrice. Values with the same superscript are not significantly different at a 5% probability level, according to DMRT.

The green compact nodular callus induced dark green shoot buds over the entire surface on medium containing BA and Kn (0.1-3.0 mg/L) either alone or in combination with IAA or IBA or NAA (0.5 mg/L) after 4 weeks of subculture (Figs. **2d-f**). Initially, the shoot buds appeared tiny and light green in colour, they eventually became healthy shoots after two weeks of the culture period. Shoot regeneration frequency and shoot number were comparatively low on Kn than BA treatments, owing to browning and turning of compact callus into friable nature within three weeks of culture. Of the synergistic treatments, BA 1.0 mg/L with IBA 0.5 mg/L combination exhibited the highest shoot regeneration of 100% for internode, 97.3% for leaf and 94.6% for petiole within 25 days of culture. The number of shoots were of 29.8, 22.0 and 19.7 from the respective explants (Table **6**) and Figs. (**2g-i**).

Fig. (2). Organogenesis from internode, leaf and petiole explants of *Oxalis corniculata* L. (**a-c**) Callus initiation from internode, leaf and petiole explants after 25 days (1.5x), (**d-f**) Callus proliferation from internode, leaf and petiole explants after 25 days (2.0x), (**g-i**) Shoot bud initiation from callus derived from internode, leaf and petiole explants after 15 days (2.0x), (**j**) Multiple shoot formation after 25 days (1.5x), (k) Root formation after 25 days (1.0x), (**l**) Hardened plantlets after 30 days (0.5x).

Table 6. Shoot regeneration response from callus induced from internode, leaf and petiole explants of *O. corniculata* grown on MS medium supplemented with auxin and cytokinins after 4 weeks.

PGRs (mg/L)	Internode		Leaf		Petiole	
	Shoot bud induction (%)	Shoot number	Shoot bud induction (%)	Shoot number	Shoot bud induction (%)	Shoot number
Control	69.3^{m}	5.9^{no}	57.3^{op}	4.3^{s}	48.0^{qr}	4.0^{q}
BA	-	-	-	-	-	-

(Table 6) cont.....

PGRs (mg/L)	Internode		Leaf		Petiole	
	Shoot bud induction (%)	Shoot number	Shoot bud induction (%)	Shoot number	Shoot bud induction (%)	Shoot number
0.1	84.0^{gh}	11.8^{j}	77.3i^{j}	9.2^{l}	78.6^{fg}	8.9^{i}
0.5	90.0^{e}	14.0^{g}	88.0^{c}	11.3^{ij}	84.0^{de}	9.5^{hi}
1.0	97.3^{b}	16.6^{d}	94.6^{b}	12.5^{h}	90.6^{b}	10.4^{gh}
2.0	85.3^{gh}	15.0^{f}	78.6^{i}	10.4^{k}	73.3^{i}	8.0^{j}
3.0	82.6^{h}	9.8^{kl}	72.0^{l}	9.01^{m}	70.6^{j}	6.9^{kl}
Kn	-	-	-	-	-	-
0.1	65.3^{op}	7.1^{n}	50.6r	6.1^{q}	42.6^{t}	4.8^{p}
0.5	68.0^{mn}	8.4^{m}	58.6o	6.8^{p}	49.3^{qr}	6.0^{mn}
1.0	64.0^{op}	6.5^{no}	62.6^{n}	5.7^{r}	56.0^{p}	5.2^{o}
2.0	58.6^{q}	5.2^{op}	57.3^{op}	4.3^{s}	50.6^{q}	5.0^{op}
3.0	53.3^{r}	4.1^{q}	54.6^{q}	3.8^{st}	45.3^{s}	3.4^{r}
BA + IBA	-	-	-	-	-	-
0.1 + 0.5	88.0^{f}	16.5^{de}	81.3^{g}	13.2^{fg}	80.0^{f}	12.5^{de}
0.5 + 0.5	96.0^{bc}	19.4^{c}	90.6^{cd}	18.1^{b}	85.3^{d}	15.4^{b}
1.0 + 0.5	100.0^{a}	29.8^{a}	97.3^{a}	22.0^{a}	94.6^{a}	19.7^{a}
2.0 + 0.5	93.3^{d}	21.6^{b}	92.0^{c}	17.5^{bc}	89.3^{bc}	14.3^{c}
3.0 + 0.5	89.3^{ef}	18.0^{d}	86.6^{ef}	15.6^{de}	80.0^{f}	11.8^{f}
BA + IAA	-	-	-	-	-	-
0.1 + 0.5	72.0^{l}	12.8^{hi}	74.6^{k}	12.0^{hi}	70.6^{j}	9.6^{hi}
0.5 + 0.5	78.6^{i}	14.1^{fg}	77.3^{ij}	13.6^{f}	76.0^{h}	10.5^{gh}
1.0 + 0.5	86.6^{g}	16.3^{e}	80.0^{gh}	15.8^{d}	78.6^{fg}	12.6^{de}
2.0 + 0.5	76.0^{j}	13.0^{h}	70.6^{lm}	11.0^{j}	69.3^{jk}	9.2^{hi}
3.0 + 0.5	70.6^{lm}	9.8^{kl}	68.0^{m}	7.9^{no}	65.3^{l}	6.1^{l}
BA + NAA	-	-	-	-	-	-
0.1 + 0.5	69.3^{m}	11.8^{j}	65.3^{n}	10.5^{jk}	61.3^{n}	9.6^{hi}
0.5 + 0.5	70.6^{lm}	13.4^{gh}	69.3^{m}	11.8^{i}	64.0^{lm}	10^{gh}
1.0 + 0.5	74.6^{k}	10.3^{k}	72.0^{l}	8.2^{n}	68.0^{jk}	7.0^{kl}
2.0 + 0.5	66.6^{o}	7.8^{mn}	68.0^{m}	6.5^{pq}	60.0^{no}	5.8^{mn}
3.0 + 0.5	64.0^{op}	5.6^{o}	62.6^{n}	6.0^{qr}	56.0^{p}	5.1^{op}

Values are a mean of 25 replicates per treatment and repeated thrice. Values with the same superscript are not significantly different at a 5% probability level, according to DMRT.

Shoot Elongation

The regenerated microshoots transferred to MS medium consisting of BA (1.0 mg/L) and different concentrations of GA_3 showed shoot elongation in varying responses. Of the treatments, 0.5 mg/L GA_3 responded to the highest shoot elongation of 7.2 cm from shoots obtained from internode callus within 20 days of culture (Table **7**) and Fig. (**2j**). Further increase in GA_3 and the shoot obtained from the other two explants showed a decrease in shoot length.

Table 7. Shoot elongation response of multiple shoots raised from callus induced from internode, leaf and petiole explants of *O. corniculata* grown on MS medium supplemented with BA (1.0 mg/L) and GA_3 after 20 days.

GA_3 (mg/L)	Internode		Leaf		Petiole	
	Freq. of elongated shoots (%)	Shoot length (cm)	Freq. of elongated shoots (%)	Shoot length (cm)	Freq. of elongated shoots (%)	Shoot length (cm)
0.1	86.6^{ab}	6.9^{b}	66.6^{bc}	6.1^{ab}	64.0^{ab}	5.0^{c}
0.5	92.0^{a}	7.2^{a}	73.3^{a}	6.4^{a}	65.3^{a}	5.7^{a}
1.0	82.6^{b}	6.4^{c}	68.0^{b}	5.9^{b}	57.3^{c}	5.4^{ab}
1.5	76.0^{c}	6.0^{d}	65.3^{c}	5.3^{c}	54.6^{cd}	4.7^{cd}
2.0	65.3^{d}	5.8^{de}	57.3^{d}	4.8^{d}	46.6^{e}	4.0^{e}

Values are a mean of 25 replicates per treatment and repeated thrice. Values with the same superscript are not significantly different at a 5% probability level, according to DMRT.

Rooting

The elongated shoots cultured on the rooting medium consisting of IAA or IBA with Kn produced better rooting than the individual treatments. MS basal medium devoid of PGRs failed to root on shoots. IBA (3.0 mg/L) responded better than IAA both at individual and combination treatments. The highest of 97.3%, 89.3% and 81.3% rooting was recorded on IBA 3.0 mg/L with Kn 0.5 mg/L, within 25 days of culture from internode, leaf, and petiole explants, respectively (Table **8**) and Fig. (**2k**). The highest root number and root length observed on the treatment were between 7.2-10.7 and 6.8-7.4 cm.

Hardening and Acclimatization

All the rooted plantlets from micropropagation and indirect organogenesis experiments transferred to the potting mixtures grew well under culture room conditions for up to one month. Then they were successfully transferred to the greenhouse condition with the survival rate of 95 and 90%, from the respective experiments. All the plants established in the greenhouse appeared morphologically uniform with normal leaf form, shape and growth pattern (Figs. **1k & 2l**).

Table 8. Rooting response of multiple shoots raised from callus induced from internode, leaf and petiole explants of *O. corniculata* grown on MS medium supplemented with auxins and cytokinins after 4 weeks.

PGRs (mg/L)	Internode			Leaf			Petiole		
	Rooting response (%)	Root number	Root length (cm)	Rooting response (%)	Root number	Root length (cm)	Rooting response (%)	Root number	Root length (cm)
IAA	-	-	-	-	-	-	-	-	-
0.1	46.6^{l}	3.9^{l}	4.0^{k}	41.3^{kl}	3.0^{i}	3.5^{n}	36.0^{m}	3.1^{k}	2.0^{n}
0.5	49.3^{l}	4.5^{jk}	4.2^{j}	44.0^{k}	3.5^{h}	3.8^{lm}	38.6^{l}	3.4^{j}	2.4^{m}
1.0	52.0^{kl}	4.8^{j}	4.9^{hi}	48.0^{j}	3.9^{fg}	3.9^{lm}	42.6^{k}	3.8^{hi}	2.7^{kl}
2.0	64.0^{i}	5.2^{j}	5.3^{g}	53.3^{i}	4.2^{f}	4.3^{k}	48.0^{j}	4.0^{hi}	2.9^{k}
3.0	66.6^{hi}	5.9^{h}	5.8^{ef}	60.0^{gh}	4.4^{f}	4.5^{j}	52.0^{ij}	4.1^{hi}	3.0^{jk}
IBA	-	-	-	-	-	-	-	-	-
0.1	54.6^{k}	6.3^{fg}	4.5^{j}	50.6^{ij}	4.0^{f}	4.3^{k}	45.3^{k}	3.8^{hi}	3.9^{gh}
0.5	61.3^{ij}	6.9^{e}	5.5^{g}	57.3^{h}	4.7^{e}	4.9^{i}	50.0^{j}	4.2^{h}	4.0^{g}
1.0	64.0^{i}	7.5^{e}	6.0^{ef}	60.0^{gh}	5.1^{de}	5.2^{g}	53.3^{i}	4.7^{ef}	4.0^{g}
2.0	68.0^{h}	7.8^{de}	6.1^{ef}	62.6^{g}	5.2^{de}	5.7^{e}	60.0^{fg}	5.0^{e}	4.6^{e}
3.0	73.3^{fg}	8.0^{d}	6.4^{cd}	69.3^{e}	5.4^{d}	6.2^{c}	64.0^{de}	5.2^{d}	5.3^{c}
IAA + Kn	-	-	-	-	-	-	-	-	-
3.0 + 0.1	76.0^{f}	5.6^{hi}	4.8^{hi}	60.0^{gh}	3.9^{fg}	4.0^{l}	56.0^{g}	3.9^{hi}	3.2^{j}
3.0 + 0.5	90.0^{c}	7.2^{e}	6.2^{e}	76.0^{cd}	5.8^{c}	5.5^{f}	72.0^{c}	6.0^{bc}	5.1^{cd}
3.0 + 1.0	86.6^{d}	6.4^{f}	5.7^{ef}	69.3^{e}	4.9^{e}	5.1^{gh}	65.3^{de}	4.5^{g}	4.4^{ef}
3.0 + 2.0	82.6^{e}	5.9^{g}	5.0^{h}	66.6^{ef}	4.4^{f}	4.5^{j}	61.3^{f}	4.0^{hi}	3.6^{i}
IBA + Kn	-	-	-	-	-	-	-	-	-
3.0 + 0.1	85.3^{de}	8.5^{c}	5.9$_{ef}$	73.3^{d}	4.6^{ef}	5.7^{e}	66.6^{d}	4.8^{e}	4.9^{cd}
3.0 + 0.5	97.3^{a}	10.7^{a}	7.4^{a}	89.3^{a}	7.8^{a}	7.0^{a}	81.3^{a}	7.2^{a}	6.8^{a}
3.0 + 1.0	94.6^{ab}	9.8^{b}	7.1^{b}	82.6^{b}	6.5^{b}	6.8^{b}	78.6^{b}	6.1^{b}	6.1^{b}
3.0 + 2.0	88.0^{cd}	8.9^{b}	6.5^{c}	78.6^{c}	5.3^{de}	6.0^{d}	72.0^{c}	5.0^{e}	5.7^{c}

Values are a mean of 25 replicates per treatment and repeated thrice. Values with the same superscript are not significantly different at a 5% probability level, according to DMRT.

DISCUSSION

The present study has demonstrated successful *in vitro* shoot proliferation and complete plant development of *O. corniculata* from node and shoot tip explants. Of the two explants used, the nodal explant was found to be better for shoot proliferation and multiplication compared to the shoot tip explant. Shoot bud induction and multiplication were highly influenced by the type of plant growth regulators and their concentration and combination.

Cytokinins are useful in stimulating cell division and releasing lateral bud dormancy in plant tissue culture. The most commonly used cytokinins in *Oxalis* species are BA and Kn. Cell division is also regulated by the synergistic action of auxins and cytokinins, each of which can influence the different phases of the cell cycle. Auxins affect DNA replication, while cytokinins control mitosis and cytokinesis. Thus, the balance between auxin and cytokinin levels can effectively control the production of plantlets in plant tissue culture [9]. In this study, the differential response of the regeneration potential seemed to be attributed to the physiological differences and genetic makeup of cells among the explants.

Shoot bud break and multiple shoots induction from both the explants were highly influenced by the function of cytokinin activity. In the present study, multiple shoot bud induction from the different explants varied with cytokinin type and concentrations.

The potential for shoot multiplication in *O. corniculata* appears to be strong in the presence of BA alone in the culture medium. The stimulatory effect of the singular supplement of BA on bud spurt and multiple shoot formation is similar to that reported in other species, such as *Biophytum sensitivum* and *O. triangularis* [10, 11]. In the previous study, Prasuna and Srinivas [12] reported that the shoot-forming capacity of the nodes was greatly influenced by the BA (93%) at 1.0 mg/L. On the whole BA was the most effective growth regulator in Oxalidaceae members, indicating cytokinin specificity for multiple shoot induction. In our study, BA at higher concentrations (2.0–3.0 mg/L) produced the best number of shoots compared to the previous literature in Oxalidaceae.

MS medium containing BA was more effective than Kn for inducing proliferation of axillary buds, as in the previous report [12]. Thus, the BA (3.0 mg/L) with Kn (1.0 mg/L) combination was found efficient for maximum responses of shoot proliferation and multiplication. The combined effect of BA and Kn on shoot development is already documented in *O. hedysaroides* [13].

The shoot multiplication potential in *O. corniculata* appeared to be very strong in the presence of cytokinin and auxin in the medium. Previous literature indicated that a low concentration of NAA assists in modifying shoot induction response in combination with cytokinins as observed in *B. sensitivum* [10]. Earlier, NAA was found to produce better shoots with cytokinins than IBA in *O. corniculata* and *O. tuberosa* [12, 14]. However, comparatively better shoot production was recorded on NAA (0.5 mg/L) with BA and Kn combination in terms of frequency of shoot proliferation (93-100%) and shoot number (28-35 shoots) from both the explants. In contrast, NAA alone was recorded as optimum for shoot proliferation and multiplication in *O. bowiei* [15]. The results indicated a strong influence of cytokinin either alone or with combinations in activating the axillary meristems to develop multiple shoot buds, while further growth and development of shoots are highly influenced by the synergistic auxin.

Shoot elongation response was best on BA with GA_3 treatments. Earlier, liquid MS medium with GA_3 alone treatment was found satisfactory to produce multiple shoots and rooting or shoot length in *O. tuberosa* and *O. erosa* [14, 16]. In our study, BA (3.0 mg/L) in combination with GA_3 produced healthy, green and long shoots in *O. corniculata*. Comparatively similar or less shoot elongation was previously recorded from the same species and *O. hedysaroides* on Kn with NAA or BA combination [12, 13].

In the present study, root induction was recorded on all three auxins (IAA, IBA and NAA) but with differential response. Nevertheless, the quality as well as quantity of roots varied with the concentration and type of auxins. Of all, IBA was found to be more potent than others. A previous report on *B. sensitivum* has recorded that IBA from 0.2-1.0 mg/L was most effective in producing multiple roots [10]. An earlier report on *O. erosa* recorded IBA as an effective auxin for rooting [16]. Similarly high number of roots was recorded in higher concentrations of both IBA and NAA in the same species [12] and in *B. sensitivum* [10].

Depending upon the concentration of auxin in the medium, there was simultaneous callusing and rooting at the thickened root base. Since the callus formed was superficial and highly friable, it could be easily removed from shoots before being transferred to the field.

The *in vitro* organogenesis governed by the balance of auxin and cytokinin in the medium cannot be demonstrated universally due to the explant sensitivity or the level of endogenous growth regulators [17]. Results obtained from this study revealed variations in response of callus induction, shoot bud regeneration, shoot elongation and root induction owing to differences in PGRs and explants in *O.*

corniculata. Of the four auxins, IBA (3.0 mg/L) performed better in producing green compact callus than 2,4-D, IAA and NAA. In *Oxalis* species, NAA has been most frequently reported for callus induction combined with either Kn or BA [18]. IBA at low concentration in combination with Kn and GA_3 was recorded only from *O. erosa* [16]. Their results showed similar dark green, compact and nodular callus as observed in the current study.

Of the explants, the internode explant showed the highest response (100%) for green, compact nodular callus on IBA treatment, followed by leaf and petiole. Similar callus nature was also observed on IAA treatments from all the explants, while 2,4-D and lower concentrations of NAA produced white and yellow friable slow growing callus. On the other hand, the red colour pigmented callus was observed on MS medium containing 2,4-D (3.0 mg/L). Interestingly, BA, 2,4-D or IAA, either singly or in combinations with other auxins or cytokinins, was recorded as suitable for callus induction and proliferation in other genera of Oxalidaceae [19].

Morphogenesis is the key to the genetic manipulation of plant cells in tissue culture, and is majorly influenced by the explant type, media formulation and growth regulators. Usually, a combination of a cytokinin with an auxin is required for the shoot organogenesis [20]. The highest shoot differentiation response was observed on BA with IBA treatments within a short period of culture in *O. corniculata*. The superiority of BA/BAP among the cytokinins on shoot bud induction has been attributed to the abilities of plant tissues to metabolize the natural growth regulators more efficiently than synthetic growth regulators. A similar response of cytokinin with auxin or GA_3 has been found suitable for shoot regeneration from callus in Oxalidaceae members, such as *O. erosa* [16], *O. glaucifolia*, *O. rhombeo-ovata* [21, 22], and *B. sensitivum* [19].

GA_3 has less explored for shoot elongation in Oxalidaceae. Very few studies have used GA_3 either singly [16] or in combination with a cytokinin for shoot differentiation from callus [21, 22]. GA_3, in addition to BA (1.0 mg/L), was observed effective for shoot elongation for *O. corniculata* in the present study, which is in accordance with the previous record on *O. erosa* [16].

The highest rooting response was observed on a medium supplemented with IBA (3.0 mg/L) with Kn (0.5 mg/L). Previously, a higher concentration of NAA with a cytokinin (BA or Kn) or GA_3 treatment has been satisfactory to produce callus or shoot with root formation in *Oxalis* species [18]. In the present study, auxins (IAA and IBA) at individual treatments and in combination were found to induce roots, preferably at higher frequencies on IBA with Kn, followed by IAA combinations.

CONCLUSION

The present study established two reproducible and efficient regeneration systems for *O. corniculata* through micropropagation and indirect organogenesis using various explants. Complete plant development following successful hardening and acclimatization with over 90% survival rate under greenhouse conditions was achieved. Both systems would serve as a model for improved biomass production and further exploration of pharmacological properties in the same and allied taxa.

ACKNOWLEDGEMENTS

The authors are thankful to the Head, Department of Botany, Bharathidasan University, Tiruchirappalli, Tamil Nadu, for providing the facilities.

REFERENCES

[1] World flora online (WFO). Oxalidaceae R.Br. Available from: http://www.worldfloraonline.org/taxon/wfo-7000000434

[2] Simpson MG. Diversity and classification of flowering plants: Eudicots. Plant systematics. 2nd. Academic press 2010; pp. 275-448.
[http://dx.doi.org/10.1016/B978-0-12-374380-0.50008-7]

[3] Oberlander KC, Dreyer LL, Bellstedt DU, Reeves G. Systematic relationships in southern African *Oxalis* L. (Oxalidaceae): Congruence between palynological and plastid *trnLF* evidence. Taxon 2004; 53(4): 977-85.
[http://dx.doi.org/10.2307/4135564]

[4] Khare CP. Indian Medicinal Plants—An illustrated dictionary First Indian reprint, Springer (India). New Delhi: Pvt Ltd 2007; pp. 717-8.

[5] Singh H, Maheshwari JK. Traditional remedies for snakebite and scorpion sting among the Bhoxas of Nainital district. Aryavaidyan 1992; 6(2): 120-3.

[6] Sarkar T, Ghosh P, Poddar S, Choudhury S, Sarkar A, Chatterjee S. *Oxalis corniculata* Linn. (Oxalidaceae): A brief review. J Pharmacogn Phytochem 2020; 9(4): 651-5.
[http://dx.doi.org/10.22271/phyto.2020.v9.i4i.11777]

[7] Murashige T, Skoog F. A revised medium for rapid growth and bioassay with tobacco tissue culture. Physiol Plant 1962; 15(3): 473-97.
[http://dx.doi.org/10.1111/j.1399-3054.1962.tb08052.x]

[8] Gomez KA, Gomez AA. Statistical procedure for agricultural research with emphasis of rice. Los Bans, Philippines: International Rice Research Institute 1976.

[9] Veselý J, Havlicek L, Strnad M, *et al.* Inhibition of cyclin-dependent kinases by purine analogues. Eur J Biochem 1994; 224(2): 771-86.
[http://dx.doi.org/10.1111/j.1432-1033.1994.00771.x] [PMID: 7925396]

[10] Chandra KS, Kokkanti M. *In vitro* propagation of *Biophytum sensitivum* Linn. using mature nodal explants. J Indian Bot Soc 2018; 97(1&2): 36-42.

[11] Rittirat S, Klaocheed S, Thammasiri K. Large scale *In vitro* micropropagation of an ornamental plant, *Oxalis triangularis* A.St.-Hil, for commercial application. In: IX International Scientific and Practical Conference on Biotechnology as an Instrument for Plant Biodiversity Conservation (physiological, biochemical, embryological, genetic and legal aspects), ISHS Acta Horticulturae pp.1339,year.2022.

[12] Prasuna SVN, Srinivas B. *In vitro* regeneration studies of *Oxalis corniculata* from nodal explants. Int J Recent Sci Res 2005; 6(9): 6216-20.

[13] Maene L, Debergh P. *In vitro* propagation and culture of *Oxalis hedysaroides* H.B.K. cv. Fire Tree. Med Fac Landbouw Rijskuniv Gent 1981; 46(4): 1201-3.

[14] Ochatt SJ, Ciai AA, Caso OH. Tissue culture techniques applied to American crops: micropropagation of Oca (*Oxalis tuberosa* Mol.), an Andean tuber-bearing species. Turrialba 1986; 36: 187-90.

[15] Wang JG, Wu D, Fu HJ, Che DD, Yang CP. Tissue culture and rapid propagation of *Oxalis bowiei.* Zhiwu Xuebao 2010; 45: 233-5.

[16] Ochatt SJ, de Azkue D. Callus proliferation and plant recovery with *Oxalis erosa* Knuth shoot tip cultures. J Plant Physiol 1984; 117(2): 143-6. [http://dx.doi.org/10.1016/S0176-1617(84)80027-9] [PMID: 23195609]

[17] Skoog F, Miller CO. Chemical regulation of growth and organ formation in plant tissues cultured in vitro. Symp Soc Exp Biol 1957; 11: 118-30. [PMID: 13486467]

[18] Van Staden J. *Oxalis* Species: *In vitro* culture, micropropagation, and the formation of anthocyanins. In: Bajaj YPS, Ed. Medicinal and aromatic plants X Biotechnology in Agriculture and Forestry. Berlin, Heidelberg: Springer 1998; 41. [http://dx.doi.org/10.1007/978-3-642-58833-4_16]

[19] Shivanna MB, Vasanthakumari MM, Mangala MC. Regeneration of *Biophytum sensitivum* (Linn.) DC. through organogenesis and somatic embryogenesis. Indian J Biotechnol 2009; 8: 127-31.

[20] Evans DA, Sharp WR, Flick CE. Growth and behaviour of cell cultures: Embryogenesis and organogenesis. In: Thorpe TA, Ed. Plant Tissue Cult Lett. Berlin: Springer Verlag 1981; pp. 345-62. [http://dx.doi.org/10.1016/B978-0-12-690680-6.50008-5]

[21] Ochatt SJ, Stampacchio ML, Escandon AS, Martinez AJ. Plant regeneration from callus of two shrubby *Oxalis* species from South America. Phyton 1988; 48: 21-6.

[22] Ochatt SJ, Escandon AS, Martinez AJ. Isolation, culture, and plant regeneration of protoplasts of two shrubby *Oxalis* species from South America. J Exp Bot 1989; 40(4): 493-6. [http://dx.doi.org/10.1093/jxb/40.4.493]

CHAPTER 21

An Effective Micropropagation Strategy for *Pseudarthria viscida* (L.) Wight & Arn.

G. Sangeetha[1] and **T. S. Swapna**[1,*]

[1] *Department of Botany, TKM College of Arts and Science, Kollam, Kerala, India*

Abstract: Habitat destruction and over-harvesting have resulted in the gradual disappearance of many medicinally important plants from their natural habitat. At present, their number is highly reduced in the wild. To conserve the genetic stocks of such plants, *in vitro* propagation can be utilized successfully. One such medicinally important plant that needs to be conserved is *Pseudarthria viscida* (L.) Wight & Arn. It is a perennial viscid pubescent semi-erect, diffuse undershrub belonging to the family Fabaceae. It is an essential component of many famous Ayurvedic formulations like Dashamoola, Mahanarayana taila, and Dhantara taila. The root is the most important part of the plant with high medicinal value. Major chemical compounds reported to be present in the roots are 1,5 dicaffeoyl quinic acid, oleic acid, tetradecanoic acid, rutin, quercetin, gallic acid, ferulic acid, and caffeic acid. The present study focused on *in vitro* regeneration and mass propagation of *P. viscida*. Fresh young leaves, nodes, and internodal segments were used as explants. Murashige and Skoog medium (MS medium), Gamborg's (B_5) medium, and White's mediums were selected for *in vitro* regeneration and mass propagation. Among the various media used, the MS medium gave a successful result in *in vitro* culture by showing a response within four weeks, and the percentage of response was also high compared to B_5 and White's medium. The leafy explant was found to be more suitable for profuse callus induction, somatic embryogenesis, and indirect organogenesis than that of internodal and nodal explants, whereas nodal explant was best for direct organogenesis in *P. viscida*. Of the different combinations tried, NAA (Naphthalene acetic acid) + BAP (6-Benzyl aminopurine) combinations were best for callus induction, somatic embryogenesis and indirect organogenesis. 2.5 mg/L BAP was best for shoot induction from nodal explants, whereas 2.5 mg/L NAA was best for root induction from *in vitro* regenerated micro shoots as explants. Well-developed plantlets were transferred to greenhouse and later to natural conditions. This study thus reports an efficient protocol for plant regeneration, and this could be vital for the multiplication and field transfer of this ethnomedicinal plant. Based on the ethnomedicinal potential, there is an urgent need for organized cultivation of this vulnerable plant for its conservation and sustainable utilization.

* **Corresponding author T. S. Swapna:** Department of Botany, TKM College of Arts and Science, Kollam, Kerala, India; Tel: 9745824670; E-mail: swapnats@yahoo.com

T. Pullaiah (Ed.)

Keywords: Conservation, Dashamoola, Dhantara taila, Ethnomedicinal, Hormonal Combinations, *In vitro* Regeneration, Mass Propagation, Mahanarayana Taila, *Pseudarthria viscida*, Sustainable Utilization.

INTRODUCTION

Medicinal plants have been emerging as a part of the modern life of man and with the greatest demand due to their nutritional, pharmaceutical, cosmetic and medicinal application without much negative impact. Hundreds of medicinal plants are at high risk of extinction due to over-exploitation and habitat destruction. This threatens the invention of future heals for diseases. For primary health care, about five billion people depend on traditional phytoremedies [1]. About 50% of prescription drugs are obtained from phytoconstituents, which were first identified in plants. Commercial and scientific attention to medicinal plants leads to high pressure on them, and continuous harvesting leads to threat and is facing a high risk of extinction. Scientists say that a minimum of one potentially important drug is lost by Earth every two years [2]. Each species lost to extinction also represents the loss of possible vitamin and protein-rich foods and stable crops. So, more attention is needed to their conservation and sustainable utilization. Conservation of medicinal plants provides sustainable livelihoods as well as the vital protection of biodiversity.

Unchecked commercialization and habitat loss of wild medicinal plants are threatening the beauty, diversity, and natural heritage, as well as the future of vital resources of our planet. This loss of diversity may also take with its important future cures for diseases as potential resources to combat poverty, hunger, and social and economic insecurity. Besides their high medicinal value, these plants have not been cultivated for commercial purposes due to low seed viability, germination rate, and high rate of mortality of seedlings in the early stages. In many threatened plants, *in vitro* propagation is successfully utilized for the conservation of the genetic stocks. *Pseudarthria viscida* is one of such medicinally important plants that need conservation [3].

Pseudarthria viscida (L.) Wight & Arn., which is commonly known as Salaparni in Sanskrit, Moovila in Malayalam, is a perennial viscid pubescent, semi-erect, diffuse undershrub belonging to the family Fabaceae and sub-family Faboideae. The plant is distributed throughout India and is especially found in river basins and in hills up to above 900m. This vulnerable plant has several medicinal uses in the indigenous system of medicine and is an essential component of many famous Ayurvedic formulations like Dashamoolarishta, Mahanarayana taila, Agastya haritaki rasayana, Brahma rasayana, Dhanuanthara ghrita, Anuthaila, Sudarshana churna and Dhantara taila [4]. The most important part is roots, which are used as digestive, astringent, anthelmintic, anti-inflammatory, thermogenic, cardiotonic, aphrodisiac, febrifuge, nervine, and rejuvenating tonic [4, 5]. It is an important remedy for blood disorders and heart diseases, as it effectively stops bleeding and alleviates edema. Major chemical compounds reported to be present in the roots are 1,5 dicaffeoyl quinic acid, oleic acid, tetradecanoic acid, rutin, quercetin, gallic acid, ferulic acid, and caffeic acid. The roots and leaves contain proteins, tannins, and flavonoids and also showed significant inhibitory activity against some fungal pathogens causing major diseases in crop plants and stored food grains [6]. A polyphenolic compound was reported from the roots of *P. viscida,* which was suggested as the reason for the antioxidant activity of this plant. The plant is used in tridoshas, cough, asthma, fever, dysentery, cardiac ailments, and rheumatoid arthritis and aids in the fast healing of fractured bone. The plant possesses antifungal, antioxidant, antitumor, anti-hypertensive, antidiarrhoeal activity, neuroprotective and anti-inflammatory activity [6, 7]. Due to its high medicinal value, the annual consumption of the root by the Ayurvedic medicine industry in Kerala is 140 tons. Home gardens of *P. viscida* were maintained by the Kani tribe of Kanyakumari Wildlife Sanctuary, Southern Western Ghats [8]. The main aim of this work is to identify effective micropropagation strategies for *P. viscida.*

IN VITRO REGENERATION AND MASS PROPAGATION OF *PSEUDARTHRIA VISCIDA*

The commercial exploitation of medicinal plants has resulted in the reduction of the population of many species in their natural habitat. Therefore, the cultivation of these plants is urgently needed to ensure their availability to the industry and the people associated with the traditional system of medicine. Consequently, it is inevitable to propagate these plants in suitable agroclimatic conditions. *In vitro* propagation is the best method for the conservation of threatening medicinally and pharmaceutically important plants. In the present study, various explants like leaf, internodal and nodal segments isolated from a young, healthy plant were selected for *in vitro* propagation of *P. viscida.* One single medium cannot be suggested for all types of plant tissue and organs. Therefore, the establishment of a new system must fulfil all the specific requirements for the proper growth of a particular tissue because nutritional requirements are essential for optimal *in vitro* growth [9]. Important media used mainly for tissue culture studies were MS medium [9], Gamborg's (B_5) medium [10] and White's mediums [11] *etc.* MS medium was originally formulated to induce organogenesis and regeneration of plants in cultured tissues, whereas, at present, it is widely used for types of culture systems. B_5 medium was originally designed for cell suspension and callus cultures. At present, with certain modifications, this medium is used for protoplast culture and other cell cultures. White's medium was one of the earliest plant tissue culture media developed for root culture. In *P. viscida*, MS medium, Gamborg's (B_5) medium and White's medium were selected for *in vitro* studies.

Different explants showed different responses in MS, B_5, and White's medium with a different hormonal combination. Phytohormones used in medium preparation were auxins like NAA (Naphthalene acetic acid), IBA (Indole butyric acid), IAA (Indole 3- acetic acid) and 2, 4-D (2,4-Dichlorophenoxy acetic acid) whereas cytokinins like BAP (6-Benzyl amino purine) and Kn (Kinetin) in either singly or in combinations. All chemicals were purchased from HIMEDIA, India. Growth characteristics and other features of the callus and organogenesis were noted. Subculturing was done at four-week intervals. All experiments were carried out thrice. The well-developed rooted shoots were removed from culture tubes and washed thoroughly in sterile distilled water to remove remnants of agar. Each plantlet was transferred to a small cup containing sterilized sand, garden soil, and farm yard manures in a 1:1:1 ratio. The plantlets were transferred to a shade net house for four weeks and eventually to the field.

Among the various media used, MS medium gave the best result by showing response within four weeks, and the percentage of response was also higher than B_5 and White's medium. Somatic embryogenesis and indirect organogenesis (rooting and shoot induction from callus) were also noticed in the MS medium. Profused callus induction, somatic embryogenesis, direct organogenesis from nodal explants, and indirect organogenesis from callus were noticed in different hormonal combinations (auxins and cytokinins) in MS medium, whereas in B_5 and White's medium showed comparatively less callus induction and organogenesis.

Callus Induction

Different hormonal combinations like NAA+BAP, NAA+Kn, IBA+BAP, IBA+Kn, IAA+BAP, IAA+Kn, 2,4-D+BAP, 2,4-D+ Kn were tested for callus induction in *P.viscida* at the range between 0.5 to 2.5 mg/L. The combinations with higher auxins (2.5 mg/L) and cytokinin (2.0 or 2.5 mg/L) were able to induce callus profusely. Callus initiation was started at the first week of inoculation and continued its growth for four weeks in MS medium and six weeks in B_5 and White's medium. After that growth rate of the callus was retarded, and the medium became dried and depleted.

MS Medium

In MS medium, maximum callus induction was noticed in NAA+BAP combinations (Fig. **1**). 98.22±0.45% callus induction was observed in 2.5 mg/L NAA + 1.5 mg/L BAP combination from the leafy explant and 96.35±1.12%

(internodal segment) in 2.5 mg/L NAA and 2.0 mg/L BAP combination (P<0.001) (Table **1**). In these combinations, light brown friable callus was found after four weeks of inoculation. In the case of NAA+ Kn combinations, maximum callus induction was noticed in 2.5 mg/L NAA+ 2.0 mg/L Kn from both leafy (87.16±1.23%) and internodal segments (84.23±0.58%) (P<0.001). Greenish white friable callus was noticed in these combinations (Table **2**). Like NAA+BAP combinations, 2, 4-D + Kn combinations also showed the best callusing in *P. viscida* (Table **3**). Maximum callus induction like 96.57± 0.98% and 95.20±0.91% were noticed on 2.5 mg/L 2,4-D + 1.5 mg/L Kn from both leaf (white compact callus) and internodal segments (white friable callus) (P<0.001) whereas in 2,4-D+BAP combinations, maximum callus induction (85.23±0.98%) was noticed from leafy explants (brownish callus) in 2.5 mg/L 2,4-D+1.5 mg/L BAP combination and 85.14±0.10% were noticed in 2.5 mg/L 2,4-D + 2.0 mg/L BAP from internodal segments (brownish callus) after four weeks of inoculation (29 days) (Table **4**).

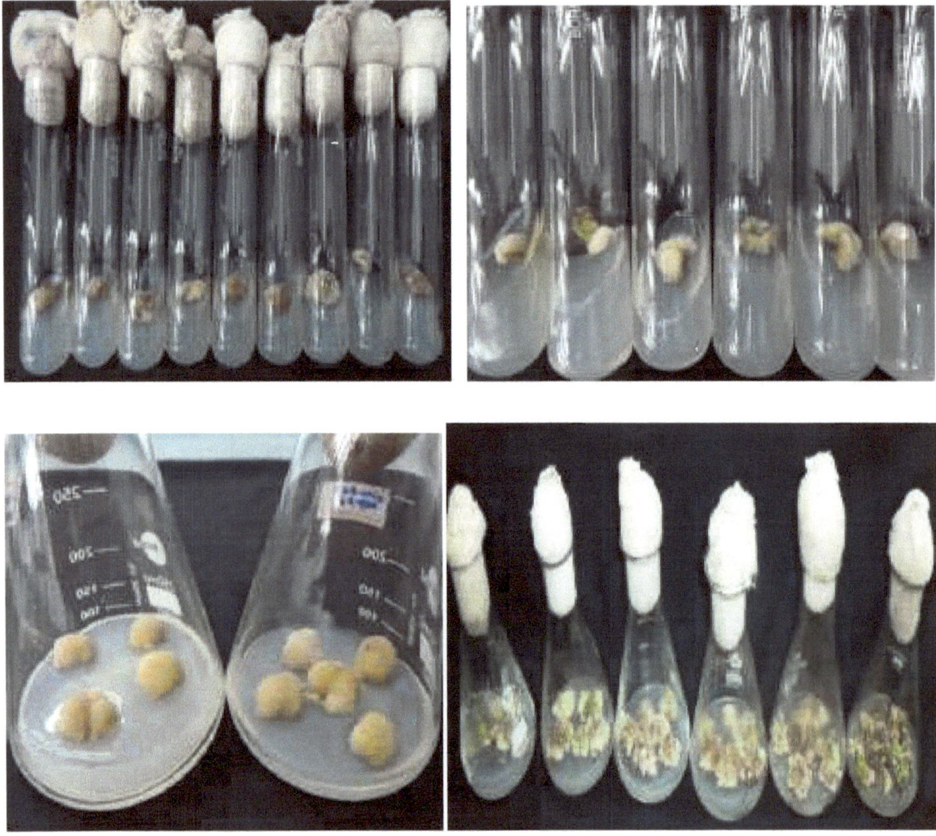

Fig. (1). Callus culturing of *Pseudarthria viscida* in MS medium.

Table 1. Callus induction of *Pseudarthria viscida* in MS medium with different combinations of NAA and BAP. Culture period: 4 weeks, Subculturing: after 4 weeks.

Hormonal Concentrations (mg/L)		Percentage of Callus Response		Nature of Callus	
NAA	BAP	L	IN	L	IN
0.5	0.5	43.20±0.34	40.33±0.12	WC	WC
1.0	0.5	58.82±2.11***	58.33±0.25***	GWC	LBF
1.5	0.5	66.00±0.92***	73.23±0.33***	LBC	LBF
2.0	0.5	79.21±0.02***	81.34±0.10***	LBC	BF
2.5	0.5	93.02±0.11***	94.23±0.26***	BF	BF
0.5	1.0	44.22±0.21ns	42.23±0.45ns	GWC	GWC
1.0	1.0	69.36±0.28***	59.22±0.15***	WC	LBF
1.5	1.0	73.54±-0.38***	65.14±0.27***	BF	BF
2.0	1.0	86.89±2.45***	83.62±0.24***	BF	F
2.5	1.0	94.00±0.09***	95.30±2.11***	LBF	F
0.5	1.5	48.13±1.17**	43.91±1.42*	WF	YWC
1.0	1.5	59.05±2.35***	59.82±2.45***	GWC	BF
1.5	1.5	76.34±0.01***	64.64±0.56***	LBF	F
2.0	1.5	90.82±0.25***	78.22±1.02***	BF	BF
2.5	**1.5**	**98.22±0.45*****	95.23±1.24***	**LBF**	BF
0.5	2.0	48.02±0.98**	42.02±1.14ns	LBF	F
1.0	2.0	62.17±0.21***	58.36±0.36***	WF	LBF
1.5	2.0	83.31±0.15***	78.63±0.26***	YWF	BF
2.0	2.0	90.67±0.05***	87.22±0.26***	GWF	BF
2.5	**2.0**	96.18±0.18***	**96.35±1.12*****	BF	**LBF**
0.5	2.5	45.13±0.12ns	44.93±2.56**	GWF	LYF
1.0	2.5	57.23±0.03***	59.23±0.58***	LBF	LBF
1.5	2.5	61.23±0.34***	64.13±1.39***	GWF	GWF
2.0	2.5	72.12±2.32***	79.56±0.78***	BF	BF
2.5	2.5	87.25±1.03***	94.23±2.11***	BF	BF

L-Leaf, IN- Inter Node, B- Brownish, C-Compact, F- Friable, G- Greenish, W-White, LB-Light Brown, LY-Light Yellow, Y- Yellowish; Each value represents the mean ± SD of triplicate measurements and superscript represent level of significance comparing to control value (0.5mg/L NAA+0.5mg/L BAP); * significant at $p<0.05$, ** significant at $p<0.01$, *** significant at $p<0.001$, ns-not significant (according to Tukey-Kramer Multiple Comparisons Test).

Table 2. Callus induction of *Pseudarthria viscida* in MS medium with different combinations of NAA and Kn. Culture period: 4 weeks, Subculturing: after 4 weeks.

Hormonal Concentrations (mg/L)		Percentage of Callus Response		Nature of Callus	
NAA	Kn	L	IN	L	IN
0.5	0.5	36.25±0.46	29.13±0.45	WC	F
1.0	0.5	59.39±0.78***	41.12±0.98***	GWC	GWC
1.5	0.5	65.31±1.25***	63.42±2.11***	BF	LBF
2.0	0.5	78.34±3.22***	76.11±1.23***	LBF	LBF
2.5	0.5	84.21±1.20***	80.11±0.48***	LBF	LY
0.5	1.0	37.03±2.14ns	35.13±0.19**	LY	BF
1.0	1.0	42.24±2.67**	48.45±1.67***	WF	LBF
1.5	1.0	56.23±0.89***	53.14±0.96***	BF	LY
2.0	1.0	69.34±0.23***	68.36±1.46***	BF	F
2.5	1.0	78.32±0.41***	75.23±0.87***	BF	BF
0.5	1.5	40.24±0.97*	35.05±2.11**	LY	BF
1.0	1.5	65.67±2.10***	58.45±1.14***	LBF	LBF
1.5	1.5	77.36±1.21***	69.71±0.64***	WC	LBF
2.0	1.5	79.25±2.34***	77.43±0.94***	LBF	BF
2.5	1.5	81.24±2.74***	83.12±1.23***	LBF	F
0.5	2.0	39.94±0.43*	31.18±2.31ns	GWC	LY
1.0	2.0	47.83±2.57***	49.12±2.16***	BF	WC
1.5	2.0	56.13±0.67***	66.54±0.12***	BF	BF
2.0	2.0	71.35±0.25***	70.29±0.84***	LBC	F
2.5	2.0	**87.16±1.23*****	**84.23±0.58*****	GWF	GWF
0.5	2.5	37.37±0.68ns	30.23±0.98ns	F	WC
1.0	2.5	48.14±0.34***	42.17±1.23***	LY	GWC
1.5	2.5	53..58±2.18***	56.23±0.57***	LBF	LBF
2.0	2.5	69.35±0.42***	64.36±3.10***	LBF	LBF
2.5	2.5	79.56±1.67***	78.34±0.24***	BF	LBF

L-Leaf, IN- Inter Node, B- Brownish, C-Compact, F- Friable, G- Greenish, W-White, LB-Light Brown, LY-Light Yellow, Y- Yellowish; Each value represents the mean ± SD of triplicate measurements and superscript represent level of significance comparing to control value (0.5mg/L NAA+0.5mg/L Kn); * significant at $p<0.05$, ** significant at $p<0.01$, *** significant at $p<0.001$, ns-not significant (according to Tukey-Kramer Multiple Comparisons Test).

Table 3. Callus induction of *Pseudarthria viscida* in MS medium with different combinations of 2,4-D and Kn. Culture period: 4 weeks, Subculturing: after 4 weeks.

Hormonal Concentrations (mg/L)		Percentage of Callus Response		Nature of Callus	
2,4-D	**Kn**	**L**	**IN**	**L**	**IN**
0.5	0.5	44.38±0.35	40.24±0.13	WF	WC
1.0	0.5	50.06±0.01**	53.89±1.26***	WF	LYF
1.5	0.5	73.34±0.98***	69.38±1.45***	LBF	LYF
2.0	0.5	85.13±0.34***	83.45±0.49***	WF	BF
2.5	0.5	91.23±1.24***	90.43±2.15***	WF	BF
0.5	1.0	47.24±2.36*	44.27±2.05**	WC	WC
1.0	1.0	50.25±3.37**	61.34±2.10***	LYF	WC
1.5	1.0	65.40±0.98***	73.98±1.79***	LBF	LYF
2.0	1.0	83.20±1.58***	81.39±0.98***	BF	BF
2.5	1.0	93.31±1.10***	93.25±2.18***	BF	WF
0.5	1.5	53.23±0.22***	46.29±1.17***	WC	WC
1.0	1.5	69.48±2.89***	69.43±2.43***	LYF	WC
1.5	1.5	75.89±3.46***	77.34±2.09***	F	LYF
2.0	1.5	87.27±2.89***	84.28±2.34***	WC	LYF
2.5	1.5	**96.57**±0.98***	**95.20±0.91*****	WC	WF
0.5	2.0	49.39±0.76**	42.23±0.19ns	LYF	LYF
1.0	2.0	58.15±2.18***	69.34±3.56***	LYF	WC
1.5	2.0	72.98±2.34***	76.23±1.34***	WF	LYC
2.0	2.0	88.00±1.23***	89.43±0.98***	LBF	LBF
2.5	2.0	94.26±0.23***	95.11±2.01***	WC	LBF
0.5	2.5	45.19±0.95ns	41.24±1.86ns	LYF	WC
1.0	2.5	59.54±2.79***	49.39±0.08***	WF	WC
1.5	2.5	64.38±0.74***	68.30±2.45***	BF	LYF
2.0	2.5	73.69±0.98***	81.23±2.59***	LBF	LBF
2.5	2.5	82.15±2.16***	90.15±1.56***	LBF	BF

L-Leaf, IN- Inter Node, B- Brownish, C-Compact, F- Friable, G- Greenish, W White, LB-Light Brown, LY-Light Yellow, Y- Yellowish; Each value represents the mean ± SD of triplicate measurements and superscript represent level of significance comparing to control value (0.5mg/L 2, 4-D + 0.5mg/L Kn); * significant at p<0.05, ** significant at p<0.01, *** significant at p<0.001, ns-not significant (according to Tukey-Kramer Multiple Comparisons Test).

Table 4. Callus induction of *Pseudarthria viscida* in MS medium with different combinations of 2,4-D and BAP. Culture period: 4 weeks, Subculturing: after 4 weeks.

Hormonal Concentrations (mg/L)		**Percentage of Callus Response**		**Nature of Callus**	
2,4-D	**BAP**	**L**	**IN**	**L**	**IN**
0.5	0.5	28.32±0.45	25.13±0.56	WF	WC
1.0	0.5	32. 23±1.12**	30.23±0.98**	WF	WF
1.5	0.5	67.23±3.56***	47.27±0.94***	LYF	WF
2.0	0.5	71.68±0.98***	59.34±2.13***	LBF	LBF
2.5	0.5	80.00±2.18***	75.23±0.95***	LBF	LBF
0.5	1.0	29.23±0.09ns	31.25±0.32**	WC	WF
1.0	1.0	33.33±0.14**	49.33±2.34***	WC	WF
1.5	1.0	56.34±1.63***	62.34±2.10***	WF	LBF
2.0	1.0	74.27±1.23***	75.23±1.86***	LYF	LYF
2.5	1.0	83.45±1.87***	81.11±2.11***	LYF	WC
0.5	1.5	30.20±0.11ns	30.65±0.23**	WF	WC
1.0	1.5	34.13±0.94**	43.20±0.23***	WF	WF
1.5	1.5	55.34±2.56***	50.32±2.15***	LYF	WF
2.0	1.5	73.45±3.12***	78.34±0.98***	BF	LYF
2.5	1.5	**85.23**±0.98***	83.65±2.64***	BF	LBF
0.5	2.0	29.23±0.93ns	27.12±1.02ns	LBF	LBF
1.0	2.0	36.67±0.85***	47.23±2.34***	WC	WC
1.5	2.0	57.23±0.34***	59.23±0.98***	WF	LYF
2.0	2.0	71.23±2.15***	70.00±2.67***	WF	BF
2.5	2.0	82.12±0.08***	**85.14**±0.10***	LYF	BF
0.5	2.5	30.34±2.89ns	29.23±0.34*	LBF	WF
1.0	2.5	47.24±0.86***	58.11±0.23***	LBF	WF
1.5	2.5	59.11±0.34***	63.00±1.25***	WF	WF
2.0	2.5	66.13±0.12***	75.04±0.45***	BF	LBF
2.5	2.5	74.34±2.12***	81.67±2.34***	BF	LBF

L-Leaf, IN- Inter Node, B- Brownish, C-Compact, F- Friable, G- Greenish, W-White, LB-Light Brown, LY-Light Yellow, Y- Yellowish; Each value represents the mean ± SD of triplicate measurements and superscript represent level of significance comparing to control value (0.5mg/L 2, 4-D + 0.5mg/L BAP); * significant at $p<0.05$, ** significant at $p<0.01$, *** significant at $p<0.001$, ns-not significant (according to Tukey-Kramer Multiple Comparisons Test).

In IBA+BAP combinations, 93.63±0.15% (leafy explants) and 91.59±0.08% (internodal explants) of optimum callus (greenish friable callus) were found to be noticed in 2.5mg/L IBA+ 1.5 mg/L BAP and 2.5mg/L IBA+2.0 mg/L BAP respectively whereas in IBA+ Kn combinations, 84.11±0.45% callus induction (brownish callus) was observed in 2.5 mg/L IBA+2.0 mg/L Kn combination from leafy explants and 83.21±0.01% in2.5 mg/L IBA+1.5mg/L Kn combination from internodal explant (Table **5-6**). IAA+ BAP combinations showed comparatively less callus response in *P. viscida*. Only 65.35±1.11% and 60.24±2.13% of callus inductions were noticed in IAA 2.5 and BAP 2.0 mg/L combinations in both leaf and internodal segments, respectively (Table **7**), whereas in IAA+ Kn combinations only 49.89±1.34% and 46.16±1.02% of callus was noticed (Table **8**). A whitish compact callus was obtained from both these IAA combinations.

Table 5. Callus induction of *Pseudarthria viscida* in MS medium with different combinations of IBA and BAP. Culture period: 4 weeks, Subculturing: after 4 weeks.

Hormonal Concentrations (mg/L)		Percentage of Callus Response		Nature of Callus	
IBA	**BAP**	**L**	**IN**	**L**	**IN**
0.5	0.5	18.31±1.40	12.47±0.94	WC	YC
1.0	0.5	22.56±1.11*	14.50±1.14ns	GWC	GWF
1.5	0.5	30.11±3.21***	27.65±4.61***	GWF	LYC
2.0	0.5	58.50±1.16***	43.00±4.22***	BF	F
2.5	0.5	74.22±0.81***	59.45±5.13***	BF	BC
0.5	1.0	21.02±2.54ns	15.02±0.99**	YC	GWF
1.0	1.0	40.00±1.56***	43.06±2.83***	LBF	BF
1.5	1.0	59.15±0.42***	54.04±3.21***	LBF	LYC
2.0	1.0	78.36±7.80***	71.35±7.10***	F	BF
2.5	0	85.21±4.50***	87.11±0.57***	BF	BF
0.5	1.5	23.43±1.20**	16.02±1.00**	GWC	WC
1.0	1.5	30.32±2.11***	26.11±2.03***	LBF	GWF
1.5	1.5	70.16±4.20***	55.0±1.90***	BF	BF
2.0	1.5	83.21±5.13***	79.81±2.01***	F	F
2.5	1.5	**93.63±0.15*****	91.17±1.22***	GF	GF
0.5	2.0	24.06±5.32**	15.18±3.55*	LBC	YC
1.0	2.0	42.03±2.56***	28.36±5.32***	BC	GWF
1.5	2.0	48.32±1.03***	39.28±4.75***	F	BC
2.0	2.0	89.25±2.12***	77.25±3.89***	F	BF
2.5	2.0	92.52±1.23***	**91.59±0.08*****	GF	GF

(Table 5) cont.....

Hormonal Concentrations (mg/L)		Percentage of Callus Response		Nature of Callus	
IBA	BAP	L	IN	L	IN
0.5	2.5	20.00±4.78ns	13.51±1.72ns	GWF	WC
1.0	2.5	37.05±6.21***	36.03±1.47***	LBC	GWF
1.5	2.5	59.12±4.56***	50.83±0.89***	F	F
2.0	2.5	72.03±3.56***	83.26±1.12***	BF	LBF
2.5	2.5	93.11±0.23***	90.28±4.78***	BF	BF

L-Leaf, IN- Inter Node, B- Brownish, C-Compact, F- Friable, G- Greenish, W-White, LB-Light Brown, LY-Light Yellow, Y- Yellowish; Each value represents the mean ± SD of triplicate measurements, and superscript represents level of significance comparing to control value (0.5mg/L IBA + 0.5mg/L BAP); * significant at $p<0.05$, ** significant at $p<0.01$, *** significant at $p<0.001$, ns-not significant (according to Tukey-Kramer Multiple Comparisons Test).

Table 6. Callus induction of *Pseudarthria viscida* in MS medium with different combinations of IBA and Kn. Culture period: 4 weeks, Subculturing: after 4 weeks.

Hormonal Concentrations (mg/L)		Percentage of Callus Response		Nature of Callus	
IBA	Kn	L	IN	L	IN
0.5	0.5	22.26±1.17	26.32±1.02	YC	WF
1.0	0.5	26.70±1.26**	31.37±0.54**	GWC	BC
1.5	0.5	45.29±0.28***	58.91±0.66***	F	F
2.0	0.5	60.11±2.11***	72.09±2.16***	BF	F
2.5	0.5	71.23±0.02***	80.11±1.45***	BF	BF
0.5	1.0	25.35±1.389*	27.34±0.62ns	WC	GWC
1.0	1.0	40.67±2.11***	45.17±0.48***	LBF	BF
1.5	1.0	51.25±0.02***	68.11±0.74***	BF	LBF
2.0	1.0	71.45±1.34***	70.34±3.86***	F	BF
2.5	0	80.23±0.94***	82.56±3.21***	BC	F
0.5	1.5	24.15±0.57ns	29.27±0.95*	WF	GWC
1.0	1.5	41.47±2.14***	58.25±0.02***	GWF	WC
1.5	1.5	61.28±0.28***	69.34±3.11***	LBF	BF
2.0	1.5	75.14±0.12***	73.10±2.46***	LBF	LBF
2.5	1.5	83.53±2.17***	**83.21±0.01****	BF	BF
0.5	2.0	24.36±0.99ns	30.10±0.21*	BC	WC
1.0	2.0	36.16±2.78***	33.26±2.37**	BC	BC
1.5	2.0	63.23±0.02***	52.16±1.67***	BF	F
2.0	2.0	74.43±0.28***	66.21±1.02***	F	LBF
2.5	2.0	**84.11±0.45****	78.12±3.13***	BF	BF

(Table 6) cont.....

Hormonal Concentrations (mg/L)		Percentage of Callus Response		Nature of Callus	
IBA	Kn	L	IN	L	IN
0.5	2.5	27.03±0.95**	28.24±0.04ns	WC	F
1.0	2.5	39.44±2.56***	32.31±1.02**	WC	BF
1.5	2.5	56.23±2.52***	66.28±3.11***	BF	F
2.0	2.5	71.56±0.72***	80.57±0.92***	LBF	LBF
2.5	2.5	79.30±2.35***	82.14±0.26***	BWF	LBF

L-Leaf, IN- Inter Node, B- Brownish, C-Compact, F- Friable, G- Greenish, W-White, LB-Light Brown, LY-Light Yellow, Y- Yellowish; Each value represents the mean ± SD of triplicate measurements and superscript represent level of significance comparing to control value (0.5mg/L IBA+0.5mg/L Kn); * significant at $p<0.05$, ** significant at $p<0.01$, *** significant at $p<0.001$, ns-not significant (according to Tukey-Kramer Multiple Comparisons Test).

Table 7. Callus induction of *Pseudarthria viscida* in MS medium with different combinations of IAA and BAP. Culture period: 4 weeks, Subculturing: after 4 weeks.

Hormonal Concentrations (mg/L)		Percentage of Callus Response		Nature of Callus	
IAA	BAP	L	IN	L	IN
0.5	0.5	8.23±0.14	7.54±0.15	WC	WC
1.0	0.5	14.35±1.23**	24.31±2.21***	GWC	WC
1.5	0.5	30.12±0.87***	32.45±2.45***	LY	LY
2.0	0.5	45.19±0.78***	44.20±0.94***	LBF	LBC
2.5	0.5	59.13±0.13***	56.29±0.89***	BF	BF
0.5	1.0	12.20±0.12*	8.35±0.13ns	WC	LY
1.0	1.0	20.38±2.65***	25.35±0.96***	LY	F
1.5	1.0	35.37±0.78***	41.21±0.45***	LBC	LY
2.0	1.0	51.31±1.46***	47.10±1.39***	LBC	WC
2.5	1.0	60.24±1.10***	59.35±1.26***	BF	BC
0.5	1.5	12.84±0.89*	11.47±3.24*	BC	LBC
1.0	1.5	29.38±0.94***	18.18±2.28***	WC	GWC
1.5	1.5	46.00±0.76***	29.19±0.09***	LY	WF
2.0	1.5	57.38±0.61***	44.36±1.56***	BF	LBC
2.5	1.5	63.23±2.45***	52.24±1.23***	LBC	BF
0.5	2.0	10.23±0.67ns	9.14±0.81ns	GWC	LY
1.0	2.0	27.32±1.02***	29.45±0.79***	F	GWC
1.5	2.0	58.23±0.98***	48.29±1.47***	F	F
2.0	2.0	62.21±0.57***	55.23±2.57***	BC	BC
2.5	2.0	**65.35±1.11*****	**60.24**±2.13***	WC	WC

(Table 7) cont.....

Hormonal Concentrations (mg/L)		Percentage of Callus Response		Nature of Callus	
IAA	BAP	L	IN	L	IN
0.5	2.5	9.23±0.23ns	6.23±0.36ns	BF	WF
1.0	2.5	25.25±2.15***	13.35±1.45**	BC	WF
1.5	2.5	49.21±0.76***	31.39±0.58***	LY	LBF
2.0	2.5	57.21±1.23***	47.35±0.68***	LBF	LBC
2.5	2.5	60.12±2.31***	58.34±0.03***	BF	LBC

L-Leaf, IN- Inter Node, B- Brownish, C-Compact, F- Friable, G- Greenish, W-White, LB-Light Brown, LY-Light Yellow, Y- Yellowish; Each value represents the mean ± SD of triplicate measurements and superscript represent level of significance comparing to control value (0.5mg/L IAA+0.5mg/L BAP); * significant at $p<0.05$, ** significant at $p<0.01$, *** significant at $p<0.001$, ns-not significant (according to Tukey-Kramer Multiple Comparisons Test).

Table 8. Callus induction of *Pseudarthria viscida* in MS medium with different combinations of IAA and Kn. Culture period: 4 weeks, Subculturing: after 4 weeks.

Hormonal Concentrations (mg/L)		Percentage of Callus Response		Nature of Callus	
IAA	Kn	L	IN	L	IN
0.5	0.5	9.58±0.49	8.15±0.15	GWC	GWC
1.0	0.5	14.00±0.96**	12.22±0.35*	WC	WC
1.5	0.5	28.38±1.47***	26.34±2.78***	GWC	LBC
2.0	0.5	35.12±0.04***	35.20±0.89***	LBF	LBF
2.5	0.5	48.23±2.17***	41.34±3.12***	BF	F
0.5	1.0	13.23±0.56*	13.24±0.10**	WC	WC
1.0	1.0	28.63±0.78***	29.34±0.98***	WC	WC
1.5	1.0	30.08±1.67***	32.58±0.29***	WC	LBF
2.0	1.0	39.18±2.73***	39.98±0.25***	LY	F
2.5	1.0	48.26±1.56***	42.55±2.46***	LY	LY
0.5	1.5	14.78±0.67**	14.67±0.57**	GWC	GWC
1.0	1.5	19.28±0.71***	21.23±0.89***	F	WC
1.5	1.5	28.16±0.95***	29.12±0.22***	LYF	BF
2.0	1.5	35.78±0.01***	37.23±0.27***	WC	LBF
2.5	1.5	**49.89**±1.34***	**46.16±1.02*****	WC	WC
0.5	2.0	16.21±0.74**	15.14±2.31***	BF	GWC
1.0	2.0	27.18±0.35***	26.25±1.07***	F	LY
1.5	2.0	38.21±0.98***	31.25±0.25***	WC	LY
2.0	2.0	47.37±0.96***	39.78±3.14***	LY	BF
2.5	2.0	51.13±3.12***	45.32±0.98***	BF	WC

(Table 8) cont.....

Hormonal Concentrations (mg/L)		Percentage of Callus Response		Nature of Callus	
IAA	Kn	L	IN	L	IN
0.5	2.5	12.19±2.12ns	10.29±0.87ns	WC	WC
1.0	2.5	28.34±2.91***	18.49±0.38***	WC	LY
1.5	2.5	31.38±2.78***	27.28±1.36***	LBC	LBF
2.0	2.5	39.93±2.11***	37.14±0.98***	LBC	LBF
2.5	2.5	43.11±0.08***	44.13±2.13***	BF	BF

L-Leaf, IN- Inter Node, B- Brownish, C-Compact, F- Friable, G- Greenish, W-White, LB-Light Brown, LY-Light Yellow, Y- Yellowish; Each value represents the mean ± SD of triplicate measurements and superscript represent the level of significance comparing to control value (0.5mg/L IAA + 0.5mg/L Kn); * significant at $p<0.05$, ** significant at $p<0.01$, *** significant at $p<0.001$, ns-not significant (according to Tukey-Kramer Multiple Comparisons Test).

Among the different media tested (MS, B5, and White's media), MS medium supplemented with NAA + BAP combinations was best for callus induction in *P. viscida* (Fig. **1**). An efficient plant regeneration system through callus for *P. viscida* was reported by Meena and Thomas [12] using cotyledonary node explants on MS medium supplemented with 1.5 mgl^{-1} 2,4-D. But in the present study, 96.57± 0.98% and 95.20±0.91% callus inductions were noticed on 2.5 mg/L 2, 4-D + 1.5 mg/L Kn from both leaf and internodal segments. Srivastava *et al.* [13] reported that the MS basal medium supplemented with different concentrations of either 2, 4-D or NAA or IBA alone and in combination with cytokinins were used for callus induction in *Desmodium gangeticum*. In this study of *D. gangeticum,* callus induction was noticed from the leafy explants on MS medium supplemented with 2.26 to 22.62 µM 2, 4-D, or 5.36 to 16.1 µM NAA. Of which 92% of callus induction was noticed on MS medium with 9.05 µM 2, 4-D whereas, in the present study of *P. viscida,* the highest frequency of callus induction was obtained from both leaf (98.22±0.45%) and internodal segments (96.35±1.12%) grown in medium supplemented with 2.5 mg/L NAA+ 1.5 mg/L BAP and 2.5 mg/L NAA and 2.0 mg/L BAP, respectively.

B_5 Medium

In the B_5 medium, maximum callus induction was noticed in the NAA + BAP combination similar to that of the MS medium, whereas the percentage of callus response was less than that of the MS medium. The highest frequency of callus induction was noticed in 2.0 mg/L NAA+1.0 mg/L BAP combination from both leafy (43.41±0.38%) and internodal segments (39.90±0.22%). Light brown friable callus was seen in leafy explants, whereas light yellow friable callus was noticed in internodal segments after six weeks of inoculation. In the case of NAA + Kn combination, 35.36±0.21% light brown friable callus was obtained from the leafy segment (2.0 mg/L NAA + 1.0 mg/L Kn), and 34.19±0.001% callus induction (light brown friable callus) was noticed in internodal explants (2.0 mg/L NAA+2.0 mg/L Kn). In 2, 4-D + BAP combinations, the highest callus induction was found to be 39.13±0.85% and 36.12±0.32% in both leaf and internodal segments (2.0 mg/L 2,4-D+2.0 mg/L BAP), respectively; whereas in 2,4-D + Kn combinations, 43.11±0.49% callus induction from leafy explants in 2.0 mg/L 2,4-D + 1.0 mg/L Kn and 44.15±0.38% callus induction from internodal explants in 2.0 mg/L 2,4-D + 2.0 mg/L Kn combination were observed. From above mentioned 2,4-D combinations, 2,4-D+ Kn was superior over 2,4-D+BAP combination in callus induction.

In IBA + BAP and IBA + Kn combinations, maximum callus induction was found to be 39.25±0.15% and 37.72±1.20% in 2.0 mg/L IBA + 1.0 mg/L BAP and 2.0 mg/L IBA + 1.0 mg/L Kn from the leafy explants, respectively, whereas 40.13±0.22% and 33.12±0.28% callus induction was noticed in 2.0 mg/L IBA+2.0 mg/L BAP and 2.0 mg/L IBA+1.0 mg/L Kn from internodal segment, respectively. From the above combinations, the internodal segment induced more callus than the leafy segment. In IAA+ BAP and IAA+ Kn combinations, comparatively, very less callus production was observed. An efficient procedure for callus induction of *Desmodium gangeticum* in a modified B_5 medium was reported [14]. In this study of *D. gangeticum,* a modified B_5 medium supplemented with 1.0 mg/L IBA was the most favourable medium for callus induction, whereas, in *P. viscida*, B_5 medium with 2.0mg/L NAA+1.0 mg/L BAP combination were best for callus induction in both leafy and internodal segments.

White's Medium

Comparatively, very little callus production was noticed in White's medium. Similar to MS and B_5 media, maximum callus productions were also noticed in NAA+BAP combinations like 29.08±0.34% and 25.51±0.67% whereas least callus productions were observed in IAA+ Kn combinations like 9.98±0.23% and 8.44±0.17%. Efficient callus induction of *Desmodium gangeticum* was reported [14]. In this study of *D. gangeticum,* better callus response was noticed in the MS medium than in White's medium, and this was similar to that of *Pseudarthria viscida.*

Somatic Embryogenesis

Somatic embryogenesis was noticed in the highest concentration of auxin and cytokinin combinations from both leafy and internodal segments in the MS medium (Table **9**, Fig. **2**). Highest frequency of somatic embryogenesis was induced in 2.5mg/L NAA +1.5 mg/L BAP combination in leafy explants (94.21±1.02%) and 2.5 mg/L NAA+ 2.0 mg/L BAP in internodal segments (95.32±0.39%). Somatic embryogenesis and germination of somatic embryos from *in vitro* grown seedling explants like cotyledon, hypocotyls, epicotyls, and leaf explants of *Desmodium gangeticum* in MS medium with various growth regulators like BAP, Kn, 2, 4-D and NAA [15].

Table 9. Effect of different hormones on somatic embryogenesis from the leaf and internodal explants of *Pseudarthria viscida* in MS medium. Culture period: 4 weeks.

Hormonal Concentrations (mg/L)						Percentage of Response in Somatic Embryo Formation	
NAA	IBA	IAA	2,4- D	BAP	Kn	L	IN
2.0	-	-	-	1.0	-	75.34±0.25	73.23±0.56
2.5	-	-	-	**1.5**	-	**94.21±1.02*****	92.15±1.34***
2.5	-	-	-	**2.0**	-	83.56±0.87***	**95.32**±0.39***
-	1.5	-	-	1.5	-	70.15±1.33	70.34±0.58
-	2.0	-	-	1.0	-	73.34±2.04*	74.04±1.20**
-	2.5	-	-	1.5	-	81.52±0.22***	80.20±1.12***
-	2.5	-	-	2.0	-	82.16±0.38***	83.23±0.22***
-	-	2.5	-	1.5	-	75.23±1.45	54.21±0.48
-	-	2.5	-	2.0	-	78.33±0.07*	62.32±1.54***
-	-	-	1.5	-	1.0	71.54±0.45	85.34±0.22
-	-	-	2.0	-	1.5	84.32±0.69***	88.32±1.21*
-	-	-	2.5	-	2.0	85.28±0.56***	86.06±1.47ns

Each value represents the mean ± SD of triplicate measurements and the superscript represents the level of significance compared to the control value (least concentrations); * significant at $p<0.05$, ** significant at $p<0.01$, *** significant at $p<0.001$, ns-not significant (according to Tukey-Kramer Multiple Comparisons Test).

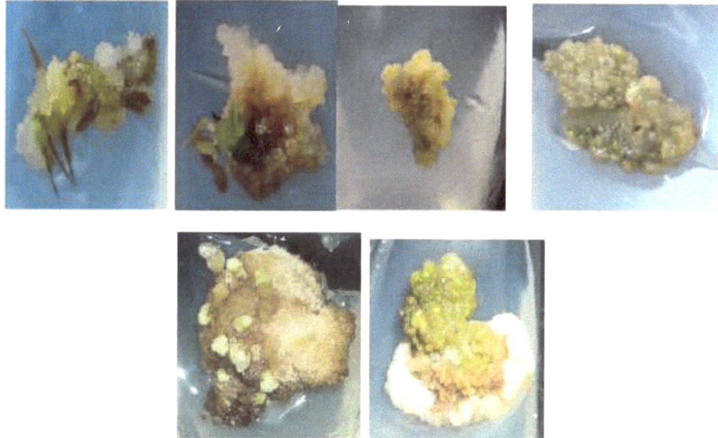

Fig. (2). Somatic embryogenesis of *Pseudarthria viscida* in MS medium.

Fig. (3). Indirect organogenesis of *Pseudarthria viscida* in MS medium.

Fig. (4). Shoot induction *of Pseudarthria viscida* in MS medium.

Shoot Induction

In MS Medium

For shoot regeneration, various concentrations of BAP or Kn alone (0.5 to 3.0 mg/L) or in combination with NAA, IBA, and IAA hormones were tried in node, leaf, and internodal segments. The percentage of regeneration was low with a low concentration of hormones. Of the two cytokinins, BAP was comparatively better for shoot initiation.

In the case of nodal explants, optimum responses (89.32±0.23%) were noticed in 2.5 mg/L BAP concentration with an average shoot number of 5.2±0.1 and average shoot length of 6.3±0.3 cm (Table **10**, Fig. **3**). Kn at 2.5 mg/L concentration produced only 79.32±1.34% shoot response with an average shoot number of 5.1±0.2 and an average shoot length of 6.0±0.3 cm (Table **11**). The presence of a low concentration of NAA/IBA/IAA with a high concentration of BAP/Kn resulted in enhancement of the response and also increased the formation of shoots during indirect organogenesis from the callus. Of the three different auxin concentrations with BA or Kn, NAA was best for shoot initiation from nodal explants. 79.34±1.23% of shoot induction was noticed in 3.0 mg/L BAP+ 0.5 mg/L NAA concentration with an average shoot number of 4.3±0.3% and average shoot length of 5.7±0.2 cm, whereas 71.43±1.45% optimal shoot induction was observed in 2.5 mg/L Kn + 1.0 mg/L NAA concentration with an average shoot number of 4.2±0.1 and average shoot length of 5.3±0.2cm. All other auxin combinations with high BA or Kn induced comparatively less shoot induction than NAA with BA or Kn.

Table 10. Effect of BAP alone or in combination with NAA, IBA and IAA on shoot induction from the nodal explants of *Pseudarthria viscida* in MS medium.

Hormonal Concentrations (mg/L)				Percentage of Response-Node as Explants	Average Number of Shoots	Average Number of Nodes/Shoots	Average Length of Shoots (cm)
BAP	NAA	IBA	IAA				
0.5	-	-	-	56.21±0.78	2.3±0.1	2.1±0.2	1.3±0.1
1.0	-	-	-	67.12±2.04***	2.3±0.3ns	2.1±0.4ns	2.4±0.2**
1.5	-	-	-	79.10±1.98***	2.3±0.1ns	2.1±0.2ns	3.3±0.4***
2.0	-	-	-	85.11±0.36***	6.4±0.4***	4.3±0.2***	4.5±0.9***
2.5	**-**	**-**	**-**	**89.34±2.23*** **	**5.2±0.1*** **	**5.1±0.3*** **	**6.3±0.3*** **
3.0	-	-	-	85.41±1.10***	5.0±0.2***	3.1±0.6**	5.6±0.3***
0.5	0.5	-	-	39.12±0.21	2.1±0.1	2.1±0.2	2.3±0.2
1.0	0.5	-	-	41.40±2.56ns	2.3±0.4**	2.5±0.3**	3.7±0.1**

(Table 10) cont.....

Hormonal Concentrations (mg/L)				Percentage of Response- Node as Explants	Average Number of Shoots	Average Number of Nodes/Shoots	Average Length of Shoots (cm)
BAP	NAA	IBA	IAA				
1.5	0.5	-	-	55.12±1.23***	3.3±0.1**	2.2±0.1ns	4.9±0.3***
2.0	0.5	-	-	63.25±2.34***	4.0±0.3***	2.3±0.2*	5.7±0.2***
2.5	0.5	-	-	71.33±1.12***	4.0±0.5***	2.5±0.3**	4.8±0.4***
3.0	**0.5**	**-**	**-**	**79.34±1.23*****	**4.3±0.3*****	**4.3±0.2*****	**5.7±0.2*****
0.5	1.0	-	-	33.16±2.49	2.3±0.1	2.1±0.2	3.1±0.1
1.0	1.0	-	-	35.03±1.34ns	2.5±0.6*	2.3±0.2*	5.1±0.2***
1.5	1.0	-	-	49.34±2.01***	2.3±0.8ns	2.2±0.2ns	4.3±0.6***
2.0	1.0	-	-	60.34±2.17***	3.4±0.6***	3.1±0.1***	5.2±0.1***
2.5	1.0	-	-	75.32±1.23***	4.4±0.1***	3.1±0.6***	6.1±0.3***
3.0	1.0	-	-	65.32±0.34***	2.3±0.1ns	2.3±0.2**	5.3±0.3***
0.5	-	0.5	-	30.43±0.78	2.1±0.1	2.1±0.2	2.1±0.1
1.0	-	0.5	-	38.53±1,34***	2.2±0.7ns	2.3±0.2**	3.7±0.3**
1.5	-	0.5	-	44.38±0.32***	2.6±0.3*	2.3±0.2**	5.4±0.1***
2.0	-	0.5	-	69.26±2.31***	3.1±0.3***	3.2±0.1***	5.7±0.2***
2.5	**-**	**0.5**	**-**	**77.38±1.20*****	**3.9±0.5*****	**3.1±0.6*****	**6.2±0.1*****
3.0	-	0.5	-	60.29±0.23***	2.1±0.4ns	2.2±0.2ns	3.6±0.2***
0.5	-	1.0	-	29.36±0.78	2.2±0.7	2.2±0.1	2.5±0.2
1.0	-	1.0	-	33.17±0.56**	2.2±0.4ns	2.2±0.1ns	4.7±0.2***
1.5	-	1.0	-	45.53±0.56***	2.6±0.7**	2.5±0.3**	4.7±0.1***
2.0	-	1.0	-	73.56±0.34***	2.4±0.3**	2.2±0.2ns	4.4±0.3***
2.5	-	1.0	-	75.89±1.03***	3.2±0.3***	2.2±0.1ns	5.9±0.2***
3.0	-	1.0	-	65.19±1.20***	2.3±0.3ns	2.2±0.2ns	4.0±0.6***
0.5	-	-	0.5	18.34±0.23	2.1±0.1	2.1±0.2	1.3±0.1
1.0	-	-	0.5	22.87±1.34**	2.1±0.1ns	2.2±0.2*	1.6±0.3**
1.5	-	-	0.5	35.42±0.88***	2.1±0.1ns	2.1±0.2ns	2.7±0.2***
2.0	-	-	0.5	41.56±0.34***	2.3±0.4*	2.1±0.2ns	2.6±0.1***
2.5	-	-	0.5	45.14±1.23***	2.1±0.1ns	2.2±0.1ns	1.7±0.3***
3.0	-	-	0.5	39.34±0.46***	2.2±0.2ns	2.2±0.1*	2.6±0.2***
0.5	-	-	1.0	15.87±0.34	2.1±0.1	2.1±0.2	1.3±0.1
1.0	-	-	1.0	27.34±1.02***	2.1±0.1ns	2.1±0.2ns	1.3±0.1ns
1.5	-	-	1.0	30.45±0.89***	2.1±0.1ns	2.2±0.1*	2.1±0.1***
2.0	-	-	1.0	33.63±2.34***	2.2±0.2ns	2.2±0.1*	1.3±0.1ns

(Table 10) cont.....

Hormonal Concentrations (mg/L)				Percentage of Response-Node as Explants	Average Number of Shoots	Average Number of Nodes/Shoots	Average Length of Shoots (cm)
BAP	**NAA**	**IBA**	**IAA**				
2.5	-	-	1.0	35.63±0.98***	2.3±0.1*	2.1±0.2ns	2.1±0.1***
3.0	-	-	1.0	35.13±1.22***	2.1±0.1ns	2.1±0.2ns	1.3±0.1ns

Each value represents the mean ± SD of triplicate measurements and the superscript represents the level of significance compared to the control value (least concentrations); * significant at $p<0.05$, ** significant at $p<0.01$, *** significant at $p<0.001$, ns-not significant (according to Tukey-Kramer Multiple Comparisons Test).

Table 11. Effect of Kn alone or in combination with NAA, IBA and IAA on shoot induction from the nodal explants of *Pseudarthria viscida* in MS medium.

Hormonal Concentrations (mg/L)				Percentage of Response-Node as Explants	Average Number of Shoots	Average Number of Nodes/Shoots	Average Length of Shoots (cm)
Kn	**NAA**	**IBA**	**IAA**				
0.5	-	-	-	40.34±0.56	2.1±0.1	2.1±0.1	2.3±0.2
1.0	-	-	-	51.09±1.67***	2.5±0.3**	2.2±0.1ns	3.1±0.1***
1.5	-	-	-	54.56±1.21***	3.1±0.4***	2.7±0.4***	4.5±0.3***
2.0	-	-	-	69.06±2.34***	4.7±0.1***	3.9±0.3***	4.8±0.9***
2.5	**-**	**-**	**-**	**79.32±1.34*** **	**5.1±0.2*** **	**3.9±0.5*** **	**6.0±0.3*** **
3.0	-	-	-	75.67±0.98***	4.8±0.3***	2.2±0.5*	4.5±0.2***
0.5	0.5	-	-	26.12±0.06	2.1±0.1	2.2±0.1	2.3±0.2
1.0	0.5	-	-	45.34±1.56***	3.4±0.3***	2.3±0.2ns	3.1±0.1***
1.5	0.5	-	-	49.95±0.68***	3.6±0.3***	3.2±0.5***	2.3±0.2ns
2.0	0.5	-	-	57.35±1.34***	4.7±0.3***	2.5±0.2**	3.1±0.1***
2.5	0.5	-	-	65.34±1.56***	4.9±0.3***	2.2±0.1ns	4.8±0.9***
3.0	0.5	-	-	69.15±2.34***	5.3±0.1***	2.4±0.3**	4.3±0.2***
0.5	1.0	-	-	40.00±0.76	2.1±0.3	2.2±0.2	2.1±0.1
1.0	1.0	-	-	43.67±2.45*	3.1±0.2***	2.2±0.8ns	2.1±0.3ns
1.5	1.0	-	-	49.34±1.56***	3.4±0.1***	2.4±0.1**	3.5±0.3***
2.0	1.0	-	-	55.11±0.78***	3.1±0.3***	2.7±0.4***	4.8±0.9***
2.5	**1.0**	**-**	**-**	**71.43±1.45*** **	**4.2±0.1*** **	3.9±0.3***	**5.3±0.2*** **
3.0	1.0	-	-	65.11±1.34***	3.9±0.2***	2.7±0.2***	4.5±0.6***
0.5	-	0.5	-	33.45±0.45	2.1±0.1	2.2±0.5	2.3±0.3
1.0	-	0.5	-	45.87±1.23***	2.4±0.3**	2.4±0.1**	3.3±0.7***
1.5	-	0.5	-	59.10±0.78***	2.3±0.1*	2.2±0.7ns	3.2±0.2***
2.0	-	0.5	-	63.45±01.4***	2.7±0.2***	2.4±0.1**	3.1±0.3***
2.5	-	0.5	-	64.26±0.45***	2.5±0.7**	2.7±0.2***	3.8±0.9***

(Table 11) cont.....

Hormonal Concentrations (mg/L)				Percentage of Response- Node as Explants	Average Number of Shoots	Average Number of Nodes/Shoots	Average Length of Shoots (cm)
Kn	**NAA**	**IBA**	**IAA**				
3.0	-	0.5	-	61.12±1.33***	2.1±0.5ns	3.9±0.3***	**3.3±0.2*****
0.5	-	1.0	-	29.23±0.56	2.1±0.3	2.4±0.4	2.2±0.5
1.0	-	1.0	-	32.12±1.45**	2.4±0.2**	2.7±0.1**	2.4±0.9*
1.5	-	1.0	-	45.98±0.34***	2.7±0.2***	3.9±0.2***	2.5±0.3**
2.0	-	1.0	-	59.34±1.34***	2.3±0.3**	2.7±0.5**	2.4±0.7*
2.5	-	1.0	-	60.34±2.12***	2.1±0.1ns	2.4±0.7ns	**2.5±0.5****
3.0	-	1.0	-	25.76±0.45***	2.4±0.1**	2.4±0.8ns	2.5±0.1**
0.5	-	-	0.5	30.65±1.19	2.1±0.1	2.2±0.4	2.2±0.2
1.0	-	-	0.5	35.35±2.11**	2.1±0.2ns	2.7±0.8**	3.1±0.5***
1.5	-	-	0.5	38.23±1.02***	2.4±0.3**	2.6±0.1**	2.5±0.6**
2.0	-	-	0.5	40.05±2.11***	3.1±0.3***	3.8±0.7***	2.2±0.3ns
2.5	-	-	0.5	48.54±1.67***	3.2±0.5***	2.7±0.2**	2.3±0.7*
3.0	-	-	0.5	43.23±1.59***	3.8±0.2***	2.2±0.5ns	2.4±0.2**
0.5	-	-	1.0	30.17±0.76	2.1±0.1	2.4±0.1	2.1±0.3
1.0	-	-	1.0	34.11±2,45**	2.1±0.3ns	2.7±0.7**	3.8±0.2***
1.5	-	-	1.0	38.34±0.23***	2.3±0.1*	2.8±0.4***	25±0.5**
2.0	-	-	1.0	40.56±2.33***	3.7±0.1***	3.6±0.3***	2.3±0.1*
2.5	-	-	1.0	45.23±0.56***	3.4±0.3***	2.4±0.2***	3.1±0.5***
3.0	-	-	1.0	47.54±1.45***	2.3±0.2*	2.4±0.5ns	2.3±0.2*

Each value represents the mean ± SD of triplicate measurements and the superscript represents level of significance compared to the control value (least concentrations); * significant at p<0.05, ** significant at p<0.01, *** significant at p<0.001, ns-not significant (according to Tukey-Kramer Multiple Comparisons Test).

Similar to that of nodal explants, maximum shoot induction was noticed in 2.5mg/L concentration of BAP from both leaf (89.00±0.27%) and internode (81.22±0.27%). In Kn, 2.5 mg/L concentration produced 77.33±0.28% and 75.39±1.20% optimal shoot induction from leaf and internodal explants (Table **12**, Fig. **4**). Node and internodal segments were used for the adventitious shoot induction from the callus. The presence of a low concentration of NAA/IBA with a high concentration of BA/Kn resulted in further increases in the response and formation of more shoots from the callus. MS medium with 1.0 mg/L of NAA+ 2.5 mg/L BAP was the most efficient hormone combination for shoot induction in *P. viscida* (Table **13**). All other combinations showed less shoot induction than that of 1.0 mg/l of NAA+ 2.5 mg/L BAP.

Table 12. Effect of BAP alone or in combination with NAA, IBA and IAA on shoot induction from the leaf and internodal explants of *Pseudarthria viscida* in MS medium.

Hormonal Concentrations (mg/L)				Percentage of Response		Average Number of Shoots	
BAP	Kn	NAA	IBA	L	IN	L	IN
0.5	-	-	-	66.20±0.21	54.30±0.27	4.1±0.1	2.1±0.2
1.0	-	-	-	69.12±0.04*	58.12±1.09**	4.2±0.3ns	2.1±0.4ns
1.5	-	-	-	72.10±0.25***	69.18±0.45***	4.3±0.1*	2.1±0.2ns
2.0	-	-	-	85.11±0.06***	72.45±0.56***	5.4±0.4***	4.3±0.2***
2.5	**-**	**-**	**-**	**89.00±0.27*****	**81.22**±0.27***	**6.4±0.3*****	**5.1±0.3*****
3.0	-	-	-	81.41±0.98***	75.34±0.04***	4.2±0.2*	3.1±0.6**
-	0.5	-	-	54.21±0.56	50.10±0.10	3.4±0.6	2.1±0.2
-	1.0	-	-	60.11±1.23***	56.05±1.11**	3.5±0.2ns	2.6±0.3**
-	1.5	-	-	69.78±1.10***	67.29±0.43***	4.4±0.6***	3.3±0.2***
-	2.0	-	-	71.20±0.89***	69.23±1.03***	4.3±0.1***	3.1±0.2***
-	**2.5**	**-**	**-**	**77.33±0.28*****	**75.39±1.20*****	**4.7±0.1*****	**4.3±0.2*****
-	3.0	-	-	68.30±1.11***	67.20±0.98***	4.1±0.1***	3.2±0.2***
0.5	-	0.5	-	45.12±0.05	41.33±0.24	4.1±0.1	2.1±0.2
1.0	-	0.5	-	51.23±1.20**	49.23±0.58***	4.1±0.4ns	2.5±0.3**
1.5	-	0.5	-	60.12±0.33***	56.23±0.73***	4.2±1.1*	2.2±0.1ns
2.0	-	0.5	-	82.25±1.34***	68.56±1.20***	4.3±1.2**	2.3±0.2*
2.5	-	0.5	-	89.13±0.12***	77.30±1.16***	4.4±0.5***	2.5±0.3**
3.0	-	0.5	-	71.29±0.23***	69.29±1.05***	5.1±0.2***	4.3±0.2***
0.5	-	1.0	-	53.16±0.49	48.26±0.43	2.3±0.5	2.1±0.2
1.0	-	1.0	-	55.03±1.12*	53.37±1.45***	3.4±0.6***	2.3±0.2*
1.5	-	1.0	-	69.34±0.98***	63.48±0.98***	3.5±0.2ns	2.2±0.2ns
2.0	-	1.0	-	72.34±1.20***	69.45±1.21***	3.4±0.6***	3.1±0.1***
2.5	**-**	**1.0**	**-**	**76.32±1.45*****	**74.56**±0.96***	**3.7±0.1*****	**3.1±0.6*****
3.0	-	1.0	-	68.32±0.11***	67.23±0.07***	4.3±0.1***	2.3±0.2**
0.5	-	-	0.5	45.43±0.18	39.17±0.73	4.1±0.1	2.1±0.2
1.0	-	-	0.5	47.53±1.34*	45.98±0.45***	4.2±0.7ns	2.3±0.2**
1.5	-	-	0.5	49.31±1.02***	46.78±1.39***	42.6±0.3ns	2.3±0.2**
2.0	-	-	0.5	59.16±2.01***	53.11±0.27***	5.1±0.3***	3.2±0.1***
2.5	**-**	**-**	**0.5**	**67.12±1.20*****	**64.34**±1.06***	**5.9±0.5*****	**3.1±0.6*****
3.0	-	-	0.5	63.24±0.23***	57.21±0.98***	2.7±0.4**	2.2±0.2ns
0.5	-	-	1.0	20.25±0.13	15.58±0.77	2.1±0.7	2.2±0.1

(Table 12) cont.....

Hormonal Concentrations (mg/L)				Percentage of Response		Average Number of Shoots	
BAP	**Kn**	**NAA**	**IBA**	**L**	**IN**	**L**	**IN**
1.0	-	-	1.0	29.11±0.46***	21.34±0.39**	2.2±0.4ns	2.2±0.1ns
1.5	-	-	1.0	37.13±0.06***	35.78±0.12***	2.4±0.6**	2.5±0.3**
2.0	-	-	1.0	43.26±0.14***	44.29±1.02***	2.5±0.1**	2.2±0.2ns
2.5	**-**	**-**	**1.0**	**55.59±1.43*** **	**49.34**±0.09***	**3.2±0.3*** **	**2.2±0.1ns**
3.0	-	-	1.0	45.39±1.20***	39.59±1.08***	2.2±0.9ns	2.2±0.2ns
-	0.5	0.5	-	40.34±0123	38.86±0.33	2.1±0.1	2.1±0.2
-	1.0	0.5		48.87±1.44***	44.40±1.35***	3.1±0.6***	2.2±0.2*
-	1.5	0.5	-	55.42±0.11***	50.00±1.24***	3.2±0.1***	2.1±0.2ns
-	2.0	0.5	-	61.46±0.34***	58.34±0.98***	3.3±0.4***	2.1±0.2ns
-	2.5	0.5	-	65.14±0.23***	67.31±0.59***	3.4±0.1***	2.2±0.1ns
-	3.0	0.5	-	46.34±0.76***	45.64±0.44***	4.2±0.2***	2.2±0.1*
-	0.5	1.0	-	35.17±0.54	32.21±0.38	3.1±0.1	2.1±0.2
-	1.0	1.0	-	47.34±1.22***	45.67±1.32***	2.1±0.3ns	2.1±0.2ns
-	1.5	1.0		50.45±0.19***	59.32±1.22***	3.1±0.1***	2.2±0.1*
-	2.0	1.0		63.63±0.34***	61.45±0.98***	3.1±0.2***	2.2±0.1*
-	**2.5**	**1.0**		**69.63±0.68*** **	**66.22±1.36*** **	**4.3±0.1*** **	**2.1±0.2ns**
-	3.0	1.0		54.13±1.22***	59.54±1.02***	3.1±0.1***	2.1±0.2ns
-	0.5	-	0.5	25.03±0.11	29.17±0.73	2.1±0.1	2.1±0.2
-	1.0	-	0.5	27.63±1.30*	35.08±0.45***	2.2±0.7ns	2.1±0.2ns
-	1.5	-	0.5	39.21±1.22***	36.71±1.39***	2.2±0.3ns	2.4±0.2**
-	2.0	-	0.5	49.56±1.05***	43.10±0.27***	3.1±0.3***	3.3±0.1***
-	2.5	-	0.5	57.32±0.20***	54.32±1.06***	3.9±0.5***	3.0±0.6***
-	3.0	-	0.5	53.44±0.43***	47.27±0.98***	2.2±0.4ns	2.1±0.2ns
-	0.5	-	1.0	25.25±0.63	19.55±0.77	2.1±0.7	2.2±0.1
-	1.0	-	1.0	30.31±0.40***	23.32±0.39**	2.1±0.4ns	2.2±0.1ns
-	1.5	-	1.0	37.23±0.46***	37.75±0.12***	2.5±0.6**	2.6±0.3**
-	2.0	-	1.0	43.46±0.34***	41.21±1.02***	2.6±0.1**	2.2±0.5ns
-	2.5	-	1.0	55.69±0.43***	50.33±0.09***	3.0±0.3***	2.2±0.2ns
-	**3.0**	**-**	**1.0**	**45.19±1.02*** **	**45.57±1.08*** **	**2.1±0.9ns**	**2.2±0.1ns**

Each value represents the mean ± SD of triplicate measurements and the superscript represents the level of significance compared to the control value (least concentrations); * significant at p<0.05, ** significant at p<0.01, *** significant at p<0.001, ns-not significant (according to Tukey-Kramer Multiple Comparisons Test).

Table 13. Effect of different hormones like NAA, 2, 4-D, IBA and IAA alone on root induction from *the in vitro* developed micro shoot of *Pseudarthria viscida* in MS medium.
Culture period: 4 weeks, Subculturing: after 4 weeks.

Hormonal concentrations (mg/L)				% of response-Micro shoots as explants	Average length of roots (cm)
NAA	2,4-D	IBA	IAA		
0.5	-	-	-	35.00±0.03	3.2±0.1
1.0	-	-	-	48.30±0.01***	4.7±0.4***
1.5	-	-	-	52.66±0.02***	4.3±0.7***
2.0	-	-	-	88.66±0.03***	6.8±0.5***
2.5	-	-	-	**91.00±0.04*****	7.5±0.2***
3.0	-	-	-	50.16±0.02***	4.2±0.1***
-	0.5	-	-	23.33±0.01	2.5±0.2
-	1.0	-	-	39.16±0.03***	2.9±0.3*
-	1.5	-	-	48.50±0.29***	3.1±0.2***
-	2.0	-	-	82.33±0.03***	6.4±0.4***
-	**2.5**	-	-	**89.32±0.02*****	7.1±0.3***
-	3.0	-	-	38.33±0.02***	4.1±0.5***
-	-	0.5	-	14.00±0.39	1.3±0.2
-	-	1.0	-	20.56±0.04**	2.2±0.2***
-	-	1.5	-	36.16±0.03***	2.5±0.3***
-	-	2.0	-	72.50±0.02***	6.3±0.4***
-	-	**2.5**	-	**75.33±0.32*****	7.0±0.1***
-	-	3.0	-	40.06±0.03***	3.5±0.1***
-	-	-	0.5	19.16±0.03	1.2±0.2
-	-	-	1.0	25.00±0.29**	3.2±0.4***
-	-	-	1.5	31.66±0.03***	3.6±0.1***
-	-	-	2.0	39.32±0.02***	4.7±0.7***
-	-	-	2.5	58.33±0.12***	5.3±0.9***
-	-	-	**3.0**	**76.06±0.03*****	6.4±0.5***

Each value represents the mean ± SD of triplicate measurements and superscript represents level of significance compared to the control value (least concentrations); * significant at $p<0.05$, ** significant at $p<0.01$, *** significant at $p<0.001$, ns-not significant (according to Tukey-Kramer Multiple Comparisons Test).

The callus cultured with a high concentration of BAP or Kn alone exhibited initial signs of shoot induction from both leaf and internodal segments after four weeks of culture, and the addition of NAA resulted in rapid and enhanced regeneration frequency than the addition of IBA. Microshoot regeneration was initiated within two weeks of culture. A large number of adventitious shoots was produced from the callus surface within two weeks of culture. Microshoots were further developed into plantlets within a period of four to six weeks of culture. These microshoots were used for root induction.

Regeneration from callus of *P. viscida* in MS medium with 0.5 to 4.0 mg/L BA or 0.5 to 4.0 mg/L Kn alone or in combination with 0.5 to 1.0 mg/L 2,4-D was already reported [12] whereas, in the present investigation of *P. viscida,* the various concentration of BAP or Kn alone (0.5 to 3.0 mg/L) or combination with NAA, IBA and IAA hormones were tried. Of which maximum shoot regeneration was noticed in 2.5 mg/L BAP, 2.5 mg/L Kn, and 3.0 mg/L BAP+ 0.5 mg/L NAA concentrations, and this study indicated that cytokinin alone or in combination with a low concentration of auxin was essential for getting a better response for shoot regeneration in *P.viscida.*

In B_5 Medium and White's Medium

In B_5 medium and White's medium, the highest three concentrations (1.0, 20.0 and 3.0mg/L) of BAP or Kn were tested in nodal segments for shoot induction. In B_5 medium, maximum shoot induction was noticed in 3.0 mg/L BAP (45.11±1.20%) and 3.0 mg/L Kn (42.29±3.23%) concentrations. Similar to that of B_5 medium, maximum shoot induction was noticed in 3.0 mg/L BAP (30.12±1.03%) and 3.0 mg/L concentration of Kn (29.23±0.32%). After the addition of 0.5 mg/L NAA or IBA with high BAP or Kn concentration, comparatively lower regeneration frequencies were observed in both the B_5 medium and White's medium. Efficient *in vitro* plant regeneration of *Desmodium gangeticum* on White's medium was reported [14].

Root Induction

In MS Medium

In *Pseudarthria viscida, in vitro* developed micro shoots (the elongated microshoots were separated and then individually inoculated in culture tubes containing auxin rich rooting media) were inoculated in a medium supplemented with different concentrations of auxins such as NAA, IBA, IAA and 2,4-D at a range of 0.5 to 3.0 mg/L for root initiation. Leaf and internodal segments were inoculated for adventitious root formation. After four weeks, the percentage of root formation, root number and length of roots were recorded regularly.

In *Pseudarthria viscida*, maximum *in vitro* root formation was noticed on MS medium supplemented with 2.5mg/L NAA concentration (91.00±0.04%) with an average root length of 7.5±0.2 cm after four weeks of inoculation (Fig. **5**). Similarly, 2.5 mg/L 2, 4-D concentration also showed higher *in vitro* root induction, like 89.32±0.02% with 7.1±0.3cm of average root length. In the case of IBA only, 2.5 mg/L concentration showed 75.33±0.32% *in vitro* root inductions with 7.0±0.1 cm average root length obtained. But in IAA, 2.5mg/L concentration showed 58.33±0.12% root induction, whereas, in 3.0 mg/L concentration, 76.06±0.03% of *in vitro* root formation with an average root length of 6.4±0.5 cm was observed after four weeks of inoculation (Table **13**).

a. Root formation in 0.25 mg/L concentration of NAA

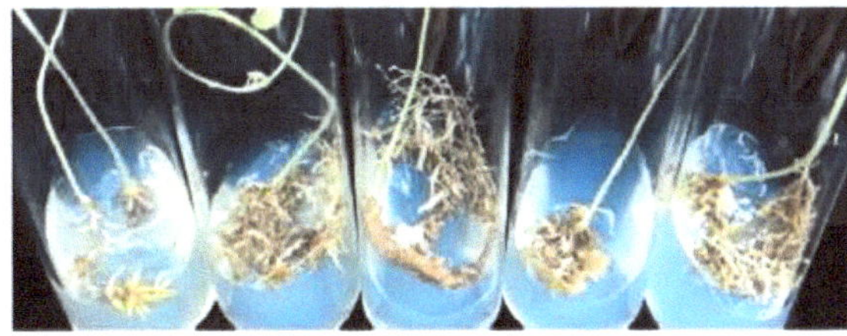

b. Rooting in 0.25 mg/L concentration of 2,4-D

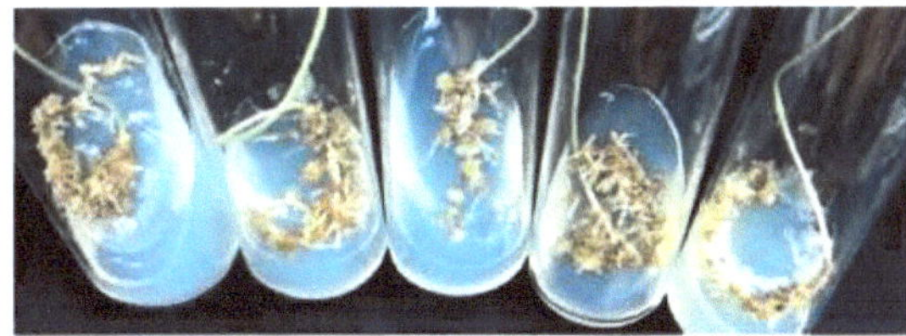

c. Rooting in 0.3mg/L concentration of IAA

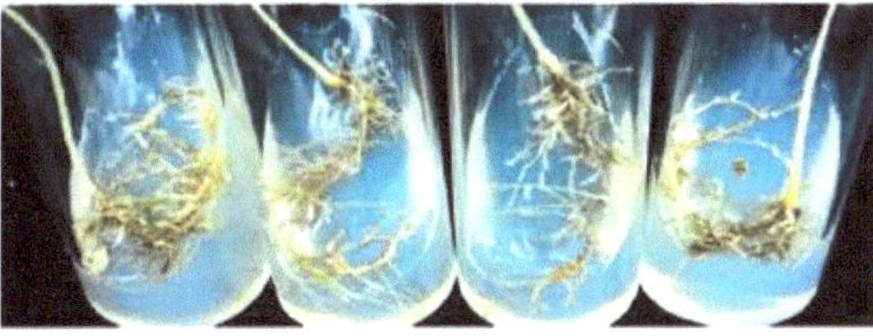

d. Rooting in 0.25 mg/L concentration of IBA

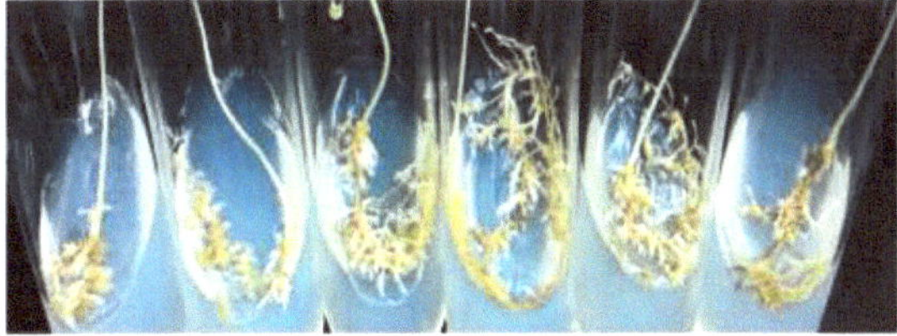

Fig. (5). *In vitro* root induction of *Pseudarthria viscida* in MS medium.

Maximum adventitious root induction was noticed at 2.5 mg/L concentration of NAA from leafy explants (54.11±0.45%) (Fig. **6**), whereas, in internodal segments, maximum adventitious root induction (48.20±1.20%) was observed at 2.0 mg/L NAA concentration. In 2,4-D concentrations, 2.5 mg/L was best for adventitious root induction in both leaf (45.23±1.21%) and internodal segments (38.28±1.07%). IBA and IAA concentrations showed comparatively less response in the induction of adventitious root (Table **14**). IBA proved to be most effective for adventitious root induction in legumes [12], whereas, in *P. viscida*, the IBA combination showed only 37.56% of adventitious root induction from leafy segments at 3.0 mg/L concentration, and a similar concentration showed only 35.32% adventitious root induction from internodal segments.

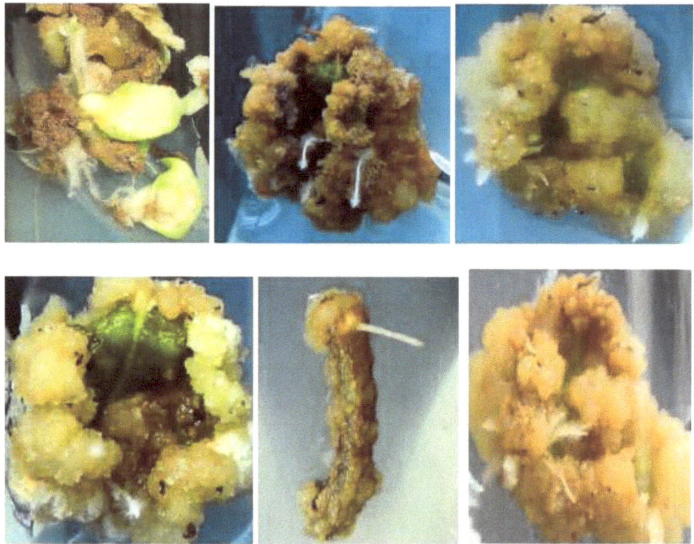

Fig. (6). Adventitious roots induction of *Pseudarthria viscida* in MS medium.

Table 14. Effect of different hormones like NAA, 2,4-D, IBA and IAA alone on adventitious root induction from the leaf and internodal explants of *Pseudarthria viscida* in MS medium. Culture period: 4 weeks.

Hormonal Concentrations (mg/L)				Percentage of Response in Adventitious Root Formation	
NAA	**2,4-D**	**IBA**	**IAA**	**L**	**IN**
1.5	-	-	-	45.73±0.11	31.24±0.43
2.0	-	-	-	49.33±1.38*	**48.20**±1.20***
2.5	-	-	-	**54.11±0.45*****	41.00±1.34***
3.0	-	-	-	46.52±1.21ns	31.29±0.78ns
-	1.5	-	-	37.23±0.15	26.84±1.26

(Table 14) cont.....

Hormonal Concentrations (mg/L)				Percentage of Response in Adventitious Root Formation	
NAA	2,4-D	IBA	IAA	L	IN
-	2.0	-	-	43.26±0.42**	27.91±0.56**
-	**2.5**	-	-	**45.23±1.21*** **	**38.28**±1.07***
-	3.0	-	-	37.29±0.32ns	27.12±1.12*
-	-	1.5	-	20.36±0.28	18.23±-0.34
-	-	2.0	-	22.21±1.16*	20.12±1.21*
-	-	**2.5**	-	**23.12**±1.47**	**22.15±0.45** **
-	-	3.0	-	20.67±1.23ns	21.03±1.64**
-	-	-	1.5	31.24±1.21	29.32±0.23
-	-	-	2.0	34.31±1.11**	30.28±0.32ns
-	-	-	2.5	34.21±0.21**	32.12±1.42**
-	-	-	**3.0**	**37.56±1.23*** **	35.32±01.56**

Each value represents the mean ± SD of triplicate measurements and the superscript represents a level of significance compared to the control value (least concentrations); * significant at p<0.05, ** significant at p<0.01, *** significant at p<0.001, ns-not significant (according to Tukey-Kramer Multiple Comparisons Test).

Leaf and internodal segments were used for root induction for secondary metabolite production. In *Pseudarthria viscida*, most of the important secondary metabolites were reported in roots. Both microshoots and adventitious roots from the callus of leaf and internodal segments were selected for secondary metabolite production experiments.

In B_5 Medium and White's Medium

In B_5 and White's medium, comparatively fewer root inductions were noticed in high concentrations of auxins. Only 35.15% and 21.36% optimal root inductions were observed in B_5 and White's Medium, respectively, after six weeks of inoculation. As previously reported, B5 and White's medium showed less root induction than MS medium [16].

Hardening and Acclimatization of Plantlets

Well-developed plantlets were transferred to *in vivo* conditions. The plantlets with roots were carefully taken out and then washed with tap water. The plants were transferred to paper cups containing sand, garden soil, and farmyard manure in the ratio of 1:1:1 and kept for one month in the culture room. After one month, plants were transferred to plastic bags, maintained at room temperature for ten days, and later moved to the greenhouse. Of the 30 plantlets transplanted into the soil, only

19 survived (Fig. 7). The *in vitro* regenerated plants were typically similar to that of the parental plant (without any visible phenotypic variations).

Fig. (7). Mass propagation and Hardening of *Pseudarthria viscida* (L.) Wight & Arn. a. Root formation in 0.25 mg/L concentration of NAA b. Rooting in 0.25 mg/L concentration of 2,4-D c. Rooting in 0.3mg/L concentration of IAA d. Rooting in 0.25 mg/L concentration of IBA.

CONCLUSION

So, the present study recommended that MS medium was better for *in vitro* propagation of *P. viscida* than that of B5 and White's medium. In the case of explants used, the leafy explant was more suitable for profused callus induction, somatic embryogenesis, and indirect organogenesis than that of internodal and nodal explants, whereas nodal explant was best for direct organogenesis in *P.*

viscida. This study thus reports an efficient protocol for plant regeneration, and this could be vital for the multiplication and field transfer of this ethnomedicinal plant. Based on the ethnomedicinal potential, there is an urgent need for organized cultivation of this vulnerable plant for its conservation and sustainable utilization.

REFERENCES

[1] Bapat VA, Yadav SR, Dixit GB. Rescue of endangered plants through biotechnological applications. Natl Acad Sci Lett 2008; 31: 201-10.

[2] Hassan RBA. Medicinal plants (Importance and Use). Pharm Anal Acta 2012; 3(10): 139.

[3] Nakka S, Devendra BN. A rapid *in vitro* propagation and estimation of secondary metabolites for *in vivo* and *in vitro* propagated *Crotalaria* species, A Fabaceae member. J Microbiol 2012; 2(3): 897-916.

[4] Jeeja G, Ansari R. Plants and plant science in Kerala Malabar Botanical Gardens. Kozhicode 2006; pp. 34-65.

[5] Sivarajan VV, Indira B. Ayurvedic drugs and their plant sources. 1st edn. Oxford and IBH publishing co. Pvt. Ltd.: NewDelhi 1994; p. 414.

[6] Deepa MA, Narmatha Bai V, Basker S. Antifungal properties of *Pseudarthria viscida.* Fitoterapia 2004; 75(6): 581-4. [http://dx.doi.org/10.1016/j.fitote.2004.04.008] [PMID: 15351113]

[7] Vijayabaskaran M, Sajeer P, Perumal P. Wound haealing activity of ethanolic extract of *Pseudarthria viscida* Linn. Inter rese. J Pharm (Cairo) 2011; 2(4): 141-4.

[8] Suba SM, Ayun VA, Kingston C. Vascular plant diversity in the tribal homegardens of Kanyakumari Wildlife Sancturary, Southern Western Ghats. Bioscience Discovery 2014; 5(1): 99-111.

[9] Murashige T, Skoog F. A revised medium for rapid growth and-bio assays with tobacco tissue cultures. J Physiol 1962; 15: 473-97.

[10] Gamborg OL, Miller RA, Ojima K. Nutrient requirements of suspension cultures of soybean root cells. Exp Cell Res 1968; 50(1): 151-8. [http://dx.doi.org/10.1016/0014-4827(68)90403-5] [PMID: 5650857]

[11] White PR. A Handbook of Plant Tissue Culture. Lancaster, Pennsylvania, USA: The Jacques Catlell Press 1943; p. 304.

[12] Cheruvathur MK, Thomas TD. An efficient plant regeneration system through callus for *Pseudarthria viscida* (L.) Wright and Arn., A rare ethnomedicinal herb. Physiol Mol Biol Plants 2011; 17(4): 395-401. [http://dx.doi.org/10.1007/s12298-011-0089-z] [PMID: 23573033]

[13] Srivastava P, Singh BD, Tiwari K, Strivastava G. High frequency shoots regeneration for mass multiplication of *Desmodium gangeticum* (L.) DC- An important anticancer, antidiabetic and hepatoprotective endangered medicinal plant. Int J Sci Res 2015; 4(8): 508-12.

[14] Kumar NG, Kumar S, Vimalan S, Prakash P, Kumar R. Optimization of growth promoters on *Desmodium gangeticum* (L.) DC using RSM-CCD and its antioxidant activity. Int J Pharm Pharm Sci 2014; 6(8): 503-7.

[15] Puhan P, Rath SP. Induction, development and germination of somatic embryos from *in vitro* grown seedling explant in *Desmodium gangeticum* (L.) DC-A medicinal plant. Res. Faslnamah-i Giyahan-i Daruyi 2012; 6(5): 346-69.

[16] Shimamoto Y, Hayakawa H, Abe J, Nakashima H, Mikami T. Callus induction and plant regeneration of *Beta* germplasm. J Sugar Beet Res 1993; 30(4): 317-9. [http://dx.doi.org/10.5274/jsbr.30.4.317]

SUBJECT INDEX

A

T. Pullaiah (Ed.)

D

E

F

G

H

R

S

T

U

V

W

Z

www.ingramcontent.com/pod-product-compliance
Lightning Source LLC
Chambersburg PA
CBHW050801220326
41598CB00006B/86

* 9 7 8 9 8 1 5 2 3 8 3 2 7 *